STUDY GUIDE
to accompany

CHEMISTRY
& Chemical Reactivity
Fourth Edition

KOTZ *&* TREICHEL

PAUL W. W. HUNTER
Michigan State University

SAUNDERS GOLDEN SUNBURST SERIES

Saunders College Publishing

Harcourt Brace College Publishers

Fort Worth   Philadelphia   San Diego   New York   Orlando   Austin
San Antonio   Toronto   Montreal   London   Sydney   Tokyo

Printed in the United States of America

ISBN 0-03-023794-7

9 0 1 2 3 4 5 6 7 8 202 11 10 9 8 7 6 5 4 3 2

CONTENTS

for

Kathryn, Lauren, and Dan

INTRODUCTION

The study of chemistry is required by almost all science majors at college. However, chemistry has the reputation of being a subject difficult to master. Introductory courses in chemistry are universally regarded by students as hard—a rite of passage that must be endured. Why is this?

Chemistry presents a considerable amount of material that needs to be mastered early in the course. It is for many students an almost overwhelming amount of material: names to be learned, formulas to write, equations to balance, stoichiometric calculations to perform, concepts to understand. And then, after the introductory topics have been dealt with, the pace never lets up; it just seems that the chemistry gets more and more difficult.

This study guide is intended to help you organize your study of chemistry so that you can learn more efficiently, understand the subject better, and ultimately achieve success in the course.

Our experience has shown that one study habit is essential for success, and another study habit helps enormously. The essential habit is a regular, preferably daily, study and *active* practice of the subject. Be prepared for the lectures and be an active participant in them; take notes and review and *rewrite* them afterward. Solve problems and answer the questions assigned as homework and *never* look at the solutions or answers until you have tried the problems yourself.

What helps in your study is a collaboration and cooperation with your fellow students. Never underestimate the power of your peers! Talking about the chemistry with other students, explaining concepts to others, reading about the chemistry in your textbook, and practicing the problems is the way to success. Chemistry is like a foreign language to many students. Like any other language, the more you use it, the easier it will become. In this study guide we hope to make your learning of chemistry easier.

We will provide suggestions on how to use and learn from the textbook, summarize and point out the important features within each chapter, and provide supplementary problems, and their solutions, on each topic.

Benjamin Franklin (1706-1790) once said that education is what remains after the information that has been taught has been forgotten. This is true of chemistry. You will probably have forgotten many of the facts you memorize in this course by the time you graduate in four years' time. However, the concepts, ideas, and principles of chemistry that you have *understood* will remain with you. You will apply this knowledge time and time again in your major studies. Let's begin.

A Note on the Organization of these Study Guides

There is a study guide for each chapter in the textbook. After a brief introduction, each study guide begins with a section-by-section summary of the corresponding chapter in the text. This summary outlines the essential points of the chapter—the content to learn and understand. Key points and comments are noted in the margins and important words or phases are written in boldface. References to tables and figures are to those in the text; page numbers are usually given.

After the summary your knowledge and understanding of the material will be tested through a series of short review questions. Answers follow these review questions.

At the end of the study guide a limited number of problems are presented. These vary in difficulty and again they test and reinforce your mastery of the subject. It is important to try all these problems, either by yourself or with your peers, before looking through the solutions that follow. Key points and comments are again noted in the margins.

Crossword and number puzzles are provided in some chapters—for both entertainment and learning. Some chapters include challenges that require more thought than typical problems—sometimes a lot more! Solutions to these puzzles and problems are provided at the end of this book.

A sample examination is presented at the end of each of the five parts of the textbook. The questions on these sample exams are intended to prepare you for your own examinations. The answers are again provided at the end of the book.

Although care has been taken to avoid errors in this study guide, undoubtedly some have slipped through. Any corrections, criticisms, or comments are most welcome and may be emailed to me at hunter@pilot.msu.edu.

CHAPTER 1

Introduction

This first chapter introduces some fundamental properties of matter. The three states of matter are described. Physical properties of matter (such as density and temperature), elements and compounds, atoms and molecules, and chemical and physical changes are discussed. The chapter presents a scheme for the classification of matter. The latter part of the chapter introduces some of the mathematics essential to introductory chemistry—the International System of units, dimensional analysis, scientific notation, significant figures, and problem-solving strategies.

Contents

1.1 Classifying Matter

Matter normally exists in one of three states—solid, liquid, or gas. A **solid** has a rigid shape and a volume that changes very little as the temperature and pressure change. A **liquid** has a fixed volume but is fluid—it takes on the shape of the lower part of its container and has a well-defined surface. A **gas** expands to fill its container and its volume varies considerably with change in temperature or pressure.

Physical properties of matter can be interpreted by the **kinetic molecular theory**—the idea that matter consists of tiny particles called molecules or atoms that are in constant motion. In a solid these particles are packed tightly in a regular pattern. In a liquid the particles are in constant motion and are not restricted to specific locations. The particles collide with each other and interact constantly. In a gas the particles are far apart and are independent of each other; they move rapidly between collisions with each other and the container in which the gas is confined. As the temperature is increased the kinetic energy of the particles increases—a solid will eventually melt and a liquid will vaporize.

Some physical properties can be observed by unaided human senses—these are macroscopic properties. Some matter is too small to be seen without a microscope and is called **microscopic**. The structure of matter at the atomic scale is called submicroscopic or **particulate**.

See Figure 1.2 on page 20 for an illustration of the three states of matter.

Understand how the behavior of particles (molecules or atoms) is different in each of the three states of matter.

Kinetic energy is the energy of motion.

Chemistry has three main components:

the macroscopic—
    tangible, visible

the particulate—
    atoms, ions, & molecules

the symbolic—
    formulas & equations.

See Figure 1.1 on page 20 for a scheme for the classification of matter.

Recognize two types of mixtures: homogeneous and heterogeneous.

A (pure) substance can be identified by a characteristic set of properties.

Figure 1.6 on page 23 illustrates the separation of iron and sulfur by physical means.

Learn the symbols and names of the first eighteen elements. When you have mastered these extend the range to the first 38. These are the elements you will encounter most frequently.

Nuclear reactions are described in Chapter 24.

There is a Periodic Table on the inside front cover of this study guide. There is also one on the inside front cover of the text.

Understand the difference between a molecule and a compound. Not all molecules are compounds, nor are all compounds molecules!

Note that some properties of substances (elements and compounds) change as the sample size decreases to just a few molecules.

Ionic compounds are described further in Chapter 3.

Be able to distinguish between compounds of two or more elements and mixtures of the same elements.

Most naturally occurring samples of matter are mixtures. There are two types of mixtures: heterogeneous and homogeneous. In a **heterogeneous mixture** the properties of the mixture vary throughout. Often the individual components of the mixture are observable (for example in a mixture of salt and pepper) but sometimes closer examination or magnification is necessary. A **homogeneous mixture** is called a **solution**—this is a mixture of components at a molecular level. No amount of optical magnification can reveal different properties in different regions of a solution.

Purification is the separation of a mixture into its components. The purity of a component can be determined by examination of its physical properties such as its color or melting point. A substance is considered pure when no observable change in its physical properties occurs with further purification.

The separation of mixtures can be quite difficult; the methods used to isolate the components of a mixture take advantage of the different physical properties of the components.

1.2 Elements and Atoms

A **substance** that cannot be decomposed further by chemical means is called an **element**. At present only 112 different elements are known, and only 90 of these are found in nature. The remainder have been created artificially by nuclear reaction. Each element has a name and a **symbol**, for example hydrogen (H), gold (Au), tin (Sn), and aluminum (Al). The symbol may consist of one or two letters. Each element has its own place in the **Periodic Table of Elements**. The smallest particle of an element is called an **atom** of that element.

1.3 Compounds and Molecules

A substance may be an element or a **compound** of two or more elements. When two or more elements combine to form a compound, the compound has properties quite distinct from the properties of the original elements. Just as elements have symbols, so compounds have **formulas**. There are two major classes of compounds: molecular and ionic. **Molecules** are small discrete units of atoms bonded together. They are the smallest units that exhibit the chemical characteristics of the substance. An example is the water molecule H_2O, composed of two hydrogen atoms and one oxygen atom. Water has chemical and physical properties quite unlike those of the constituent elements hydrogen and oxygen. **Ionic compounds** are composed of **ions**—these are electrically charged atoms or groups of atoms. An example is common salt, sodium chloride NaCl. This is a compound of sodium and chlorine in which the sodium is present as positive sodium ions Na^+ and chlorine is present as negative chloride ions Cl^-.

Compounds are quite different from mixtures of elements. Individual elements can be identified in mixtures, they may be present in any proportion, and they can be separated by physical means. A compound can be decomposed only by chemical means and the elements are always present in the same ratio by mass (the law of constant composition).

1.4 Physical Properties

All matter possesses physical properties such as melting point, boiling point, density, solubility, conductivity, viscosity, heat of fusion, and color.

Density is a physical property that relates the mass of an object and its volume. If two of the three variables, density, mass, and volume are known, the third can always be calculated.

Temperature is the property of matter that measures how hot a substance is. Since heat travels only from a hotter object to a colder one, temperature determines the direction of heat flow.

Any temperature scale can be established by assigning numbers to two reproducible events: **Celsius** chose 0° for melting ice and 100° for boiling water; **Fahrenheit** chose 0° for an ice/salt/water mixture and 100° for the body temperature of a human being—although now the Fahrenheit scale is established using 32° for melting ice and 212° for boiling water. The size of a Fahrenheit degree (one division on the scale) is 5/9 of the size of a Celsius degree and this relationship can be used to convert from one scale to the other. The **Kelvin** (or absolute) temperature scale assigns a value of 0° to the lowest temperature attainable and has a degree size the same as the Celsius scale. Zero on the Celsius scale (0°C) is equal to 273.15K. (Note the absence of the °degree symbol.)

Physical properties are often affected by a change in temperature; density, for example, decreases as temperature increases.

There are two types of physical property: extensive and intensive. **Extensive** properties, like mass and volume, depend upon the amount of substance present. **Intensive** properties, like color and density, are independent of the amount of substance.

Some physical properties are listed in Table 1.2 (page 28).

Density = mass/volume.

See study question #5.

Know how to convert a temperature on one scale into a temperature on another scale. Although the size of one °F is 5/9 the size of one °C, you cannot convert from one to the other by simply multiplying or dividing by 5/9. The zero points are different! See study question #4.

Understand the significance of zero degrees kelvin (0K).

Be able to classify properties as either extensive or intensive.

1.5 Physical and Chemical Change

In a **physical change**, the identity of all the substances remains unchanged. Examples are the conversion of liquid water into water vapor, and dissolving salt in water to make a solution. In both cases the identities of the substances remain the same.

The conversion of one or more substances into other substances is referred to as a chemical change or a **chemical reaction**. For example, the two substances hydrogen and oxygen react to form water, another substance. Atoms of the elements involved are neither created nor destroyed, they are simply rearranged. The molecules (or ionic compounds) before the reaction are changed into new molecules (or ionic compounds). A chemical reaction is represented by a **chemical equation**.

Both chemical and physical changes are almost always accompanied by a transfer of energy.

Be able to classify a change as either chemical or physical.

Equations for chemical reactions will be examined in Chapter 4.

1.6 Units of Measurement

Quantitative measurements are essential in most chemical reactions. Most measurements in science are based upon the **International System of Units** (abbreviated SI). In SI all units are derived from seven base units (of which six

SI is derived from the French: système international d'unites.

Learn the six base SI units of interest to chemists.
Table 1.3 page 37.
Learn the common prefixes for SI units.
Table 1.4 page 37.
Conversion factors for changing from non SI to SI units are listed on the inside back cover.

are of interest to chemists). Often a prefix is added to indicate a power of ten (or order of magnitude). For example, highway distances are given in kilometers where the kilo prefix indicates an order of 10^3 (i.e. 1000 meters = 1 kilometer).

Scientific, or **exponential, notation** is explained in Appendix A of the text. To use this notation a number is written as a number between 1 and 10 (but not including 10) multiplied by 10 raised to a power (which may be positive, negative, or zero).

1.7 Using Numerical Information

Manipulating numerical data is an everyday part of experimental chemistry. **Dimensional analysis** is a particularly useful approach to solving many problems in chemistry. The method involves the use of appropriate conversion factors to obtain the desired result. If the units cancel correctly, then the numerical result will be correct. Examples are presented in the text (pages 37–43) and later in this study guide.

Understand the difference between accuracy and precision. An illustration of the difference is shown in Figure 1.17 on page 43 of the text.

The **precision** of a measurement indicates how well several determinations of a quantity agree with one another (the reproducibility of the measurement). **Accuracy** on the other hand indicates how close the determinations are to the actual value. A precise series of measurements does not guarantee an accurate determination—there may be a systematic error in the experimental method.

The result of an experimental measurement is written as precisely as possible. The final digit in a number is assumed to be uncertain to the extent of ± 1. For example, the thickness of a tabletop written as 23 mm means that it might be 23+1 or 23-1 mm, that is, somewhere between 22 and 24 mm.

Be able to do calculations with proper regard for the precision of the measurements involved.

When manipulating numbers and their units in calculations, care must be taken to retain all digits until the end of the calculation and then to write the result with no more significant digits than is warranted. The guidelines for determining **significant figures** are listed on page 47 of your textbook. In short, when adding or subtracting numbers, line up the decimal points and quote the answer with no more decimal places than the number with the fewest places. When multiplying or dividing numbers, the number of significant figures in the answer should be the same as in the number with the least number of significant figures. Use common sense!

In 1000 grams of solution there is 60 grams of sugar.

Percent (%) means per centum, or the fraction of one hundred. Usually the percent is w/w or mass per mass. For example, a 6% solution of sugar in water means that in 100 g of solution there is 6 grams of sugar (and 94 g of water).

1.8 Problem Solving

Knowing how to interpret a question, and how to solve the problem, is often more difficult than the actual solution. If you misunderstand the problem, the chances of a successful solution are remote. Logical steps are:

Be sure to read the Problem-Solving Strategies listed on page 49 of the text.

1. Define the problem (understand the question!)
2. Develop a plan.
3. Execute the plan.
4. Check the answer: is it reasonable? are the units correct?

Review Questions

1. List the boldface words or phrases in the preceding pages and define each of them.

2. How do atoms or molecules behave differently in the three states of matter?

3. What is the relationship between density, volume, and mass?

4. What is the difference between the Fahrenheit and Celsius temperature scales?

5. What is the significance of zero degrees kelvin (0K)?

6. What is an element?

7. Is a compound always a molecule?

8. Is a molecule always a compound?

9. What are the two different kinds of mixtures and what is the difference between them?

10. What is the difference between a chemical change and a physical change?

11. What is the difference between an extensive property and an intensive property?

12. What is the base SI unit for mass?

13. List the six base SI units of interest to chemists.

14. Are the values of the exponents in scientific notation always positive? Can the value of the exponent be zero?

15. Is it possible for a series of measurements to be accurate but not very precise?

Answers to Review Questions

As you read the summaries in subsequent chapters in this study guide, it is a good idea to list the boldface words and phrases and their definitions.

1. All of the words and phrases that are boldface in the text are defined there. These are words and terms that you should clearly understand.

2. In a solid, atoms or molecules are packed tightly in a regular array. In a liquid the atoms or molecules are in constant motion although they interact constantly with each other. In a gas the particles are far apart, independent of each other; and move rapidly between collisions with each other and the container.

Remember this relationship; you will use it often.

3. Density = mass/volume.

The Fahrenheit scale is now established using 32° for melting ice and 212° for boiling water.

4. Celsius chose the numbers 0° for melting ice and 100° for boiling water for his temperature scale. Fahrenheit chose 0° for an ice/salt/water mixture and 100° for the body temperature of a human being. An identical event has a different number (°degrees) on the two scales. The size of a Fahrenheit degree is 5/9 of the size of a Celsius degree. However, the zero points are different for the two scales—0°C equals 32°F. The only temperature at which both scales have the same value is −40°.

Zero K is the temperature at which an ideal gas would have zero volume—see Chapter 12.

5. Zero degrees kelvin is the lowest attainable temperature.

6. An element is a substance that cannot be decomposed further by chemical means. Only 112 different elements are known, and only 90 of these are found in nature. Each element has its own name, symbol, and place in the Periodic Table.

Molecules and compounds are discussed in more detail in Chapter 3.

7. There are two principal types of compound: molecular and ionic. Compounds like sugar and water exist as molecules; compounds like salt (sodium chloride) are ionic.

There is a drawing of a C_{60} molecule in Figure 2.22 on page 89 of the text. It is also shown in Figure 11.1 on page 481 with the other allotropes of carbon.

8. A molecule is a group of atoms that act as a discrete unit and some elements exist as molecules. Examples are H_2, N_2, S_8, P_4, and C_{60}. Most molecules, however, are compounds of two or more elements.

9. There are two types of mixtures: heterogeneous and homogeneous. In a heterogeneous mixture the properties of the mixture vary throughout (although this may be difficult to see). A homogeneous mixture is called a solution—this is a mixture of components at a particulate level and the properties of a solution are constant throughout.

10. A chemical change is the conversion of one or more substances into other substances. Atoms of the elements involved are neither created nor destroyed, they are simply rearranged. In a physical change, the identities of the substances remain unchanged.

11. An extensive property depends upon the amount of substance present; an obvious example is the mass of the substance. An intensive property is independent of the amount present; color is an example.

12. The base SI unit for mass is the kilogram (kg)—not the gram.

13. mass(kg); length(m); time(s); temperature(K); current(A); amount(mol).

14. No, the exponent in scientific notation can be positive (for numbers 10 or greater) or negative (for numbers less than 1) and it can be zero—although 10^0 equals 1 and is usually omitted.

15. Yes, if there is considerable random error in the series of measurements, the precision will be poor. However, providing the errors are truly random, the average will be accurate.

Study Questions and Problems

One must learn by doing the thing; though you think you know it, you have no certainty until you try.

Sophocles
(496-406BC)

1. Identify the following changes as physical or chemical changes:
   a. Baking soda reacts with vinegar to produce carbon dioxide.
   b. The copper sheath on the Statue of Liberty turns green.
   c. Addition of salt melts ice on the highway.
   d. Steam condenses on the windowpane.
   e. Epoxy resin cures and hardens.
   f. Sugar dissolves in a cup of coffee.
   g. Natural gas burns in a furnace.

2. a. Calculate the density of lead if a 10 kg block has a volume of 885 cm^3.
   b. What is the volume of a 100 g bar of aluminum if its density is 2.70 g cm^{-3}?
   c. Calculate the mass of 100 cm^3 of uranium (density 19.07 g cm^{-3}).

3. At what temperature do the Fahrenheit and Celsius scales have the same value?

4. Convert
   a. 12°F to °C
   b. 300K to °F
   c. 25°C to K

5. Suppose that you developed your own new temperature scale (°N) in which you assigned a value of 20°N to the melting point of ice and a value of 220°N to the boiling point of water. At what temperature on your scale (°N) would sulfur melt (m.pt.113°C)?

You can develop any temperature scale you wish. Many scientists had their own scale: Newton, Rankine, and Réamur, for example.

6. Which of the following physical properties are extensive?
    a. heat of fusion   d. viscosity
    b. melting point   e. conductivity
    c. color   f. density

7. Write the names of the following elements:
    a. N
    b. Ca
    c. K
    d. Cu
    e. P
    f. V

8. Write the symbols for the following elements:
    a. silicon
    b. chlorine
    c. iron
    d. sodium
    e. silver
    f. sulfur

9. Convert
    a. 1342 mL into L
    b. 3.26×10^{-6} km into mm
    c. 8,768 mg into g
    d. 400 cm^3 into m^3
    e. 3600 sq.in. into sq.ft.

Guidelines for determining significant figures are described on page 47 of the text.

10. Write the following numbers in scientific notation with the correct number of significant figures:
    a. 1,327
    b. 0.00562
    c. 2.76
    d. 0.166
    e. 0.09911

11. Measurements of the boiling point of a liquid were taken by two laboratory technicians (A and B). The actual boiling point was 92.3. Which technician achieved the most accurate result and which technician was most precise?
    A: 92.0   92.1   92.4   92.2
    B: 91.9   92.5   92.6   92.0

12. Match the prefix with the correct multiplier:
    milli   micro 10^3 10^{-6}
    kilo   mega 10^{-12} 10^{-2}
    centi   pico 10^{-3} 10^6

13. Evaluate the following expressions. Express the answers in scientific notation with the correct number of significant figures and the correct units.

    a. 0.0045 in + 1.0098 in + 0.987 in + 23.08 in

    b. $(3.45 \text{ cm}^3 \times 2.70 \text{ g cm}^{-3}) + (7.433 \text{ cm}^3 \times 1.677 \text{ g cm}^{-3})$

    c. 2.703 g/(1.376 cm × 2.45 cm × 3.78 cm)

14. A 12.3 g block of an unknown metal is immersed in water in a graduated cylinder. The level of water in the cylinder rose. The level of water in the cylinder rose exactly the same distance when 17.4 grams of aluminum (density 2.70 g cm^{-3}) was added to the same cylinder. What is the density of the unknown metal?

15. Use the outline described in Section 1.8 to plan your approach to solving the following problem. Describe the strategy you use.
    The level of water in a graduated cylinder is at the 100 mL mark. When a platinum crucible floats on the surface of the water, the level reads 157.9 mL. When the crucible is totally immersed in the same cylinder, the level reads 102.70 mL. What is the density of platinum? The density of water is 0.997 g/mL at 25°C.

16. If one pound is 453.59 grams, how many grams are there in one ounce? How many ounces are there in one kilogram?

17. A sample of a gold alloy contains 5.6% silver by mass. How many grams of silver are there in 1 kilogram of the alloy?

Answers to Study Questions and Problems

1. a. chemical
    b. chemical; the green color is caused by the reaction of the copper with the atmosphere.
    c. physical; no chemical change takes places.
    d. physical
    e. chemical; the epoxy resin is a polymeric ether and its formation involves a chemical reaction.
    f. physical
    g. chemical

The green compound is $Cu(OH)_2CO_3$

2. a. Density = mass/volume = $10,000\text{g}/885\text{cm}^3 = 11.3 \text{ g cm}^{-3}$

    b. Volume = mass/density = $100\text{g}/2.70 \text{ g cm}^{-3} = 37.0\text{cm}^3$

    c. Mass = volume × density = $100 \text{ cm}^3 \times 19.07 \text{ g cm}^{-3} = 1910 \text{ g}$.

These expressions are all equivalent representations of

Density = mass/volume.

If two of the variables are known, the third can always be calculated.

3.   The gradients of the two scales are different. The Fahrenheit scale rises 180° between the melting point of ice and the boiling point of water. The Celsius scale rises only 100°. The ratio between the gradients is 180/100 or 9/5.

Using the conversion method presented in the text:
If the value on the two scales is equal, and x is that value, then
$x°C = (5/9)(x°F - 32)$, so $9x = 5x - 160$, and $x = -40°F$ or $-40°C$.

A convenient method for converting °F to °C and *vice versa* is to add 40; multiply by 5/9 (or 9/5); and then subtract 40.

To convert °F to K, first convert to °C, and then convert °C to K.

This is because the temperature at which both scales have the same value is −40°, so add 40 to bring the common point to 0°, multiply by 9/5 or 5/9, then subtract the 40°.

Multiply by 5/9 because a Fahrenheit degree is smaller than a Celsius degree, i.e. the factor 5°C/9°F is equal to 1.

4.   a. $12°F + 40° = 52°$
         $52° × 5/9 = 28.9°$
         $28.9° - 40° = -11°C$

     b. $300K = 300° - 273.15° = 26.9°C$
         $26.9°C + 40° = 66.9°$
         $66.9° × 9/5 = 120.33°$
         $120.33° - 40° = 80°F$

     c. $25°C + 273.15° = 298K$

5.   The difference between the melting point of ice and the boiling point of water is 220–20°N = 200°N. This is equivalent to 100°C. So one °N equals 0.5°C. The melting point of sulfur is 113°C above zero°C. This is equivalent to 226°N above 20°N (the melting point of ice). So sulfur melts at 20+226°N = 246°N.

The heat of fusion is usually expressed in units of kJ/mol. (It does take more heat to melt more material.)

6.   None; all are intensive (see note at left).

The symbols for some elements that have been known for a long time are based upon names other than their English names.

7.   a. N   nitrogen
     b. Ca   calcium
     c. K   potassium
     d. Cu   copper
     e. P   phosphorus
     f. V   vanadium

8.   a. silicon   Si
     b. chlorine   Cl
     c. iron   Fe
     d. sodium   Na
     e. silver   Ag
     f. sulfur   S

9. a. $1342 \text{ mL} \times (1\text{L}/1000\text{mL}) = 1.342 \text{ L}$

    b. $3.26 \times 10^{-6} \text{ km} \times (10^{6}\text{mm}/1\text{km}) = 3.26 \text{ mm}$

    c. $8{,}768 \text{ mg} \times (1\text{g}/10^{3}\text{mg}) = 8.768 \text{ g}$

    d. $400 \text{ cm}^3 \times (1\text{m}/100\text{cm})^3 = 4 \times 10^{-4} \text{ m}^3$

    e. $3600 \text{ sq.in.} \times (1\text{ft}/12\text{in})^2 = 25 \text{ sq.ft.}$

10. a. $1{,}327 \qquad = 1.327 \times 10^3$
    b. $0.00562 \quad = 5.62 \times 10^{-3}$
    c. $2.76 \qquad = 2.76$
    d. $0.166 \qquad = 1.66 \times 10^{-1}$
    e. $0.09911 \quad = 9.911 \times 10^{-2}$

11. A: more precise   average $= 92.1_8$
    B: less precise   average $= 92.2_5$ slightly more accurate

12. milli 10^{-3}
    micro 10^{-6}
    kilo 10^{3}
    mega 10^{6}
    centi 10^{-2}
    pico 10^{-12}

13. a. $0.0045 \text{ in} + 1.0098 \text{ in} + 0.987 \text{ in} + 23.08 \text{ in} = 2.508 \times 10^{1} \text{ in}$

    b. $(3.45 \text{ cm}^3 \times 2.70 \text{ g cm}^{-3}) + (7.433 \text{ cm}^3 \times 1.677 \text{ g cm}^{-3})$
        $= 9.32 \text{ g} + 12.47 \text{ g} = 2.178 \times 10^{1} \text{ g}$

    c. $2.703 \text{ g}/(1.376 \text{ cm} \times 2.45 \text{ cm} \times 3.78 \text{ cm}) = 2.12 \times 10^{-1} \text{ g cm}^{-3}$

14. The 17.4 grams of aluminum (density 2.70 g cm^{-3}) has a volume of 6.44 cm^3. The 12.3 g block of the unknown metal has the same volume, so its density must be 12.3 g/6.44 cm^3 = 1.91 g cm^{-3}.

15. Define the problem: Need to calculate the density of platinum.

    Develop a plan: Density is mass/volume; so determine the mass and the volume perhaps? How can the mass be determined? How can the volume be determined? When something floats on the surface of the water, it displaces a mass of water equal to its own mass. So the mass can be determined by the quantity of water displaced when the crucible floats. When an object is totally immersed, it displaces a volume of water equal to its own volume, so the volume can be determined by the change in the volume when the crucible is totally immersed.

Note that the 1 is *exactly* 1; it is an integer and has infinite precision (so has the 1000). The number of significant figures in the answer is 4.

Note that the conversion factor (both the number and the units) is cubed.

There are 144 sq.in. per sq.ft.

The number of significant figures in numbers like 400 and 3600 is often in doubt. They could be exactly 400 and 3600 or just approximately 400 or 3600—the context sometimes indicates which is correct.

There is considerable advantage in writing numbers in scientific notation—when the number of significant figures is known precisely.

The first measurement (0.0045) is insignificant but must be included in the sum.

Note the answer has 4 sig. fig. even though some numbers have only 3 sig. fig.

Density = mass/volume.

Execute the plan: The mass of the crucible equals the volume of water displaced × the density of water = 57.9mL × 0.997g/mL = 57.7 g. The volume of the crucible equals 2.70 cm³. The density of the platinum metal is therefore 57.7g/2.70cm³ = 21.4 g cm⁻³.

Check the answer: The magnitude of the density is about right and the units are correct.

16. One pound is equal to 16 ounces:
(453.59 grams/1 lb) × (1 lb/16 oz) = 28.35 g/oz
(16 oz/1 lb) ×)1 lb/453.59 grams) × (1000 grams/1 kg) = 35.27 oz/kg

17. The 5.6% indicates that in every 100 grams of alloy, there are 5.6 grams of silver. The % is usually a mass/mass percent. In 1000 grams of alloy there are 56 grams of silver.

CHAPTER 2

Introduction

The history of the way in which our view of matter at the atomic level has developed is fascinating. In some cases simple experiments with rudimentary apparatus led to startling results and great steps forward in our understanding of things too small to see. This chapter describes the origins of atomic theory and how the theory has grown.

Contents

2.1 Origins of Atomic Theory

Democritus (460–370BC) compared the sand on a beach—which looks continuous when viewed from a distance—with sea water. He reasoned that just as the sand consisted of small particles, so perhaps did the water, and indeed all matter. He called these small fundamental particles of matter **atoms**. Unfortunately this idea of the atomic nature of matter was largely ignored for the next 2200 years.

In 1803, John Dalton (1766–1844) revived the **atomic theory** and used it to explain the Law of Conservation of Matter and the Law of Constant Composition. He proposed a new law, the Law of Multiple Proportions, in support of the theory.

2.2 Protons, Neutrons, and Electrons

Dalton's postulate that the atoms of an element are indivisible turned out to be incorrect. Atoms have a subatomic structure; they are composed in turn of even smaller particles.

The existence of electric charge (positive and negative) was one of the first indications that particles smaller than atoms existed. In 1833, Michael Faraday (1791–1867) performed experiments in electrolysis that led to the idea of a fundamental particle of electricity called an **electron**. Joseph J. Thompson (1856–1940) experimented with cathode ray tubes and proved the existence of the

The word *atom* means *indivisible*.

Understand the postulates of Dalton's atomic theory.

Understand these three laws:

The law of conservation of mass (Lavoisier).

The law of constant composition (Proust).

The law of multiple proportions (Dalton).

For descriptions of these three laws, see pages 61–62 of the text or the answer to the review question #2 in this study guide.

Some knowledge of the history of the development of atomic theory will help you understand it.

Radioactivity and nuclear reactions are examined further in Chapter 24.

electron. He announced his discovery in 1897, only 100 years ago. Early experiments in radioactivity and the discovery of three types of radiation, α, ß, and γ, led Marie Curie (1867–1934) to suggest in 1899 that atoms could indeed disintegrate into smaller particles. The actual electrical charge on an electron, and its mass, were finally established by Robert Millikan (1868–1953) in his oil-drop experiment.

The **proton**, the fundamental positive particle, was observed in cathode ray tubes containing hydrogen gas.

The **neutron**, the third subatomic particle, was discovered by James Chadwick (1891–1974) in 1932. This particle, as the name suggests, has no charge. It has a mass slightly greater than a proton.

Because negative and positive charges attract one another, and like charges repel each other, Thompson envisioned the atom as a homogeneous mix of positive and negative matter. Ernest Rutherford (1871–1937) tested this model by directing a beam of α particles at a thin sheet of gold. Some α particles were bounced back—which could only mean that they had encountered something in the atom more massive than themselves. So he proposed a **nuclear model** for the atom: a very small positively charged nucleus at the center responsible for virtually the entire mass of the atom surrounded by the negative electrons taking up almost the entire volume of the atom.

An α particle is a helium nucleus. A helium nucleus contains 2 protons and 2 neutrons. A diagram of the apparatus used is shown in Figure 2.8 on page 69 of the text.

2.3 Atomic Structure

Three primary particles, the proton, the neutron, and the electron, make up all atoms. The protons and neutrons exist in the nucleus; the electrons surround the **nucleus**. In a neutral atom, the number of electrons equals the number of protons.

Learn the characteristics (mass and charge) of the three primary subatomic particles.

proton:
mass 1.673×10^{-24} gram
charge +1

neutron:
mass 1.675×10^{-24} gram
no charge

electron:
mass 9.109×10^{-28} gram
charge −1

2.4 Atomic Number and the Mass of an Atom

All atoms of an element have the same number of protons in the nucleus. This number is called the **atomic number** (Z). It is the atomic number that identifies an element. For example, an atom that has 6 protons in its nucleus, and therefore an atomic number Z equal to 6, is always an atom of carbon.

Historically, the relative masses of the atoms of the elements were determined by careful experiments. For example, a nitrogen atom is fourteen times heavier than a hydrogen atom. To develop a absolute mass scale, an arbitrary standard is required. This standard has been chosen to be an atom of carbon with six protons, six neutrons, and six electrons—it is assigned a mass of exactly 12 units.

Originally, atomic masses were based upon a mass of exactly 1.000 for hydrogen. Then upon a mass of exactly 16.000 for oxygen. Unfortunately, this meant the mass of oxygen−16 for physicists, but the atomic mass of oxygen (weighted average of all isotopes) for chemists. To resolve the conflict, the new standard, carbon−12 was chosen in 1961.

One atomic mass unit (one amu or one dalton) is defined as $1/12^{\text{th}}$ the mass of a carbon atom with six protons, six neutrons, and six electrons.

One amu equals 1.661×10^{-24}g.

The sum of the number of protons and the number of neutrons in the nucleus of an atom is called the **mass number** (A). For example the mass number of a carbon atom with six protons, six neutrons, and six electrons is 12. This atom of carbon is often referred to as carbon−12.

The symbol for carbon−12 is

$^{12}_{6}$C.

2.5 Isotopes

Although atoms of the same element must, by definition, have the same number of protons in the nucleus, they may have different numbers of neutrons. Therefore the atomic number for an element is always the same, but the mass number may vary. Dalton's postulate that all atoms of an element are identical is incorrect. Atoms of the same element that are different are called **isotopes**.

Some elements have only one stable (nonradioactive) isotope, other elements have many isotopes. Hydrogen has two stable isotopes (hydrogen and deuterium) and one radioactive isotope (tritium). These three isotopes have mass numbers 1, 2, and 3 respectively. When deuterium is substituted for hydrogen in water, the result is "heavy water", a molecule with a mass of 20 amu instead of 18 amu.

The masses and the relative abundances of the various isotopes of an element can be obtained from a mass spectrometer. For example, boron has two isotopes: boron-10 with a mass of 10.0129 amu and an abundance of 19.91% and boron-11 with a mass of 11.0093 amu and an abundance of 80.09%.

Be able to write the symbol for an isotope: $^A_Z X$.

The hydrogen–1 isotope is called protium.

2.6 Atomic weight

The **atomic mass** (or more commonly **atomic weight**) of an element is a weighted average of all naturally occurring isotopes of an element. For example, boron has an atomic mass somewhere between 10.0129 and 11.0093, but nearer to 11.0093 because the boron-11 isotope is the more abundant. Its actual atomic mass is 10.81 amu.

Know how to calculate the atomic mass from the masses of the isotopes and their abundances, and *vice versa*.

2.7 The Periodic Table

The **Periodic Table of Elements** is an arrangement of the elements so that those with similar properties lie in vertical columns called groups. The **groups** are numbered 1 through 8, with a letter A or B. The A groups are called the main groups and the B groups contain the transition elements. The horizontal rows are called **periods**.

The Periodic Table is divided into two main regions: the **metals** (to the left) and the **nonmetals** (to the upper right). On the boundary between these two regions lie the **metalloids**. Most of the elements are metals. All metals except mercury are solid at room temperature. Only some nonmetals (e.g. carbon, phosphorus, sulfur, iodine) are solids; one (bromine) is a liquid; and several (e.g. hydrogen, helium, oxygen, nitrogen, chlorine, xenon) are gases. The metalloids are the elements germanium, arsenic, selenium, antimony, and tellurium that lie on the diagonal line separating the metals and nonmetals.

Dmitri Mendeleev (1834–1907) was the first to develop the Periodic Table (1869) as it is known today. He noticed a **periodicity** in the chemical and physical properties of the elements as their atomic weights increased and arranged the elements so that those elements with similar properties lay in the same group. He left spaces for unknown elements and predicted their existence. Many (e.g. gallium, germanium, scandium) were soon discovered.

The Periodic Table is the chemist's most valuable reference.

The proposed scheme for renumbering the groups in the Periodic Table starts with 1 at the left and then numbers each group (including the transition elements) sequentially through to 18 for the noble gases.

Mendeleev's Periodic Table is illustrated in Figure 2.15 on page 82.

Henry G.J.Moseley (1888–1915) determined that it is not the atomic mass that governs periodicity in properties but in fact the atomic number. The atomic number of an element was unknown to Mendeleev.

2.8 The Elements, their Chemistry, and the Periodic Table.

Note that the elements in the Periodic Table are not always in order of increasing mass. For example:

Argon and potassium

Cobalt and nickel

Tellurium and iodine

Group 1A elements are called the **alkali metals**. They are very reactive and are found in nature only in compounds. They react with water to produce alkaline solutions and they form compounds A_2O with oxygen (where A is the alkali metal). For the want of any better place, hydrogen is often placed in Group 1A although in most of its properties it is quite unlike the alkali metals.

Group 2A elements are called the **alkaline earth metals**. Not quite as reactive as the alkali metals, they still are found in nature only in compounds. They too form alkaline solutions. The general formula of their oxides (compounds with oxygen) is EO, where E is the alkaline earth metal.

Become familiar with the arrangement of the elements in the Periodic Table. You will quite quickly associate a different chemical behavior of an element with its position in the table.

The **transition elements** lie between groups 2A and 3A. This block of elements is ten columns wide and contains many common elements (e.g. iron, titanium, and manganese). Some transition metals that are less reactive occur in nature as the element (e.g. silver, copper, gold, platinum). Many of the transition metals have commercial uses.

The **lanthanides** and **actinides** are usually drawn at the bottom of the table to save space. These two rows are fourteen elements wide.

Group 3A contains one nonmetal at the top (boron). The next element (aluminum) is the third most abundant element in the earth's crust.

Group 4A contains two nonmetals at the top (carbon and silicon), one metalloid (germanium), and two metals (tin and lead). Carbon is the basis for the great variety of compounds (called organic compounds) that make up living things. The element exists in three forms (called **allotropes**): graphite, diamond, and the fullerenes. Silicon is the second most abundant element on earth and occurs with aluminum and oxygen in many minerals (aluminosilicates). An alloy of tin and copper is called bronze—after which an entire age was named. Lead, at the bottom, is a toxic metal.

Allotropes are different forms of the same element that exist in the same physical state.

Group 5A contains a gas (nitrogen N_2) at the top, followed by phosphorus, (another nonmetal), arsenic and antimony (two metalloids), and finally bismuth (a metal). Nitrogen is a diatomic molecule and is essential to life. Fixing nitrogen (forming compounds of the element) is difficult. Phosphorus is also essential to life. It exists in two forms: red and white. The white form consists of P_4 molecules. In the red form these P_4 tetrahedra are joined together. Bismuth is the last element in the Periodic Table that is not radioactive.

Group 6A begins with oxygen, a diatomic gas like nitrogen. Unlike nitrogen, oxygen readily combines with most other elements. Ozone O_3 is an allotrope of oxygen. Sulfur exists commonly as S_8 molecules and can be found in its elemental form. It is used primarily in the manufacture of sulfuric acid—a chemical manufactured in greater quantity than any other (largely for use in the agricultural fertilizer industry). Polonium at the bottom of the group is radioactive. It was isolated by Marie and Pierre Curie in 1898.

Group 7A are the **halogens**; they are all nonmetals and are all diatomic molecules. They are all reactive; they form salts with metals and they form molecular compounds with most nonmetals including themselves. The reactivity decreases down the group. Fluorine and chlorine are gases, bromine is a liquid, and iodine is a solid.

Group 8A consists of the monatomic **noble gases**; these are the least reactive of all the elements. It is, however, possible to make compounds of xenon and krypton. Radon is radioactive.

Review Questions

1. List the postulates of Dalton's atomic theory. Which turned out to be untrue?

2. Write statements of the Law of Conservation of Matter, the Law of Constant Composition, and Law of Multiple Proportions.

3. What is the multiple proportion referred in Dalton's law of multiple proportions?

4. List the three primary subatomic particles, their masses, and their charges.

5. What is an isotope?

6. List the three isotopes of hydrogen and describe how they differ.

7. Define the atomic mass unit.

8. Identify the isotope with twenty protons, twenty neutrons, and twenty electrons. Write its symbol.

9. What is a group in the Periodic Table? What is a period?

10. What is periodicity? What is the law of chemical periodicity?

11. Sketch the Periodic Table without looking at a printed copy. Fill in the table with as much detail as you can remember or deduce.

12. Describe the three allotropes of carbon.

13. List the nonmetals; state whether each is a solid, liquid, or gas.

14. List the elements essential to life. Which are "bulk elements" and which are "trace elements"?

Answers to Review Questions

1. The postulates of Dalton's atomic theory are:

    a. All matter is made up of indestructible atoms.
    b. All atoms of an element are identical in mass and property.
    c. Compounds are formed by the combination of atoms in small whole number ratios.
    d. A chemical reaction involves a only a rearrangement of atoms; atoms are neither created nor destroyed.

    The first two postulates are partially incorrect. Atoms are composed of smaller particles and all atoms of an element are not identical—isotopes do exist.

2. Law of Conservation of Matter (Antoine Lavoisier):
    Mass is conserved in a chemical reaction (*cf.* Dalton's postulate #4).

    Law of Constant Composition (Joseph Proust):
    Any compound always has the same composition, regardless of its source (*cf.* Dalton's postulate #3).

    Law of Multiple Proportions (John Dalton):
    If two elements form two different compounds, the mass ratio of the elements in one compound is always a whole number multiple of the mass ratio of the same elements in the other compound.

3. The multiple referred in Dalton's law is best explained by an example:

    Consider three oxides of nitrogen: N_2O, NO, and NO_2. The ratios of the mass of nitrogen to the mass of oxygen in these three compounds are 28/16, 14/16, and 7/16 respectively.

    The 28, 14, and 7 are in a simple whole number ratio: 4:2:1—this is the multiple proportion referred to in Dalton's law.

4. The three primary subatomic particles are:

proton	mass 1.672623×10^{-24} gram	charge +1
neutron	mass 1.674929×10^{-24} gram	no charge
electron	mass 9.109389×10^{-28} gram	charge −1

5. An isotope of an element consists of atoms with a specific number of neutrons (a specific mass number). For example, three isotopes of carbon are:

    carbon-12 with 6 protons and 6 neutrons,
    carbon-13 with 6 protons and 7 neutrons,
    carbon-14 with 6 protons and 8 neutrons.

    All are isotopes of the same element.

6. The three isotopes of hydrogen (all have an atomic number = 1) are:

hydrogen: ^1_1H 1 proton   0 neutrons   mass number 1
(protium)
deuterium: ^2_1H 1 proton   1 neutron   mass number 2

tritium: ^3_1H 1 proton   2 neutrons   mass number 3
(tritium is radioactive)

7. The atomic mass unit (amu) is the mass unit of a scale in which one amu is equal to $1/12^{th}$ of the mass of a carbon-12 atom.

8. There are 20 protons—the atomic number Z is equal to 20. There are 20 neutrons—the mass number must be 40. The atomic number identifies the element as calcium. The mass number identifies the isotope. The symbol is $^{40}_{20}Ca$. The number of electrons equals the number of protons, so the atom has no charge.

Note that the subscript $_{20}$ can be omitted from the symbol ^{40}Ca. The element is calcium (symbol Ca) so the subscript $_{20}$ is in fact redundant.

The subscript is included in nuclear reactions to make balancing the equations easier.

9. A group in the Periodic Table is a column of elements having similar chemical properties. A period is a horizontal row in the table.

10. Periodicity is a regular and repetitive cycling of similar chemical and physical properties as the atomic number of the elements increase. The law of chemical periodicity states that the properties of the elements are periodic functions of atomic number. Note that Mendeleev originally stated the law as a function of atomic mass.

11. As your course in chemistry progresses, you will learn more and more about the Periodic Table. Some features of the table may already be familiar to you. For example,

   • the two groups on the left containing the alkali and alkaline earth metals,
   • the six-column block on the right with the diagonal line running through it that separates the metals and nonmetals,
   • the 4 × 10 block of transition elements with the last row occupied by elements only recently detected in nuclear reactions,
   • and the two rows at the bottom of the table.

   You should perhaps, at this stage, be able to place 20 to 30 elements in the table. There is no need to memorize atomic masses; a table will almost always be provided for you.

12. There are three allotropes of carbon: diamond with a three-dimensional structure, graphite consisting of two-dimensional sheets of carbon atoms, and the fullerenes of which the most common is the C_{60} molecule shaped like a soccer ball.

13. There are 19 nonmetals, including two that are radioactive:

Abbreviations are:
(s) solid
(l) liquid
(g) gas

Group 3A: boron(s)
Group 4A: carbon(s), silicon(s)
Group 5A: nitrogen(g), phosphorus(s)
Group 6A: oxygen(g), sulfur(s), selenium(s)
Group 7A: fluorine(g), chlorine(g), bromine(l), iodine(s), astatine(s)
Group 8A: helium(g), neon(g), argon(g), krypton(g), xenon(g), radon(g)

14. Of the 112 elements, only about 20 are essential to life. See the current perspective on page 92 of the text.

Study Questions and Problems

The one quality which sets one man apart from another—the key which lifts one to every aspiration while others are caught up in the mire of mediocrity—is not talent, formal education, nor brightness —it is self-discipline. With self-discipline, all things are possible. Without it, even the simplest goal can seem like the impossible dream.

Theodore Roosevelt
(1858-1919)

1. Explain, at an atomic or molecular level, what happens when
   a. water freezes to form ice
   b. copper and tin combine to form bronze
   c. rainwater evaporates from the pavement

2. Two compounds containing only copper and oxygen have 79.89% and 88.82% copper respectively. Using these data, demonstrate Dalton's law of multiple proportions.

3. Which of the following atoms are isotopes of the same element? Identify the elements of these isotopes and describe the number of protons and neutrons in the nucleus of them all.

$$^{15}_{7}X \quad ^{12}_{6}X \quad ^{13}_{7}X \quad ^{18}_{8}X \quad ^{14}_{7}X \quad ^{14}_{6}X \quad ^{16}_{8}X \quad ^{13}_{6}X \quad ^{17}_{8}X$$

4. There are three naturally occurring isotopes of neon:

   neon–20   mass 19.9924 amu   abundance 90.84%
   neon–21   mass 20.9940 amu   abundance 0.260%
   neon–22   mass 21.9914 amu   abundance 8.90%

   a. Without calculation, what is the approximate atomic mass of neon?
   b. Calculate the actual atomic mass.

A small quantity (0.005%) of uranium–234 also occurs in nature.

5. Uranium has an atomic mass equal to 238.0289. It consists of two isotopes: uranium–235 with an isotopic mass of 235.044 amu and uranium–238 with an isotopic mass of 238.051. Calculate the % abundance of the uranium–235 isotope.

6. From amongst the elements sodium, chlorine, nickel, argon, calcium, uranium, and oxygen, select the alkali metal, the alkaline earth metal, the transition metal, the actinide, the halogen, the noble gas, and the chalcogen (Group 6A).

7. To illustrate Robert Millikan's determination of the charge on an electron, suppose that you were given the task of determining the mass of a single jelly bean given the following experimental data:

   Various scoops of jelly beans were weighed and the following masses determined. The number of jelly beans in each scoop was not known.

   Masses (in grams) of ten different scoops:
   4.96, 8.68, 13.64, 7.44, 21.08, 16.12, 9.92, 19.84, 6.20, 12.40.

8. Reorder this list to match the name of the scientist with his or her contribution to our understanding of the nature of matter:

Joseph Thompson	developed the idea of the atomic nature of matter
James Chadwick	established the law of conservation of matter
Robert Millikan	characterized positive and negative electrical charges
Henry Moseley	suggested that atoms could disintegrate
Michael Faraday	experimented with electrolysis
Dmitri Mendeleev	proved the existence of the electron
John Dalton	developed the idea of a nuclear atom
Henri Becquerel	discovered the neutron
Democritus	developed the first periodic table of elements
Joseph Proust	showed that periodicity depended upon atomic number
Antoine Lavoisier	formulated the law of constant composition
Ernest Rutherford	determined the charge on a single electron
Marie Curie	revived the atomic theory
Benjamin Franklin	discovered radioactivity

9. Identify the following elements:
   a. The most abundant metal in the earth's crust.
   b. Combined with chlorine, it produces a compound essential to life.
   c. A metal that occurs in vast limestone deposits and combines with oxygen to form an oxide with a formula MO.
   d. The transition element at the center of hemoglobin.
   e. Used in smoke detectors and named for the United States.
   f. A component of washing powder mined in Death Valley.
   g. The basis for the compounds that make up all living things.
   h. Primary constituent of pencil lead.
   i. The last element in the Periodic Table that is not radioactive.
   j. Exists as X_4 molecules.
   k. The element named after the sun, where it was first detected.

10. Imagine yourself living in the time of John Dalton. You have an unknown element that combines with sulfur to produce two different compounds. One of the compounds contains 47.50% sulfur by mass and the second compound contains 31.15% sulfur by mass. What atomic mass would you assign to the unknown element?

Answers to Study Questions and Problems

Energy is removed to cause the water to freeze.

1.   a.   In the liquid state the water molecules are in constant motion and are not restricted to specific locations. The molecules collide and interact with each other constantly. When the water freezes to form ice, the water molecules become locked in place in a regular crystalline lattice. Their translational motion ceases. The water molecules are held together by a relatively strong intermolecular attraction.

These intermolecular forces of attraction are called hydrogen bonds—see page 188 in this study guide.

b.   A solution (alloy) of copper and tin is a mixture at the atomic level. No amount of optical magnification can reveal the different metals in the alloy. The mixture is a random arrangement of copper and tin atoms.

c.   In a liquid there is a distribution of molecular energies—some molecules moving quite fast with high kinetic energies and others moving slowly with low kinetic energies. Those molecules with high energies may have enough energy to overcome the intermolecular attraction and escape into the vapor state. They evaporate. As the rainwater absorbs more energy, so more molecules are able to escape. Eventually all the rainwater evaporates from the pavement.

2.   Compound A: 79.89% copper;   oxygen = 20.11% by mass
Compound B: 88.82% copper;   oxygen = 11.18% by mass

Compound A, assume you have a 100 amu sample:
atomic mass of copper = 63.546 amu;   79.89/63.546 = 1.257 atoms
atomic mass of oxygen = 15.9994 amu;   20.11/15.9994 = 1.257 atoms
a ratio of one copper atom to every oxygen atom.

Compound B, assume you have a 100 amu sample:
atomic mass of copper = 63.546 amu;   88.82/63.546 = 1.398 atoms
atomic mass of oxygen = 15.9994 amu;   11.18/15.9994 = 0.699 atoms
a ratio of two copper atoms to every oxygen atom.

These ratios demonstrate Dalton's law of multiple proportions: the ratio 1/1 and the ratio 2/1 are themselves in the ratio (proportion) 1 to 2.

3. $^{12}_{6}X$ $^{13}_{6}X$ $^{14}_{6}X$ —   all have an atomic number = 6 — carbon, with 6, 7, and 8 neutrons respectively.

$^{13}_{7}X$ $^{14}_{7}X$ $^{15}_{7}X$ —   all have an atomic number = 7 — nitrogen, with 6, 7, and 8 neutrons respectively.

$^{16}_{8}X$ $^{17}_{8}X$ $^{18}_{8}X$ —   all have an atomic number = 8 — oxygen, with 8, 9, and 10 neutrons respectively.

4. There are three naturally occurring isotopes of neon:

    neon–20   mass 19.9924 amu   abundance 90.84%
    neon–21   mass 20.9940 amu   abundance 0.260%
    neon–22   mass 21.9914 amu   abundance 8.90%

    a. The most abundant isotope is neon–20 (90.84%) so the atomic mass
        will be near to the mass of this isotope, i.e. about 20 amu.

    b. The actual atomic mass is a weighted average:
        [90.84 (19.9924) + 0.260 (20.9940) + 8.90 (21.9914)]/100
        = 20.17 amu.

        Divide by 100 because the abundances are expressed as percentages not fractions.

5. The atomic mass (238.0289) is a weighted average of the two isotopes:
    uranium–235 with an isotopic mass of 235.044 amu and uranium–238
    with an isotopic mass of 238.051. Let x = the fractional abundance of the
    uranium–235 isotope.

    Uranium–235 is the isotope used in nuclear fission reactors. It must be separated from the uranium–238. See Chapter 24.

    [235.044x + 238.051(1–x)] = 238.0289
    235.044x + 238.051 – 238.051x = 238.0289   therefore x = 0.0073
    % abundance of uranium–235 = 0.73%

6. sodium   alkali metal
    chlorine   halogen
    nickel   transition element
    argon   noble gas
    calcium   alkaline earth metal
    uranium   actinide
    oxygen   chalcogen

7. The difference in mass between two scoops, and the masses of the scoops
    themselves, must be a whole number multiple of the mass of one bean.
    In the data provided, the smallest difference is 1.24g (e.g. 8.68–7.44 or
    21.08–19.84). If this was indeed the mass of one bean, the numbers of
    beans in each scoop would be: 4, 7, 11, 6, 17, 13, 8, 16, 5, and 10.

    But perhaps 1.24g could be the mass of 2 beans, or 3 beans? Unlikely—
    the chances of random scoops all being a multiple of 2 (or 3) beans is
    remote.

8. John Dalton   revived the atomic theory
    Democritus   developed the idea of the atomic nature of matter
    Antoine Lavoisier   established the law of conservation of matter
    Joseph Proust   formulated the law of constant composition
    Benjamin Franklin   characterized positive and negative electrical charges
    Henri Becquerel   discovered radioactivity
    Marie Curie   suggested that atoms could disintegrate
    Michael Faraday   experimented with electrolysis

Joseph Thompson   proved the existence of the electron
Robert Millikan   determined the charge on a single electron
Ernest Rutherford developed the idea of a nuclear atom
James Chadwick   discovered the neutron
Dmitri Mendeleev   developed the first periodic table of elements
Henry Moseley   showed that periodicity depended upon atomic
                   number

9.   a.   aluminum Al
      b.   sodium Na
      c.   calcium Ca
      d.   iron Fe
      e.   americium Am
      f.   boron B
      g.   carbon C
      h.   carbon (graphite) C
      i.   bismuth Bi
      j.   phosphorus (white) P
      k.   helium He

10.   This problem is typical of the problems encountered by John Dalton and
      his contemporaries—it wasn't so easy to establish the table of atomic
      masses known today.

      If you know the formula of a compound, and its composition by mass,
      you can calculate the relative masses of the atoms. Similarly, if you know
      the atomic masses, and the composition by mass, you can establish the
      formula of the compound. But what do you do if you know neither the
      atomic masses nor the formula? You have to make some educated as-
      sumptions and see how these assumptions fit other data.

      It took 47 years (from 1811 to 1858) to establish a consistent table of
      atomic masses—and one of the problems was that Avogadro's hypothesis
      in 1811 that nitrogen and oxygen were diatomic molecules was largely
      ignored.

      In this problem the law of multiple proportions can be applied:

      Compound A: 47.50% sulfur;   unknown = 52.50% by mass
      Compound B: 31.15% sulfur;   unknown = 68.85% by mass

      Compound A, assume you have a 100 amu sample:
            atomic mass of sulfur = 32.066 amu
            number of atoms = 47.50/32.066 = 1.481 atoms
            atomic mass of unknown = x amu
            number of atoms = 52.50/x atoms
            a ratio of one sulfur atom to 35.45/x unknown atoms.

It was Stanislao Cannizzaro in
1858 who solved the dilemma
by cleverly combining Dalton's
law and Avogadro's hypothesis.

Compound B, assume you have a 100 amu sample:
   atomic mass of sulfur = 32.066 amu
   number of atoms = 31.15/32.066 = 0.9714 atoms
   atomic mass of unknown = x amu
   number of atoms = 68.85/x atoms
   a ratio of one sulfur atom to 70.87/x unknown atoms.

According to Dalton's law of multiple proportions, the ratios 35.45/x and 70.87/x should be in a whole number ratio—they are, it's 1 to 2. They must also be whole numbers, or at least simple fractions.
The value of x, the atomic mass of the unknown element must be a simple fraction or multiple of 35.45 and 70.87. An obvious choice for x, the atomic mass of the unknown element, is 35.45 (chlorine).

The two compounds might then be A: SCl or S_2Cl_2, and B: SCl_2.

Challenge 1

This question was asked in *another* text:

Suppose the atomic mass unit had been defined as 1/10th of the mass of an atom of phosphorus. What would the atomic mass of carbon be on this scale?

Instead of 1/12th the mass of a carbon−12 atom.

This answer was provided in the solution manual:

Regardless of the starting definition of a mass scale , we should find that the relative masses remain constant from one scale to another. The following conversion factor therefore gives us the ratio of mass of a carbon atom to a phosphorus atom:

$$\frac{12.00 \text{ g C}}{30.97 \text{ g P}} = 0.3875 \frac{\text{g C}}{\text{g P}}$$

This conversion factor is correct; the mass of a carbon atom is indeed 0.3875 of the mass of a phosphorus atom.

If we start with 1/10th of the value assigned to phosphorus, namely 3.097 g, then the mass of carbon is obtained by using the above conversion factor:

$$3.097 \text{ g P} \times 0.3875 \frac{\text{g C}}{\text{g P}} = 1.20 \text{ g carbon}$$

Explain what is *wrong* with this answer. The answer to this and the challenges in other chapters are provided at the end of the study guide.

28 Elements

—a crossword puzzle

Across:

1. "Acid former" discovered by Joseph Priestley.
7. Halogen used in water purification.
8. Brimstone.
11. Produces iridescent colors.
12. Below 4 down, its lamps are yellow.
14. The milk of its hydroxide settles the stomach.
16. A colored sign and below it...
17. ...another noble gas.
19. A ferrous metal first cast as pigs.
22. Noble origin of the man of steel?
24. Its tincture is antiseptic.
25. But there's no W anywhere in the name!
26. The most abundant metal in the earth's crust.

Down:

2. The goal of the alchemists.
3. An alkaline earth metal, bachelor of arts?
4. Named for greek stone.
5. With or without the T, it's found where 23 down is.
6. Rutherford fired its nuclei at 2 down.
7. Basis of compounds in living things and below it...
9. ...a semiconductor from the valley?
10. From a mineral hauled by 20 mules.
13. A transition metal sometimes mistaken for 14 across.
15. For the site of Seaborg's laboratory.
17. Is it the first of the actinides?
18. Named for the old devil himself.
20. A penny from Cyprus...
21. ...with this it's bronze.
23. Source and symbol initially Ytterby in Sweden.

CHAPTER 3

Introduction

In this chapter chemical compounds and their properties are examined. There are two major classes of compounds: molecular and ionic—you will learn how to distinguish them, write formulas for them, and name them. The concept of the mole will be introduced and explained. Finally, calculations to determine the formula of a compound will be described.

Contents

3.1 Molecules and Compounds

Compounds are substances that can be decomposed into two or more different substances by chemical means. The **chemical formula** for a compound expresses the number, or relative number, of atoms of each element present in the compound. For example, the formula for a molecule of sucrose is $C_{12}H_{22}O_{11}$ indicating that one molecule of sucrose contains 12 carbon atoms, 22 hydrogen atoms, and 11 oxygen atoms. The formula for sodium chloride, NaCl, indicates that the number of sodium ions present equals the number of chloride ions.

As mentioned in Chapter 1, the properties of a compound are usually quite unlike the properties of its constituent elements. For example, aluminum, a silver colored metal, reacts with bromine, a red-orange liquid, to produce aluminum bromide, a white powder. The formula for aluminum bromide is Al_2Br_6.

There are different ways to write a chemical formula. An **empirical formula** indicates the simplest possible integer ratio of elements in a compound. For example, the empirical formula for aluminum bromide is $AlBr_3$—indicating that there are three bromine atoms for each aluminum atom. A **molecular formula** simply expresses the composition of the molecule—how many, and what, elements are present. A **structural formula** provides additional information about the way in which the atoms of the molecule are grouped together. For example, ethanol has the molecular formula C_2H_6O and a structural formula CH_3CH_2OH. The structural formula can be drawn in even greater detail showing how the atoms are connected together.

Equations for chemical reactions will be examined in Chapter 4.

Be able to write the molecular formula for a compound.

Empirical formulas will be discussed later in this chapter (section 3.7).

Two different substances may have the same molecular formula but will have different structural formulas.

See the drawings of structural formulas at the bottom of page 102 of the text.

3.2 Molecular Models

The purchase of a molecular model set is highly recommended.

Computer resources for molecular modelling are now readily available. See page 105 in the text.

Knowledge of the molecular structure of a substance is often essential in explaining its physical and chemical properties. It is, however, difficult to draw three-dimensional structures on paper even though certain conventions have been developed to help. Molecular models are able to give a better sense of three-dimensional structure.

3.3 Ions

There are two major classes of compounds: **molecular compounds** and **ionic compounds**. They are quite different. Molecular compounds, like ethanol (CH_3CH_2OH) and water (H_2O) mentioned earlier, consist of discrete groups of atoms bonded together at the particulate level. These groups of atoms act as a unit and have properties characteristic of the unit. Ionic compounds, as the name suggests, consist of ions. These ions may be single atoms (monatomic) or groups of atoms (polyatomic) with negative or positive charges.

There is no such thing as an ionic molecule.

Metal atoms generally lose electrons and form positive ions (**cations**). The positive charge on the ion depends upon the number of electrons lost. Nonmetals generally gain electrons and form negative ions (**anions**). The negative charge on the ion depends upon the number of electrons gained.

For example, a sodium atom, in virtually all its reactions, loses one electron. This loss results in an imbalance in the number of protons and electrons in the ion (11 protons but now only 10 electrons). The sodium ion therefore has a charge of +1. Magnesium, in Group 2A, loses two electrons and forms the Mg^{2+} ion. Transition metals characteristically lose a variable number of electrons to form different ions. For example, iron commonly forms Fe^{2+} or Fe^{3+} by the loss of 2 electrons or 3 electrons respectively.

Group 1A metals form +1 ions.
Group 2A metals form +2 ions.
Group 3A metals form +3 ions.

Group 5A elements from −3 ions.
Group 6A elements form −2 ions.
Group 7A elements form −1 ions.

It is important to become familiar with the ions formed by the elements in Figure 3.6 on page 107 of the text.

Nonmetals often form ions with a negative charge equal to the 8 minus the group number. For example, the halogens form −1 anions, oxygen and sulfur form −2 anions, nitrogen and phosphorus form −3 anions. Hydrogen either loses its electron to form the H^+ ion, or gains an additional electron to form the H^- ion.

The outermost shell of electrons in an atom is called the valence shell. The electrons in the valence shell are called the valence electrons. It is these electrons that determine the chemical properties of the element.

The loss or gain of electrons by the main group elements results in the number of electrons remaining on the ions being equal to the number of electrons on the nearest noble gas. For example, by gaining two electrons, oxygen attains the same number of electrons as neon. By losing two electrons, calcium is left with the same number of electrons as argon. Noble gas electron configurations are particularly stable.

Commit to memory the 21 polyatomic ions listed in Table 3.1 on page 110 of the text.

A **polyatomic ion** is a group of atoms bonded together, just as in a molecule, with the entire unit bearing a charge. A common positive polyatomic ion is the ammonium ion NH_4^+. A common negative polyatomic ion is the carbonate ion CO_3^{2-}.

3.4 Ionic Compounds

Ionic compounds are compounds in which the component particles are ions (monatomic or polyatomic). The net charge of the compound is neutral; the positive and negative charges must balance. For example, in the ionic compound aluminum nitrate, the charge on the aluminum ion is 3+ and the charge on the polyatomic nitrate ion is 1−. The formula requires three nitrate ions for every aluminum ion: $Al(NO_3)_3$. Note that the positive ion (the cation) is written first in the formula.

Ions are held together in ionic compound by the electrostatic force of attraction between opposite charges. The magnitude of the force depends upon the charges and the distance between them. The higher the charges, and the smaller the ions, the greater the force of attraction.

Note that the formula for the polyatomic ion is placed in parentheses when more than one occurs in the formula.

Note that the charge on the cation equals the subscript of the anion (and vice versa).

The smaller the ions, the smaller the distance between them, and the greater the force.

The strength of the force of attraction between the ions in an ionic compound determines its physical properties.

3.5 Names of Compounds

The name of an ionic compound is built from the names of the positive and negative ions in the compound. The positive ion, normally a metal ion, is simply the name of the metal. For example, Na^+ is called the sodium ion. If the metal can form more than one ion, the charge is indicated by a Roman numeral in parentheses following the name of the metal. For example, iron forms two common ions, Fe^{2+} and Fe^{3+} that are referred to as iron(II) and iron(III). There are some nonmetallic positive ions. The example you will encounter most often is the ammonium ion NH_4^+.

A monatomic negative ion is named by changing the ending of the name of the element to −ide. For example the anion of chlorine Cl is called the chloride ion Cl^-. Polyatomic negative ions have their own names. Some are listed in Table 3.1 on page 110 of the text. Many of the polyatomic ions contain oxygen and their names are derived systematically. Note the use of the endings −ite and −ate, and the prefixes hypo− and per−, explained on page 116 of the text.

In naming an ionic compound, the positive ion is named first, followed by the negative ion. The charge on the cation is specified only if more than one exists. For example, copper(I) bromide *vs.* copper(II) bromide. Likewise, prefixes denoting the stoichiometry are not necessary; $CaCl_2$ is calcium chloride, not calcium dichloride.

The binary molecules of the nonmetals are named following a different procedure. When the molecule contains hydrogen and an element from Groups 6A or 7A, the hydrogen is named first, followed by the other element with the ending -ide. For example, H_2S hydrogen sulfide, HF hydrogen fluoride. When the molecule contains hydrogen and an element from Groups 4A or 5A, the molecules have special names. For example, CH_4 methane, SiH_4 silane (note the ending −ane), NH_3 ammonia, PH_3 phosphine, AsH_3 arsine (note the ending −ine). The names of other binary compounds are more straightforward: The elements are named in order of increasing group number and the number of atoms is indicated with a prefix such as mono−, di−, tri−, tetra−, penta−, hexa−... For example: CO_2 carbon dioxide, SO_3 sulfur trioxide, N_2O_4 dinitrogen tetroxide, P_4O_{10} tetraphosphorus decoxide, SF_2 sulfur difluoride, CCl_4 carbon tetrachloride.

Do not confuse the ammonium ion NH_4^+ with the ammonia molecule NH_3.

A binary compound is a compound containing just two elements. It more often than not contains more than two atoms.

The hydrocarbons, compounds such as methane, will be discussed further in Chapter 11.

Note that the -a- preceding the -o- of oxide is often omitted to make pronunciation easier. For example, dinitrogen tetroxide instead of dinitrogen tetraoxide.

Note that the mono− prefix is omitted for the first element in a name.

3.6 Atoms, Molecules, and the Mole

Chemistry has three components
–the particulate (individual
atoms, molecules and ions,
–the macroscopic (the real
world that can be touched and
seen)
–the symbolic (the symbols,
formulas, and equations used to
represent substances and their
reactions).

Recall the definition of the
atomic mass unit.

More precisely, the number is
6.022136736 × 10^{23}.

Although it's simple enough to
determine the value of a dozen
by counting, it is impossible to
use a counting method for a
mole–it would take too long. A
reliable method for establishing
the value of Avogadro's number
involves determining the
density of a crystal, its molar
mass, and the interatomic
spacing by X-ray crystallography.

Understand the difference
between one mole and
Avogadro's Number.

One amu equals 1.661× 10^{-24}g.

You must be able to convert
between mass and moles with
ease; practice!

#moles = mass/molar mass

Check the Problem Solving Tips
and Ideas 3.3 on page 124 of
the text.

Atoms are extremely small—far too small to be seen. The number of atoms or molecules within a relatively small amount of matter, one or two grams, is unimaginably large. So, although chemists often think in terms of individual atoms and molecules, their experiments usually involve vastly larger numbers. The way in which the particulate world of individual atoms is related to the world we see is through the **mole**.

The mole is the SI unit for an **amount of substance**. It is defined as the amount of substance that contains as many elementary entities (atoms or molecules, for example) as there are atoms in exactly 12 grams of carbon–12.

Eggs are counted by the dozen, pencils by the gross, shoes by the pair, sheets of paper by the ream, and substances by the mole. What do these terms represent? One dozen eggs means 12 eggs; one gross of pencils means 144 pencils; one pair of shoes means 2 shoes; one ream of paper means 500 sheets; one mole of a substance means 6.022×10^{23} particles of that substance. There are two differences between terms such as dozen or score and the mole. The mole is a far larger number and it is not an integer. The number of particles in a mole is determined experimentally, and the actual number is given a name. It is called **Avogadro's Number**. The reason it has a name is that it's not so easy to say "six point zero two two one three six....times ten to the twenty third" as it is to say "twelve". However, the relationship between one dozen and twelve is exactly the same as the relationship between one mole and Avogadro's Number.

The actual value of Avogadro's Number depends upon the definition chosen for one mole. If a standard other than carbon–12 had been chosen, the value would be different.

The convenience of the mole as a unit results from the fact that the number of entities in one mole of anything is always the same—Avogadro's Number. (Similarly, the number of entities—eggs, donuts, cans of beer, etc.—in one dozen is always 12.) Thus, just as there are two hydrogen atoms and one oxygen atom in one molecule of water, so there are two moles of hydrogen atoms and one mole of oxygen atoms in one mole of water molecules.

The mass in grams of one mole (1 mol) of atoms of any element is called the **molar mass** of that element. It is numerically equal to the atomic mass in amu. For example, the atomic mass of sulfur is 32.07 amu. This is the average mass of one atom. The molar mass of sulfur is 32.07 grams. This is the mass of one mole of sulfur atoms.

The mole is the cornerstone of quantitative chemistry. The number of moles of a substance within a sample is the mass of the sample divided by the molar mass of the substance. For example, an aluminum can weighs 9.0 grams. How many moles of aluminum are there in the can? The molar mass of aluminum is 26.98 g/mol. Dividing the mass (9.0 grams) by the molar mass (26.98 g/mol) yields 0.33 mol. Notice how the units cancel to produce the correct answer.

The mole applies to compounds just as it does to elements. The **molar mass of a compound** is the sum of the molar masses of the constituent elements. For example, ammonia has the formula NH_3. This means that in one molecule of ammonia there is one nitrogen atom and three hydrogen atoms. It also means that in one mole of ammonia molecules, there is one mole of nitrogen atoms and three moles of hydrogen atoms. The molar mass of ammonia is 3.02 + 14.01 = 17.03 g/mol. Remembering the relationship between one mole and Avogadro's Number, this means that in 17.03 grams (one mole) of ammonia, there are 6.022×10^{23} molecules of ammonia composed of 6.022×10^{23} atoms of nitrogen and $3 \times 6.022 \times 10^{23}$ atoms of hydrogen.

The molar mass of a molecular compound is sometimes referred to as the **molecular mass** or **molecular weight** of the compound.

Since ionic compounds do not exist as molecules, they cannot have a molecular mass. Instead, the simplest (empirical) formula is written and the molar mass calculated for that formula. This is referred to as the **formula mass** or **formula weight**.

An empirical formula indicates the simplest possible integer ratio of elements in a compound.

3.7 Describing Compound Formulas

According to the Law of Constant Composition, a pure compound always contains the same elements in the same proportion by mass. This is because the compound has the fixed composition described by its formula. How can the formula be obtained from data describing the mass composition, and *vice versa*?

You should be able to convert mass data (ratio of elements present by mass) into mole data (ratio of elements present by mole—the formula).

The **percent composition** is the way by which the mass composition of a substance is often stated. It is the fraction of one hundred grams that is a particular element.

For example, the mass percent of nitrogen in ammonia (NH_3) is the molar mass of nitrogen (14.007g) divided by the molar mass of ammonia (17.030g) multiplied by 100. This equals 82.25%.

If the percent composition data for a substance is known, then the formula for the substance can be calculated. For example, suppose a compound of copper and chlorine is analyzed and found to contain 47.26% copper by mass. In other words, 100 grams of the substance contains 47.26 grams of copper. The remainder is chlorine (52.74 grams).

Conversion from a mass ratio to a mole ratio requires dividing by the molar mass:

Remember:

#moles = mass/molar mass

	47.26 g Cu	52.74 g Cl
mass ratio:		
divide by molar mass:	/63.546	/35.453
	= 0.7437	= 1.488
convert to integers by dividing by the smaller:	= 1	= 2

therefore $CuCl_2$

Read the Problem Solving Tips and Ideas 3.4 & 3.5 on finding empirical and molecular formulas on pages 131 and 134 of the text.

Only the ratio of elements in the compound can be established. This is called the **empirical formula**. To determine the molecular formula for a molecular compound, the molar mass must be obtained by further experiment.

3.8 Hydrated Compounds.

Very often ionic salts crystallize with molecules of water included in the crystal lattice. The water molecules are often associated with ions in the crystal. These compounds are called **hydrated compounds**. When a hydrated salt is heated, it loses the water to form the **anhydrous** salt.

Anhydrous means "without water".

Review Questions

1. List the boldface words or phrases in the preceding pages and define or explain each of them.

2. What is the difference between molecular and structural formulas?

3. What is the difference between a molecular compound and an ionic compound?

4. Draw representations of water at the particulate, the macroscopic, and the symbolic levels.

5. How can you determine whether a compound is ionic or molecular?

6. How can you predict the charge on the monatomic ion of a main group element?

7. What is a polyatomic ion? Is the central atom in a polyatomic ion always a nonmetal?

8. Explain the naming procedure for the polyatomic oxoanions.

9. Summarize what you know about the procedure for naming binary compounds of the nonmetals.

10. Why is the formula of methane written CH_4, and phosphine written PH_3, whereas the formula of hydrogen sulfide is written H_2S, and hydrogen chloride written HCl? (i.e. why does the hydrogen follow the C or P, but precede the S or Cl?)

11. What is the difference between "one mole" and "Avogadro's Number"?

12. Why isn't Avogadro's number an integer, like 12 (dozen) or 20 (score)?

13. Compare the definition of an atomic mass unit and the definition of one mole. What is the relationship between them?

14. What does the formula H_2O represent? What does the formula NaCl represent?

15. How can mass composition data be converted into a mole ratio? In other words, how can the formula of a compound be established from knowledge of the percent masses of each element present in the compound?

16. How does an empirical formula differ from a molecular formula?

17. What is a hydrated compound?

Answers to Review Questions

1. All the words and phrases that are boldface in the text are defined in the text. You should have a clear understanding of all these terms. Write definitions or explanations in your own words.

2. The molecular formula simply lists how many atoms of which elements are present in one molecule of the substance. The structural formula also indicates, sometimes in considerable detail, the arrangement of the atoms in the molecule. An example is the molecular formula for ethanol (C_2H_6O) and its structural formula (CH_3CH_2OH).

3. There are two major classes of compounds: molecular compounds and ionic compounds. They are quite different. Molecular compounds, like water (H_2O) consist of discrete groups of atoms bonded together at the particulate level. Ionic compounds, as the name suggests, consist of monatomic or polyatomic ions arrayed in an almost infinite three-dimensional lattice.

4. Particulate: macroscopic: 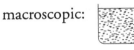   symbolic: H_2O

5. In order to recognize ionic compounds, it is necessary to know the formulas and names of the common ions. In general, an ionic compound will consist of a metal cation with either a monatomic nonmetal anion or a polyatomic anion. Almost always, if the compound contains no metal then it is molecular.

   See the Problem Solving Tips and Ideas 3.2 on page 113 of the text.

   Note that some cations are non–metallic.

6. The positive charge on a metal ion from the left of the Periodic Table is equal to the group number. For example, 1+ for sodium in group 1A. The negative charge on the nonmetals from the right side of the Periodic Table is equal to the group number – 8. For example, 2– for sulfur from group 6A.

7. A polyatomic ion is a molecule with a charge; it is an ion with two or more atoms. The central atom is very often a nonmetal, but need not be. For example, permanganate MnO_4^-, dichromate $Cr_2O_7^{2-}$.

8. Although the names of some polyatomic ions just have to be memorized (for example, dichromate $Cr_2O_7^{2-}$, acetate $CH_3CO_2^-$, cyanide CN^-), it is possible to derive the names of many other oxygen containing anions logically. There are two suffixes, –ate and –ite. The oxoanion with the greater number of oxygen atoms has the suffix –ate. There are two prefixes per– and hypo– that are used when there are more than just two oxoanions. The stem of the name is derived from the central atom, for example: nitro– for nitrogen, sulf– for sulfur, phosph– for phosphorus, etc.

   See the discussion on page 116 of the text, and Table 3.1 on page 110.

9. The elements are named in order of increasing group number and the number of atoms is indicated with a prefix such as di–, tri–, tetra–, penta– hexa–... Examples are: SO_2 sulfur dioxide, SiO_2 silicon dioxide, N_2O dinitrogen monoxide, SF_4 sulfur tetrafluoride, CCl_4 carbon tetrachloride. When the binary compound contains hydrogen and an element from Groups 6A or 7A, the hydrogen is named first, followed by the other element with the ending –ide. For example, H_2S hydrogen sulfide, HBr hydrogen bromide. When the molecule contains hydrogen and an element from Groups 4A or 5A, the molecules have other names. For example, CH_4 methane, SiH_4 silane (note the ending –ane), PH_3 phosphine, AsH_3 arsine (note the ending –ine).

10. The position of hydrogen in the Periodic Table is a problem. Very often it will be placed at the top of Group 1A, or at the top of Group 7A, or as in the front of the text, in both places. In some respects, hydrogen behaves as if it belongs somewhere above carbon. The convention is that binary compounds are named with the element further to the lower left in the Periodic Table first, followed by the element further to the upper right. The same applies to the way the formula is written. For example: LiH, CaH_2, BH_3, but H_2O, HCl. Where does the change-over take place? Between Groups 4A and 5A. So CH_4 and SiH_4 but NH_3 and PH_3.

11. A mole is the SI unit for an amount, whereas Avogadro's number is a number of entities within that amount. For example, Avogadro's number is always the same (6.022×10^{23}) whereas one mole of carbon atoms is 12.011 grams of carbon (which is the amount of carbon that contains Avogadro's number of atoms) but one mole of sulfur atoms is 32.066 grams and one mole of helium atoms is 4.0026 grams. The masses differ but the number stays the same.

12. Avogadro's number is so large that it cannot be determined by counting. It would take far too long. If it could be determined by counting then it would indeed be useful to set it equal to a simple integer (for example 1×10^{23}). The basis for Avogadro's number is arbitrary. It is currently equal to the number of atoms of carbon in exactly 12 grams of carbon–12.

13. The atomic mass unit is defined as 1/12th the mass of a single atom of carbon–12. The mole is defined as the amount of substance that contains the same number of particles as there are atoms in exactly 12 grams of carbon–12. The amu is the mass unit used at the particulate level; the gram is the mass unit used at the macroscopic (real world) level. The relationship between them is Avogadro's number. For example:
One carbon–12 atom has a mass of 12 amu (exactly).
One mole of carbon–12 atoms has a mass of 12 grams (exactly).
One gram = 6.022×10^{23} amu.

14. The formula H_2O represents either one molecule of water, or one mole of water molecules. The context determines which. The formula NaCl represents the composition of the salt sodium chloride. There is no molecule of sodium chloride; there is one sodium ion for each chloride ion however. NaCl represents one formula unit or one mole of formula units.

15. The relationship between the mass of a substance and the number of moles of the substance is fundamental. The two are related by the molar mass of the substance:

    Number of moles = mass of the substance / molar mass of the substance.

    If the mass composition data (mass ratio) is known, then dividing by the molar masses of the constituent elements will produce the mole ratio. The mole ratio is the formula of the substance.

16. An empirical formula indicates the simplest possible integer ratio of elements in a compound. The molecular formula indicates how many atoms of which elements there are in one molecule of the compound. For example, the molecular formula of ethylene is C_2H_4, but the empirical formula is CH_2—a ratio of 1 atom of C to 2 atoms of H.

17. Hydrated compounds are ionic salts with molecules of water included in the crystal lattice. When a hydrated salt is heated, it loses the water to form the anhydrous salt.

Study Questions and Problems

1. a. The structural formula for acetic acid is CH_3CO_2H. What is its empirical formula; what is its molecular formula?
    b. The molecular formula of acrylonitrile is C_3H_3N. Look up in the text, and draw, its structural formula.
    c. The molecular formula of aspartame (nutrasweet) is $C_{14}H_{18}O_5N_2$. Look up in the text, and draw, its structural formula.

2. The formulas for ethanol and ammonium nitrate are C_2H_5OH and NH_4NO_3. In what respects are these formulas and compounds different?

3. The molecular formula for both butanol and diethylether is $C_4H_{10}O$. Write structural formulas for both and show how they are different. Are any other structures possible?

4. Name the polyatomic ions:
    $CH_3CO_2^-$ $\qquad$ HCO_3^-
    $H_2PO_4^-$ $\qquad$ $Cr_2O_7^{2-}$
    SO_3^{2-} $\qquad$ ClO_4^-

Matter exists in the form of atoms and combinations of atoms. Void and matter are mutually exclusive; matter is solid and eternal.

Lucretius
(c.94-55BC)

5. What are the formulas of the polyatomic ions:
   phosphate
   sulfate
   bisulfite
   nitrite
   cyanide
   chlorite

6. Write the ions present in the following salts and predict their formulas:
   potassium bromide
   calcium carbonate
   magnesium iodide
   lithium oxide
   aluminum sulfate
   ammonium chlorate
   beryllium phosphate

7. Name the following ionic salts:
   $(NH_4)_2SO_4$
   $KHCO_3$
   $Ca(NO_3)_2$
   $Co_2(SO_4)_3$
   $NiSO_4$
   $AlPO_4$

8. Name the following binary compounds of the nonmetals:

CS_2	$SiCl_4$
SF_6	GeH_4
IF_5	P_4O_{10}
N_2H_4	S_4N_4
PCl_5	OF_2
Cl_2O_7	IF_7

9. What are the formulas for the following binary compounds?
   silicon dioxide
   boron trifluoride
   xenon tetroxide
   dinitrogen pentoxide
   bromine trifluoride
   carbon tetrachloride
   phosphine
   silicon carbide
   phosphorus tribromide
   disulfur dichloride
   hydrogen selenide

10. a. How many moles are present in 128 grams of sulfur dioxide?

You must be able to convert between mass and moles with ease; practice!

#moles = mass/molar mass

   b. What is the mass of 3 moles of oxygen molecules?
   c. If 5 moles of a metallic element has a mass of 200 grams, which element is it?
   d. What is the molar mass of methane CH_4?
   e. What is the mass of 9 moles of fluorine molecules?
   f. 102 grams of a gas contains 6 moles. What is its molar mass?
   g. How many grams are there in one mole of benzene C_6H_6?
   h. How many moles of nitrogen atoms are there in 6 moles of TNT (trinitrotoluene $CH_3C_6H_2(NO_2)_3$)?
   i. What is the molar mass of TNT?

11. What is the percent by mass of nitrogen in ammonium nitrate?

12. The hydrocarbons ethylene (molar mass 28 g/mol), cyclobutane (molar mass 56 g/mol), pentene (molar mass 70 g/mol), and cyclohexane (molar mass 84 g/mol), all have the same empirical formula. What is it? Write the molecular formulas for these four compounds.

   A hydrocarbon is a binary compound of carbon and hydrogen.

13. A compound was analyzed and found to contain 76.57% carbon, 6.43% hydrogen, and 17.00% oxygen by mass. Calculate the empirical formula of the compound. If the molar mass of the compound is 94.11 g/mol, what is the molecular formula of the compound?

14. A compound was analyzed and found to contain 53.30% carbon, 11.19% hydrogen, and 35.51% oxygen by mass. Calculate the empirical formula of the compound. If the molar mass of the compound is 90.12 g/mol, what is the molecular formula of the compound?

15. A 15.67 g sample of a hydrate of magnesium carbonate was carefully heated, without decomposing the carbonate, to drive off the water. The mass was reduced to 7.58 g. What is the formula of the hydrate?

16. Anhydrous lithium perchlorate (4.78 g) was dissolved in water and recrystallized. Care was taken to isolate all the lithium perchlorate as its hydrate. The mass of the hydrated salt obtained was 7.21 g. What hydrate is it?

Answers to Study Questions and Problems

1. a. The empirical formula for acetic acid is CH_2O; its molecular formula is $C_2H_4O_2$.

   b. The structural formula of acrylonitrile is $CH_2=CHCN$.

   c. The structural formula of aspartame (nutrasweet) is $NH_2CH(CH_2CO_2H)CONHCH(CH_2C_6H_5)CO_2CH_3$ or

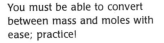

The formula can also represent one *mole* of ethanol molecules.

2. The formula for ethanol, C_2H_5OH, is the formula for one molecule of ethanol. One molecule of ethanol contains two carbon atoms, six hydrogen atoms, and one oxygen atom. Ammonium nitrate, NH_4NO_3, on the other hand, is an ionic compound. It consists of the ions NH_4^+ and NO_3^- in the 1:1 ratio indicated by the formula. There is no such entity as an ammonium nitrate molecule.

3. Butanol is an alcohol; it contains the –OH functional group. Diethylether is an ether; it contains the –O– functional group.
Butanol is $CH_3CH_2CH_2CH_2OH$.
Diethylether is $CH_3CH_2OCH_2CH_3$.
Both have the molecular formula $C_4H_{10}O$.

See Problem 3 on page 164 of this study guide.

There are two possible structures for the propyl group $CH_3CH_2CH_2-$ and $(CH_3)_2CH-$.

There are other structural isomers of butanol: one, for example, is $CH_3CH_2CH(OH)CH_3$. And another ether with the same molecular formula is methylpropylether $CH_3CH_2CH_2OCH_3$, and this ether itself has two structural possibilities.

4.
$CH_3CO_2^-$	acetate
HCO_3^-	bicarbonate or hydrogen carbonate
$H_2PO_4^-$	dihydrogen phosphate
$Cr_2O_7^{2-}$	dichromate
SO_3^{2-}	sulfite
ClO_4^-	perchlorate

5.
phosphate	PO_4^{3-}
sulfate	SO_4^{2-}
bisulfite	HSO_3^-
nitrite	NO_2^-
cyanide	CN^-
chlorite	ClO_2^-

The ratio is such that the positive and negative charges balance. For example, for beryllium phosphate: 3 × 2+ (for the cation) = 2 × 3– (for the anion).

6.
potassium bromide	K^+	Br^-	ratio 1:1	KBr
calcium carbonate	Ca^{2+}	CO_3^{2-}	1:1	$CaCO_3$
magnesium iodide	Mg^{2+}	I^-	1:2	MgI_2
lithium oxide	Li^+	O^{2-}	2:1	Li_2O
aluminum sulfate	Al^{3+}	SO_4^{2-}	2:3	$Al_2(SO_4)_3$
ammonium chlorate	NH_4^+	ClO_3^-	1:1	NH_4ClO_3
beryllium phosphate	Be^{2+}	PO_4^{3-}	3:2	$Be_3(PO_4)_2$

7.
$(NH_4)_2SO_4$	ammonium sulfate
$KHCO_3$	potassium bicarbonate
$Ca(NO_3)_2$	calcium nitrate
$Co_2(SO_4)_3$	cobalt(III) sulfate
$NiSO_4$	nickel(II) sulfate
$AlPO_4$	aluminum phosphate

8. CS_2 carbon disulfide $SiCl_4$ silicon tetrachloride
     SF_6 sulfur hexafluoride GeH_4 germane
     IF_5 iodine pentafluoride P_4O_{10} tetraphosphorus decoxide
     N_2H_4 hydrazine S_4N_4 tetrasulfur tetranitride
     PCl_5 phosphorus pentachloride OF_2 oxygen difluoride
     Cl_2O_7 dichlorine heptoxide IF_7 iodine heptafluoride

9.   silicon dioxide SiO_2
     boron trifluoride BF_3
     xenon tetroxide XeO_4
     dinitrogen pentoxide N_2O_5
     bromine trifluoride BrF_3
     carbon tetrachloride CCl_4
     phosphine PH_3
     silicon carbide SiC (note the absence of the prefix mono–)
     phosphorus tribromide PBr_3
     disulfur dichloride S_2Cl_2 (sometimes called sulfur monochloride)
     hydrogen selenide H_2Se (note the absence of the prefix di-)

These are all equivalent expressions:

#moles = mass/molar mass

10. a.   The molar mass of sulfur dioxide SO_2 is 64.06 g/mol.
          In 128 g there must be 128/64.06 moles = 2.00 moles.

     b.   The molar mass of an oxygen molecule O_2 is 2 × 16.00 = 32 g/mol.
          (This is the mass of one mole of oxygen molecules.)
          The mass of 3 moles must be 3 × 32.00 = 96 grams.

mass = #moles × molar mass

     c.   The mass is 200 grams; the number of moles is 5.
          So the molar mass is 200 g / 5 mol = 40 g/mol (calcium).

molar mass = mass/# moles

     d.   Methane (CH_4) has a molar mass =12.011 + 4×1.008 = 16.04 g/mol.

     e.   Fluorine is diatomic; its molar mass is 2 × 19.00 = 38 g/mol.
          The mass of 9 moles of fluorine = 9 moles × 38 g/mol = 342 grams.

     f.   The mass is 102 grams; the number of moles is 6.
          So the molar mass is 102 g / 6 mol = 17 g/mol (ammonia NH_3).

     g.   One mole of benzene C_6H_6 has a molar mass = 6 × 12.011 + 6 × 1.008 = 78.11 g/mol.

     h.   The number of nitrogen atoms in one molecule of trinitrotoluene $CH_3C_6H_2(NO_2)_3$ is 3, one in each NO_2 group. In 6 moles of TNT there must be 6 × 3 = 18 moles of nitrogen atoms.

i.   The molar mass of TNT ($CH_3C_6H_2(NO_2)_3$):
     carbon: $7 \times 12.011 = 84.08$
     hydrogen: $5 \times 1.008 = 5.04$
     nitrogen: $3 \times 14.007 = 42.02$
     oxygen: $6 \times 16.00 = 96.00$
     total: 227.14 g/mol

11. Ammonium nitrate, NH_4NO_3, has a molar mass of 80.04 g/mol. There are 2 nitrogen atoms for every formula unit of ammonium nitrate; so the mass of nitrogen in every mole of ammonium nitrate is 2×14.007 g. The percent mass of nitrogen is (28.01 g / 80.04 g) $\times$ 100% = 35.00%.

12. The highest common factor is 14, which corresponds to one carbon atom and two hydrogen atoms. The empirical formula of all these hydrocarbons is therefore CH_2. The hydrocarbons are:

All of these hydrocarbons contain a ratio of two hydrogen atoms for every carbon atom.

ethylene	molar mass 28 g/mol	molecular formula C_2H_4
cyclobutane	molar mass 56 g/mol	molecular formula C_4H_8
pentene	molar mass 70 g/mol	molecular formula C_5H_{10}
cyclohexane	molar mass 84 g/mol	molecular formula C_6H_{12}

13. The data provided is mass composition data. A formula for a compound is a mole ratio—the relative numbers of the various constituent elements. To convert from mass to moles divide by the molar mass. So the mass percent for each element is divided by its molar mass as follows:

Setting up the problem in a table like this is highly recommended.

	carbon	hydrogen	oxygen
mass % ratio:	76.57%	6.43%	17.00%
divide by molar mass:	/12.011	/1.008	/16.00
	= 6.37	= 6.38	= 1.06
convert to integers by dividing by the smallest:	= 6	= 6	= 1

The empirical formula is C_6H_6O
The empirical mass (the mass of this formula) is 94.11 g
This equals the molar mass, so this empirical formula is also the molecular formula.

14. Again, to convert from mass to moles divide by the molar mass. So divide the mass percent for each element by its molar mass:

	carbon	hydrogen	oxygen
mass % ratio:	53.30%	11.19%	35.51%
divide by molar mass:	/12.011	/1.008	/16.00
	= 4.44	= 11.1	= 2.22
convert to integers by dividing by the smallest:	= 2	= 5	= 1

The empirical formula is C_2H_5O
The empirical mass is 45.06 g
This is only one-half of the molar mass (90.12 g/mol), so this empirical formula is one-half the molecular formula.
The molecular formula is $C_4H_{10}O_2$.

15. The molar mass of magnesium carbonate $MgCO_3$ is 84.31 g/mol. Set up the problem as in the previous two questions. Again, to convert from mass to moles divide by the molar mass. The mass of water lost is the original mass (15.67 g) – the final mass (7.58 g) = 8.09 g.

	$MgCO_3$	H_2O
mass ratio:	7.58 g	8.09 g
divide by molar mass:	/84.31	/18.02
	= 0.0899	= 0.449
convert to integers by dividing by the smaller:	= 1	= 5

The formula of the hydrate is $MgCO_3 \cdot 5H_2O$.

16. The molar mass of lithium perchlorate $LiClO_4$ is 106.39 g/mol. Set up the problem as in the previous problem. To convert from mass to moles divide by the molar mass. The mass of water gained is 2.43 g.

	$LiClO_4$	H_2O
mass ratio:	4.78 g	2.43 g
divide by molar mass:	/106.39	/18.02
	= 0.0449	= 0.135
convert to integers by dividing by the smaller:	= 1	= 3

The formula of the hydrate is $LiClO_4 \cdot 3H_2O$.

Numbers

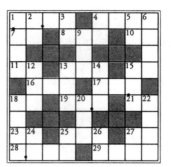

Across:

1.   Molar mass of copper.
4.   The year Antoine Lavoisier lost his head.
7   Neptune's number.
8.   The most recently detected element (Feb '96).
10. The exponent of Avogadro's number.
11. The number of inches in one foot...
13. ...and the number of sq. in in one sq.ft.
15. The molar mass of oxygen atoms.
16. One atmosphere pressure in torr.
17. 273°C in K.
18. The mass of the empirical formula for 9 down.
19. The volume of one mole of a gas at 0°C and 1 atm.
21. Niobe's number.
23. Three moles of carbon.
25. The ignition temperature of paper in °F.
27. The molar mass of methane CH_4.
28. The first four digits of Avogadro's number.
29. The year John Dalton announced his atomic theory.

Down:

1.   Lithium's mass.
2.   The atomic number of arsenic...
3.   ...and that of the element under it.
4.   The isotope of carbon upon which the amu is based.
5.   The atomic number of 20 down.
6.   The group numbers of silicon, aluminum, nitrogen, and sulfur, respectively.
9.   The mass of four moles of acetylene C_2H_2.
12. It's freezing in K.
13. A number for the prize-giver.
14. The number of grams in one pound.
15. The number of cubic cm in one cubic inch.
18. The mass of a proton divided by the mass of an electron.
20. The isotope of uranium used in nuclear reactors.
22. The year Linus Pauling won his second Nobel Prize.
24. The number of carbon atoms in a common fullerene.
25. The number of neutrons in 2 down's isotope–75.
26. The number of electrons in a sodium atom...
27. ...and after it has lost one to form the sodium ion.

CHAPTER 4

Introduction

Our objective in this chapter is to begin a study of reaction stoichiometry—the quantitative treatment of chemical reactions. Chemical equations are described—you will learn how to write them, balance them, and use them to determine the mass relationships in a chemical reaction. The concept of a limiting reactant and the calculation of the yield of a reaction are introduced.

Contents

4.1 Chemical Equations

A **chemical equation** is a representation of a chemical reaction. Symbols and formulas for the **reactants** (the substances that are going to react) are written on the left and symbols and formulas for the **products** (the substances that are produced) are written on the right. The arrow ($\rightarrow$) represents the reaction.

A **chemical reaction** is only a rearrangement of atoms—atoms are neither created nor destroyed. Therefore, in a balanced chemical equation, the number of atoms on both sides of the arrow must be the same.

This is the law of conservation of matter (Lavoisier) later incorporated into the atomic theory of matter (Dalton).

$$\textbf{Reactants} \quad \rightarrow \quad \textbf{Products}$$

In a chemical reaction, bonds between atoms are broken, new bonds between atoms are formed, and new substances are produced. The physical states of the reactants and products are sometimes denoted by *(s)* solid, *(l)* liquid, *(g)* gas, and *(aq)* solution. For example:

The symbol *(aq)* is used for substances in solution (see Chapter 5).

$$P_4(s) + 6Cl_2(g) \rightarrow 4PCl_3(l)$$

The numbers in front of the symbols and formulas in the equation are called **coefficients** and are required to balance the equation. They reflect the principle of conservation of matter: four phosphorus atoms in one molecule of P_4 on the left are balanced by the four phosphorus atoms in the four molecules of PCl_3 on the right.

The number of phosphorus atoms and the number of chlorine atoms on each side of the equation are the same.

The equation can be interpreted equally well in terms of moles. Thus one mole of P_4 molecules produces four moles of PCl_3 molecules. The ratio of one to four molecules, or one to four moles of molecules, in this equation is called the **mole ratio** or **stoichiometry** of the reaction.

A chemical equation doesn't indicate how long the reaction will take, or if energy is released or required, or even if the reaction will happen.

Learn how to balance chemical equations. Note that, by definition, an equation must be balanced—there really is no such thing as an unbalanced equation.

For a combustion reaction:

1:   Write the correct formulas
2:   Balance the C atoms
3:   Balance the H atoms
4:   Balance the O atoms

See the examples on pages 152 and 153 of the text.

4.2 Balancing Chemical Equations

Using chemical equations for quantitative calculations requires that the equations be balanced. This balancing can often be done by trial and error although a systematic approach is useful:

Step 1: Write the correct formulas for the reactants and products.
Step 2: Look for the most complicated looking formula and balance the atoms in that formula first.
Step 3: Balance atoms that occur only in one reactant and one product.
Step 4: Balance the remaining atoms.

Remember that you cannot change the formulas for the reactants and the products in an equation in order to balance the equation. Doing so would change the identity of the substances involved in the reaction.

4.3 Mass Relationships in Chemical Reactions: Stoichiometry

Given quantitative data in mass (grams), *always* convert this to moles for stoichiometric calculations.

A balanced chemical reaction illustrates the quantitative relationship between the reactants and products in a chemical reaction. The equation expresses this quantitative relationship in terms of **moles**—it expresses the mole ratio, or stoichiometry, required. There are four steps in a typical quantitative calculation for a chemical reaction:

Step 1: Write the balanced chemical equation.
Step 2: Convert mass data to moles.
Step 3: Use the stoichiometric relationship illustrated by the equation.
Step 4: If required, convert mole data back to mass data.

You must be able to convert between mass and moles!

#moles = mass/molar mass

Any stoichiometric relationship of the chemical reaction is represented by a **stoichiometric factor** or mole ratio. This is an application of the dimensional analysis discussed in Chapter 1. For example, in the equation for the combustion of benzene:

$$2C_6H_6(l) \; + \; 15O_2(g) \; \rightarrow \; 12CO_2(g) \; + \; 6H_2O(l)$$

Read the Problem-Solving Tips and Ideas 4.1 on page 157 of the text.

The stoichiometric coefficients in the equation indicate that 2 moles of benzene require 15 moles of oxygen to produce 12 moles of carbon dioxide and 6 moles of water. What mass of water is produced when 39 grams of benzene is burned?

Step 1: The balanced chemical equation is already written.
Step 2: 39 grams of benzene is 0.50 mole of benzene.
Step 3: The equation provides the stoichiometric relationship between the moles of benzene burned and the moles of water produced:

The molar mass of benzene C_6H_6 is 78 g/mol.

39g / 78 g mol^{-1} = 0.50 mol.

The mole ratio, or stoichiometry, is $2C_6H_6$ to $6H_2O$. The stoichiometric factor is $(6H_2O/2C_6H_6)$. So,

$$0.5 \text{ mol } C_6H_6 \times (6 \text{ mol } H_2O/2 \text{ mol } C_6H_6) = 1.5 \text{ mol } H_2O$$

Notice how the units cancel to give moles of H_2O.

Step 4: 1.5 moles of water = 1.5 mol × 18 g/mol = 27 grams.

4.4 Reactions in which one reactant is present in Limited Supply

In practice, reactants are rarely present in the correct stoichiometric ratio. This is often intentional; it can, for example, be done to ensure that the more expensive of two reactants is completely used. The reactant that is *not* in excess determines the amount of product that can be made and is, for that reason, called the **limiting reactant**. Some of the non-limiting reactant (the reactant in excess) will remain unused at the end of the reaction.

Suppose zinc metal is to be oxidized by hydrochloric acid according to the chemical equation:

$$Zn(s) + 2HCl(aq) \rightarrow ZnCl_2(aq) + H_2(g)$$

Suppose further that the mass of zinc is 6.54 grams and the mass of hydrochloric acid in the solution is 5.47 grams. What mass of hydrogen gas will be produced? Approach the problem as before:

Step 1: Write the balanced chemical equation—done.

Step 2: Convert mass data to moles:
6.54 g of Zn = 6.54g / 65.39g mol^{-1}= 0.100 mol
5.47 g of HCl = 5.47g / 36.46g mol^{-1} = 0.150 mol

Step 3: Determine the limiting reactant:
0.100 mol of Zn requires 0.200 mol of HCl; but there is only 0.150 mol of HCl. So HCl is the limiting reactant.

Step 4: Use the stoichiometric relationship illustrated by the equation to determine the amount of product that can be formed:
0.150 mol HCl × (1 mol H$_2$/2 mol HCl) = 0.075 mol H$_2$

Step 5: Change back to mass data:
0.075 mol H$_2$ = 0.075 mol × 2.016 g/mol = 0.15 grams.

4.5 Percent Yield

The maximum quantity of product that can be obtained in a chemical reaction is called the **theoretical yield**. In practice some waste occurs, other products may be formed, or the reaction may not even go to completion. The quantity of the desired product obtained is often less than the theoretical yield. The ratio of the **actual yield** to the theoretical yield multiplied by 100 is called the **percent yield**.

4.6 Chemical Equations and Chemical Analysis

Chemical analysis depends upon the idea that the quantity of product obtained in a reaction is related exactly to an unknown quantity of the reactant. This relationship is the stoichiometric factor derived from the stoichiometry of the appropriate balanced chemical equation.

For example, nickel is often determined by precipitating it as a very insoluble red salt of dimethylglyoxime. If the mass of the salt is determined, the

---

A useful analogy is the production of bicycles from frames and wheels. Each bicycle requires one frame and two wheels. How many bicycles can you produce from 7 frames and 10 wheels and which component is the limiting reactant?
You have enough frames for 7 bicycles, but only enough wheels for 5 bicycles; thus the wheels *limit* the number of bicycles that can be made. Two frames remain unused at the end.

The limiting reactant can often be established using common sense as in this example. The stoichiometry of the equation must be taken into account! Even though there are more moles of HCl than Zn, the HCl is the limiting reactant.

The limiting reactant can be established more formally by dividing the number of moles of each reactant present by its stoichiometric coefficient. The reactant with the lowest result is the limiting reactant.

For example, in this case, for Zn, 0.100 divided by 1 = 0.100; and for HCl, 0.150 divided by 2 = 0.075; so the HCl is the limiting reactant.

You should be able to calculate the % yield in a chemical reaction.

% yield = $\frac{\text{actual}}{\text{theoretical}}$ × 100%

See page 163 in the text.

See problem number 4.8 in this study guide.

An organic compound is a compound of carbon and hydrogen, often with oxygen and nitrogen, and sometimes with other elements such as the halogens, sulfur, and phosphorus.

A carbohydrate is a compound of carbon, hydrogen, and oxygen.

mass of nickel originally present can be calculated. This type of analysis is called **gravimetric analysis** and the stoichiometric factor in this case is called the gravimetric factor. Other methods of chemical analysis depend upon the same principle.

Another method of analysis often used for organic compounds is **combustion analysis**. In this method the substance is burned in oxygen and the quantity of carbon dioxide and water produced is measured:

Every carbon atom produces one molecule of carbon dioxide.
Every two hydrogen atoms produce one molecule of water.

For example, suppose 0.290 gram of a carbohydrate is analyzed by combustion. The analysis produces 0.440 gram of carbon dioxide and 0.090 gram of water. What is the empirical formula of the compound?

Step 1: What is the stoichiometry? —this is described above.

Step 2: Convert mass data to moles:
0.440 g of CO_2 = 0.440g / 44.01g mol^{-1}= 0.0100 mol
0.090 g of H_2O = 0.090g / 18.02g mol^{-1} = 0.0050 mol

It's useful to set up a table to organize these calculations. See problems 4.9 and 4.10 in this study guide.

You must be able to convert between mass and moles!

#moles = mass/molar mass

Step 3: Use the stoichiometric relationships above:
Number of moles of carbon C = 0.0100 mol CO_2 = 0.0100 mol
Number of moles of hydrogen H = 2×0.005 mol H_2O = 0.0100 mol

There are an equal number of moles of carbon and hydrogen in the compound:

Mass of hydrogen = 0.0100 mol × 1.008 g/mol = 0.010 g
Mass of carbon = 0.0100 mol × 12.01 g/mol = 0.120 g
Total mass of carbon and hydrogen = 0.130 g

Remainder must be oxygen = 0.290 − 0.130 = 0.160 g
Moles of oxygen = 0.160g / 16.0g mol^{-1}= 0.0100 mol

An empirical formula indicates the simplest possible integer ratio of elements in a compound.

Step 4: Write the empirical formula:
The number of moles of carbon, hydrogen, and oxygen are all the same; the empirical formula is CHO.

Review Questions

1. What is the difference between a chemical reaction and a chemical equation?

2. Why must a chemical equation be balanced?

3. List the steps in a systematic approach to balancing a chemical equation.

4. What is the stoichiometry of a reaction?

5. What is a stoichiometric factor?

6. What is a limiting reactant and how can you establish which reactant is the limiting reactant?

7. How is the percent yield for a reaction calculated?

8. Explain how a mixture can be analyzed to determine what, and how much, of a substance is in the mixture.

9. Summarize the procedure used to establish the formula of a carbohydrate by combustion analysis.

10. How can a molecular formula be established once the empirical formula is known?

Answers to Review Questions

1. A chemical reaction is a process in which substances are changed into other substances by a rearrangement of atoms. A chemical equation is a symbolic representation of a chemical reaction. The substances involved in the chemical reaction are represented by symbols and formulas. The physical states of the substances are defined by symbols and the direction of the reaction is denoted by an arrow.

2. A chemical equation must be balanced because of the law of conservation of mass. Atoms cannot be created nor destroyed in a chemical reaction. Therefore there have to be equal numbers of atoms on the two sides of an equation.

3. Step 1: Write the correct formulas for the reactants and products.
   Step 2: Look for the most complicated looking formula and balance the atoms in that formula first.
   Step 3: Balance atoms that occur only in one reactant and one product.
   Step 4: Balance the remaining atoms.

   Practice balancing equations; some examples are provided in question 1 in the next section.

4. The stoichiometry of a reaction is the quantitative relationship between the amounts of reactants and products in the reaction. Note that the principles of stoichiometry are also applied to the composition of compounds and the formulas used to represent them.

5. A stoichiometric factor is the mole ratio relating the number of moles of one reactant to the number of moles of another reactant or to the number of moles of product. The factor is obtained directly from the coefficients of the balanced chemical equation.

   Stoichiometric factors are the subject of question 2 in the next section.

6. The limiting reactant is the reactant that is used up first and therefore is the reactant that determines, or limits, the amount of product that can be formed.

Dividing by the stoichiometric coefficients takes account of the stoichiometry of the reaction automatically.

The limiting reactant can often be established using common sense although it can be established more formally by dividing the number of moles of each reactant present by its stoichiometric coefficient. The reactant with the lowest result is the limiting reactant.

$$\% \text{ yield} = \frac{\text{actual}}{\text{theoretical}} \times 100\%$$

7.   The theoretical yield is calculated from the stoichiometry of the balanced chemical equation. The actual yield is the amount of product obtained experimentally. The actual yield divided by the theoretical yield multiplied by 100 is the percent yield.

8.   Chemical analysis involves a chemical reaction. (There are other ways to analyze matter—for example spectroscopic analysis.) A reaction is chosen for the component (the analyte) within the mixture that needs to be determined; the other components in the mixture are supposed not to react. A positive reaction is all that is necessary for a qualitative test. For a quantitative determination, the quantity of product from the chemical reaction is measured, which in turn allows the quantity of the analyte in the original mixture to be determined.

9.   If a compound burns in oxygen (typically an organic molecule containing carbon and hydrogen) then each element (except oxygen) in the compound will combine with oxygen to produce the appropriate oxide. The quantity of carbon dioxide is determined by absorbing it in sodium hydroxide. The quantity of water produced is determined by absorbing it in magnesium perchlorate. The mass changes in the absorbents yield the amounts of carbon dioxide and water produced for a known mass of compound. Every mole of carbon in the original material yields one mole of carbon dioxide. Every two moles of hydrogen in the original material yields one mole of water.

10.   To establish the molecular formula from the empirical formula, a determination of the molecular mass of the compound must be done.

Most people would sooner die than think; in fact, they do so.

Bertrand Russell
(1872-1970)

Study Questions and Problems

1.   Balance the following equations:

With practice, you will develop an instinctive approach to balancing a chemical equation.

a. $\_C_4H_6(g) + \_O_2(g) \rightarrow \_CO_2(g) + \_H_2O(l)$

b. $\_NH_3(g) + \_O_2(g) \rightarrow \_NO_2(g) + \_H_2O(l)$

c. $\_PCl_3(l) + \_H_2O(l) \rightarrow \_H_3PO_3(aq) + \_HCl(aq)$

d. $\_Ca_3P_2(s) + \_H_2O(l) \rightarrow \_Ca(OH)_2(aq) + \_PH_3(g)$

e. $\_C_4H_8(OH)_2(l) + \_O_2(g) \rightarrow \_CO_2(g) + \_H_2O(l)$

f. $\_NH_3(g) + \_NO(g) \rightarrow \_N_2(g) + \_H_2O(l)$

g. $\_KClO_3(s) \rightarrow \_KCl(s) + \_O_2(g)$

h. $\_Ca(OH)_2(s) + \_H_3PO_4(aq) \rightarrow \_Ca_3(PO_4)_2(s) + \_H_2O(l)$

i. $\_C_3H_8(g) + \_O_2(g) \rightarrow \_CO_2(g) + \_H_2O(l)$

j. $\_N_2O(g) + \_O_2(g) \rightarrow \_NO_2(g)$

k. $\_Al_4C_3(s) + \_H_2O(l) \rightarrow \_Al(OH)_3(aq) + \_CH_4(g)$

l. $\_CS_2(l) + \_Cl_2(g) \rightarrow \_CCl_4(l) + \_S_2Cl_2(l)$

m. $\_C_2H_5OH(l) + \_PCl_3(l) \rightarrow \_C_2H_5Cl(l) + \_H_3PO_3(l)$

n. $\_ZnS(s) + \_O_2(g) \rightarrow \_ZnO(s) + \_SO_2(g)$

o. $\_Ag(s) + \_H_2S(g) + \_O_2(g) \rightarrow \_Ag_2S(s) + \_H_2O(l)$

2. What are the stoichiometric factors between the first product and first reactant written in the equations in question 1?

3. When asked to balance the equation:

$$C_2H_6(g) + O_2(g) \rightarrow CO_2(g) + H_2O(l)$$

the following suggestions were made:

$$C_2H_6(g) + 5O_2(g) \rightarrow 2CO_2(g) + 3H_2O(l)$$

$$C_2H_6(g) + 5O(g) \rightarrow 2CO(g) + 3H_2O(l)$$

$$2C_2H_6(g) + 7O_2(g) \rightarrow 4CO_2(g) + 6H_2O(l)$$

Which answer is correct and what is wrong with the others?

4. Write balanced chemical equations for the following reactions:

a. the decomposition of ammonium nitrate to nitrogen gas, oxygen gas, and water vapor.
b. the reaction of sodium bicarbonate with sulfuric acid to produce sodium sulfate, water, and carbon dioxide.
c. the treatment of phosphorus pentachloride with water to produce phosphoric acid and hydrogen chloride.

5. If the maximum amount of product possible is formed in the following reactions, what mass of the specified product would you obtain?

a. 10 grams of sodium chloride is treated with excess silver nitrate:

$$AgNO_3(aq) + NaCl(aq) \rightarrow AgCl(s) + NaNO_3(aq)$$

How much silver chloride is precipitated?

b. 12 grams copper metal is treated with excess dilute nitric acid:

$$3Cu(s) + 8HNO_3(aq) \rightarrow 3Cu(NO_3)_2(aq) + 2NO(g) + 4H_2O(l)$$

How much nitric oxide gas (NO) is produced?

c. 60 grams propane gas is burned in excess oxygen:

$$C_3H_8(g) + 5O_2(g) \rightarrow 3CO_2(g) + 4H_2O(l)$$

How much water is produced?

6. A furniture dealer put together a special deal for the annual sale—an entire dining room set comprising a table, six dining chairs, two bookshelves, a china cabinet, and a sideboard for $999. The dealer had in stock 280 tables, 1750 chairs, 550 bookshelves, 300 china cabinets, and 325 sideboards. He asked his assistant to figure out how many dining room sets they could sell, how much money they would make if they sold all the sets possible, and what they would have left that could not be sold as part of the deal.

7. Hydrazine reacts with dinitrogen tetroxide according to the equation:

$$2N_2H_4(g) + N_2O_4(g) \rightarrow 3N_2(g) + 4H_2O(g)$$

50.0 grams of hydrazine is mixed with 100.0 grams of dinitrogen tetroxide. How much nitrogen gas was produced?

Molar mass of $[Ni(NH_3)_6](NO_3)_2$ is 284.887 g/mol.

Molar mass of $[Ni(DMG)_2]$ is 288.917 g/mol.

Molar mass of Ni is 58.693 g/mol.

8. A nickel(II) ammonia compound was prepared in the laboratory. Its formula was supposed to be $[Ni(NH_3)_6](NO_3)_2$. In order to verify the composition of the compound, a solution of the compound was treated with the reagent dimethylglyoxime. This reagent reacts with nickel(II) to produce an insoluble red compound $Ni(DMG)_2$.
0.1324 gram of the nickel ammonia compound was dissolved in water and treated with an excess of the dimethylglyoxime reagent. The mass of red $Ni(DMG)_2$ produced was 0.1343 gram.

a. Calculate the mass of nickel in the nickel(II) ammonia compound.
b. Calculate the %mass of nickel in the compound.
c. What %mass of nickel was expected?
d. What is the gravimetric factor for this analysis?

9. 7.321 mg of an organic compound containing carbon, hydrogen, and oxygen was analyzed by combustion. The amount of carbon dioxide produced was 17.873 mg and the amount of water produced was 7.316 mg. Determine the empirical formula of the compound.

10. 0.1101 gram of an organic compound containing carbon, hydrogen, and oxygen was analyzed by combustion. The amount of carbon dioxide produced was 0.2503 gram and the amount of water produced was 0.1025

gram. A determination of the molar mass of the compound indicated a value of approximately 115 grams/mol. Determine the empirical formula, and the molecular formula, of the compound.

11. The reaction of hydrogen iodide and potassium bicarbonate produces potassium iodide according to the equation:

$$HI(aq) + KHCO_3(s) \rightarrow KI(aq) + H_2O + CO_2(g)$$

If 32 grams of potassium bicarbonate is treated with 48 grams of hydrogen iodide, what is the maximum amount of potassium iodide that can be produced?

12. TNT (trinitrotoluene) is prepared by treating toluene with nitric acid according to the equation:

$$C_7H_8(l) + 3HNO_3(aq) \rightarrow C_7H_5(NO_2)_3(s) + 3H_2O(l)$$

What mass of TNT can be made when 450 grams of toluene is treated with 1000 grams of nitric acid?

13. Sodium metal reacts vigorously with water to produce a solution of sodium hydroxide and hydrogen gas:

$$2Na(s) + 2H_2O(l) \rightarrow 2NaOH(aq) + H_2(g)$$

What mass of hydrogen gas can be produced when 10 grams of sodium is added to 15 grams of water?

14. Nitrous oxide reacts with oxygen to produce nitrogen dioxide according to the equation:

$$2N_2O(g) + 3O_2(g) \rightarrow 4NO_2(g)$$

What mass of nitrogen dioxide can be made from 42 grams of nitrous oxide and 42 grams of oxygen?

15. If only 75 grams of nitrogen dioxide was produced in the reaction described in the previous question, what was the %yield?

16. When copper was treated with dilute nitric acid as described in question 5b, the quantity of nitric oxide gas (NO) collected was only 3.0 grams. What was the %yield of this product?

Answers to Study Questions and Problems

1. Follow the logical procedure outlined earlier. Examine the equation for any obvious stoichiometric relationships between the reactants and products. Remember that the number of atoms of each element on each side of the equation must be the same. For example, one C_4H_6 on the reactant side requires four CO_2 on the product side, or one Ca_3P_2 on the reactant side requires three $Ca(OH)_2$ on the product side—such relationships are not a bad place to start.

    a. $2C_4H_6(g) + 11O_2(g) \rightarrow 8CO_2(g) + 6H_2O(l)$

    b. $4NH_3(g) + 7O_2(g) \rightarrow 4NO_2(g) + 6H_2O(l)$

    c. $PCl_3(l) + 3H_2O(l) \rightarrow H_3PO_3(aq) + 3HCl(aq)$

    d. $Ca_3P_2(s) + 6H_2O(l) \rightarrow 3Ca(OH)_2(s) + 2PH_3(g)$

    e. $2C_4H_8(OH)_2(l) + 11O_2(g) \rightarrow 8CO_2(g) + 10H_2O(l)$

    f. $4NH_3(g) + 6NO(g) \rightarrow 5N_2(g) + 6H_2O(l)$

    g. $2KClO_3(s) \rightarrow 2KCl(s) + 3O_2(g)$

    h. $3Ca(OH)_2(s) + 2H_3PO_4(aq) \rightarrow Ca_3(PO_4)_2(s) + 6H_2O(l)$

    i. $C_3H_8(g) + 5O_2(g) \rightarrow 3CO_2(g) + 4H_2O(l)$

    j. $2N_2O(g) + 3O_2(g) \rightarrow 4NO_2(g)$

    k. $Al_4C_3(s) + 12H_2O(l) \rightarrow 4Al(OH)_3(aq) + 3CH_4(g)$

    l. $CS_2(l) + 3Cl_2(g) \rightarrow CCl_4(l) + S_2Cl_2(l)$

    m. $3C_2H_5OH(l) + PCl_3(l) \rightarrow 3C_2H_5Cl(l) + H_3PO_3(l)$

    n. $2ZnS(s) + 3O_2(g) \rightarrow 2ZnO(s) + 2SO_2(g)$

    o. $4Ag(s) + 2H_2S(g) + O_2(g) \rightarrow 2Ag_2S(s) + 2H_2O(l)$

2. Knowledge of the stoichiometric factors derived from a balanced chemical equation is essential in quantitative (stoichiometric) calculations.

    a. 2 moles C_4H_6 / 8 moles CO_2

    b. 4 moles NH_3 / 4 moles NO_2

    c. 1 mole PCl_3 / 1 mole H_3PO_3

d.   1 mole Ca_3P_2 / 3 moles $Ca(OH)_2$

e.   2 moles $C_4H_8(OH)_2$ / 8 moles CO_2

f.   4 moles NH_3 / 5 moles N_2

g.   2 moles $KClO_3$ / 2 moles KCl

h.   3 moles $Ca(OH)_2$ / 1 mole $Ca_3(PO_4)_2$

i.   1 mole C_3H_8 / 3 moles CO_2

j.   2 moles N_2O / 4 moles NO_2

k.   1 mole Al_4C_3 / 4 moles $Al(OH)_3$

l.   1 mole CS_2 / 1 mole CCl_4

m.   3 moles C_2H_5OH / 3 moles C_2H_5Cl

n.   2 moles ZnS / 2 moles ZnO

o.   4 moles Ag / 2 moles Ag_2S

3. $C_2H_6(g) + 5O_2(g) \rightarrow 2CO_2(g) + 3H_2O(l)$
Not balanced—check the oxygen atoms.
$C_3H_8(g) + 5O(g) \rightarrow 2CO(g) + 3H_2O(l)$
Cannot change the formulas! Oxygen is O_2 and carbon dioxide is CO_2.
$2C_3H_8(g) + 7O_2(g) \rightarrow 4CO_2(g) + 6H_2O(l)$
This answer is correct.

4.   a.   the decomposition of ammonium nitrate to nitrogen gas, oxygen gas, and water vapor.

Write the formulas for all the reactants and products and then balance the equation.

$2NH_4NO_3(s) \rightarrow 2N_2(g) + O_2(g) + 4H_2O(g)$

b.   the reaction of sodium bicarbonate with sulfuric acid to produce sodium sulfate, water, and carbon dioxide.

$2NaHCO_3(s) + H_2SO_4(aq) \rightarrow Na_2SO_4(aq) + 2H_2O + 2CO_2(g)$

c.   the treatment of phosphorus pentachloride with water to produce phosphoric acid and hydrogen chloride.

$PCl_5(s) + 4H_2O(l) \rightarrow H_3PO_4(aq) + 5HCl(g)$

5. a. 10 grams of sodium chloride is treated with excess silver nitrate:

$$AgNO_3(aq) + NaCl(aq) \rightarrow AgCl(s) + NaNO_3(aq)$$

Notice the procedure in these problems:

1. Convert mass to moles.
2. Use the stoichiometric factor.
3. Convert from moles back to mass.

Convert mass quantities to moles:
10 grams of NaCl is $10g/58.44 \text{ g mol}^{-1} = 0.171$ mole.

The stoichiometric factor (mole ratio) is 1 mol AgCl/1 mol NaCl. One silver chloride unit is produced for every sodium chloride used. So the number of moles of silver chloride precipitated = 0.171 mole.

Convert back to mass:
0.171 mole of AgCl = 0.171 mole × 143.3 g mol^{-1} = 24.5 grams.

b. 12 grams copper metal is treated with excess dilute nitric acid:

$$3Cu(s) + 8HNO_3(aq) \rightarrow 3Cu(NO_3)_2(aq) + 2NO(g) + 4H_2O$$

Convert mass quantities to moles:
12 grams of copper is $12g/63.546 \text{ g mol}^{-1} = 0.189$ mole.

The stoichiometric factor (mole ratio) is 2 mol NO/3 mol Cu.
2 moles NO are produced for every 3 moles copper used.

Notice how the units cancel to give moles of NO.

So the NO produced = 0.189 mol Cu × (2 mol NO/3 mol Cu)
                    = 0.126 mol NO.

Convert back to mass:
0.126 mole of nitric oxide = 0.126 mole × 30.0 g mol^{-1} = 3.8 grams.

c. Propane gas is burned in excess oxygen:

$$C_3H_8(g) + 5O_2(g) \rightarrow 3CO_2(g) + 4H_2O(l)$$

Convert mass quantities to moles:
60 grams of propane is $60g/44.097 \text{ g mol}^{-1} = 1.36$ moles.

The stoichiometric factor (mole ratio) is 4 mol H_2O/1 mol C_3H_8. 4 moles of water are produced for every mole of propane burned.

Intermediate calculations should be done with sufficient precision to ensure no loss of information. The final answer should be quoted with no greater precision (in this case 2 significant figures) than the the initial data.

So the water produced = 1.36 mol C_3H_8 × (4 mol H_2O/1 mol C_3H_8)
                       = 5.44 moles water.

Convert back to mass:
5.44 moles water = 5.44 mole × 18.015 g mol^{-1} = 98 grams.

See Problem-Solving Tips and Ideas 4.2 on page 162 in the text.

6. An alternative way to determine the limiting reactant in a reaction is to calculate the quantity of product each reactant can make. The reactant that produces the least product is the limiting reactant.

The number of sets that can be produced, and the number of non-limiting components used, depends upon the limiting reactant.

tables:	280 tables	→ 280 sets	(5 remaining)
chairs:	1750 chairs	→ 291 sets	(100 remaining)
bookshelves:	550 shelves	→ 275 sets — limiting reactant	
china cabinets:	300 cabinets	→ 300 sets	(25 remaining)
sideboards:	325 sideboards	→ 325 sets	(50 remaining)

At $999 per set, the total sale is 275 × $999 = $274,725.

7. Hydrazine reacts with dinitrogen tetroxide according to the equation:

$$2N_2H_4(g) \;+\; N_2O_4(g) \;\rightarrow\; 3N_2(g) \;+\; 4H_2O(g)$$

Convert mass quantities to moles:
50.0 grams of hydrazine is 50.0g/32.045 g mol^{-1} = 1.56 moles.
100.0 grams of N_2O_4 is 100.0g/92.011 g mol^{-1} = 1.087 moles.
The limiting reactant is hydrazine (it is used up first).
The quantity of N_2 formed will depend upon the amount of N_2H_4 used.

The stoichiometric factor is (3 mol N_2/2 mol N_2H_4).
3 moles N_2 are produced for every 2 moles hydrazine used.
So the N_2 produced = 1.56 moles hydrazine × (3N_2/2N_2H_4)
= 2.34 moles N_2.

Convert back to mass:
2.34 moles of nitrogen = 2.34 mol × 28.01 g mol^{-1} = 65.6 grams.

> Remember that the limiting reactant can be established by dividing the number of moles of each reactant by its stoichiometric coefficient. The reactant with the lowest result is the limiting reactant.

8. The mass of red $Ni(DMG)_2$ produced was 0.1343 gram.
The number of moles = 0.1343 g/288.917 g mol^{-1} = 4.648×10^{-4} mole.
The number of moles of nickel must be the same, therefore
the mass of nickel = 4.648×10^{-4} mole × 58.693 g mol^{-1} = 0.02728 g.

The %mass of nickel = (0.02728g/0.1324) × 100% = 20.61%

The %mass expected = (58.693/284.887) × 100% = 20.60%

> Molar mass of [$Ni(DMG)_2$] is 288.917 g/mol.
>
> Molar mass of Ni is 58.693 g/mol.
>
> Molar mass of [$Ni(NH_3)_6$](NO_3)$_2$ is 284.887 g/mol.

In terms of moles, the gravimetric factor is 1:1. There is one mole of nickel in every mole of both the [$Ni(NH_3)_6$](NO_3)$_2$ and the $Ni(DMG)_2$. What is useful in gravimetric analyses like this is to calculate a gravimetric factor in terms of mass, so that from a mass of $Ni(DMG)_2$ the mass of nickel can be calculated directly. The mass gravimetric factor in this case is simply the ratio of the molar masses of the nickel and the $Ni(DMG)_2$.

9. Setting up a systematic table for the various steps is a useful way to keep track of the calculations:

	carbon	hydrogen	oxygen
mass in mg:	17.873 mg CO_2	7.316 mg H_2O	?
divide by molar masses to get mmol CO_2 & H_2O:	/44.010 = 0.406	/18.015 = 0.406	?
	0.406 mmol C	0.812 mmol H	?
multiply by molar masses to get masses of C and H:	4.876 mg C	0.818 mg H	1.622 mg O
back to mmol:	0.406 mmol C	0.812 mmol H	0.101 mmol O
convert to integers:	4C	8H	1O

> The molar mass of carbon dioxide is 44.010 g/mol. The molar mass of water is 18.015 g/mol.
>
> Sometimes it is more convenient to work in millimoles (mmol) rather than moles.
>
> There are 2 moles of H in every mole of water.
>
> The only reason to convert to mass is to calculate the mass of oxygen in the compound by difference. The mass of oxygen is the mass of the compound less the mass of carbon and the mass of hydrogen.

The empirical formula of the compound is C_4H_8O.

10. A similar problem:

	carbon	hydrogen	oxygen
mass in mg:	250.3 mg CO_2	102.5 mg H_2O	?
divide by molar masses to get mmol CO_2 & H_2O:	/44.010 = 5.69	/18.015 = 5.69	?
	5.69 mmol C	11.38 mmol H	?
multiply by molar masses to get masses of C and H:	68.31 mg C	11.47 mg H	30.32 mg O
back to mmol:	5.69 mmol C	11.38 mmol H	1.89 mmol O
convert to integers:	3C	6H	1O

The empirical formula of the compound is C_3H_6O. The empirical mass (the molar mass of this empirical formula) is 58 g mol^{-1}. This is one-half the experimentally determined molecular mass which means that there must be two empirical units in one molecular unit. The molecular formula of the compound is therefore $C_6H_{12}O_2$.

11. The reaction of hydrogen iodide and potassium bicarbonate:

$$HI + KHCO_3 \rightarrow KI + H_2O + CO_2$$

Convert mass quantities to moles:
32 grams of $KHCO_3$ is 32g/100.12 g mol^{-1} = 0.320 mole.
48 grams of HI is 48g/127.9 g mol^{-1} = 0.375 mole.
The potassium bicarbonate is the limiting reactant.

The stoichiometric factor (mole ratio) is 1 mol KI/1 mol $KHCO_3$.
One KI unit is produced for every $KHCO_3$ used.
So the number of moles KI produced = 0.320 mole.

Convert back to mass:
0.320 mole of KI = 0.320 mole × 166.0 g mol^{-1} = 53 grams.

12. Trinitrotoluene is prepared by treating toluene with nitric acid:

$$C_7H_8(l) + 3HNO_3(aq) \rightarrow C_7H_5(NO_2)_3(s) + 3H_2O(l)$$

Remember that the limiting reactant can be established by dividing the number of moles of each reactant by its stoichiometric coefficient. The reactant with the lowest result is the limiting reactant.

Convert mass quantities to moles:
450 grams of toluene is 450g/92.14 g mol^{-1} = 4.884 moles.
1000 grams of nitric acid is 1000g/63.013 g mol^{-1} = 15.87 moles.
The toluene is the limiting reactant.

The stoichiometric factor (mole ratio) is 1 mol TNT/1 mol toluene.
One TNT molecule is produced for every toluene molecule used.
So the number of moles TNT produced = 4.884 moles.

Convert back to mass:
4.884 moles of TNT = 4.884 moles × 227.1 g mol^{-1} = 1109 grams.

13. Sodium metal reacts vigorously with water:

$$2Na(s) + 2H_2O(l) \rightarrow 2NaOH(aq) + H_2(g)$$

Convert mass quantities to moles:
   10 grams of sodium is $10g/23 \text{ g mol}^{-1}$ = 0.435 mole.
   15 grams of water is $15g/18.02 \text{ g mol}^{-1}$ = 0.833 mole.
   The sodium is the limiting reactant.

The stoichiometric factor (mole ratio) is 1 mol H_2/2 mol Na.
   One hydrogen molecule is produced for every 2 atoms of sodium used.
   So the number of moles H_2 produced = $(1H_2/2Na) \times 0.435$ mole.

Convert back to mass:
   0.217 mole of hydrogen = 0.217 mole × 2.016 g mol^{-1} = 0.44 gram.

14. Nitrous oxide reacts with oxygen to produce nitrogen dioxide according to the equation:

$$2N_2O(g) + 3O_2(g) \rightarrow 4NO_2(g)$$

Convert mass quantities to moles:
   42 grams of N_2O is $42g/44.01 \text{ g mol}^{-1}$ = 0.954 mole.
   42 grams of O_2 is $42g/32.0 \text{ g mol}^{-1}$ = 1.313 mole.
   The oxygen is the limiting reactant.

0.954 divided by 2 = 0.477
1.313 divided by 3 = 0.438

—so oxygen is the limiting reactant.

The stoichiometric factor (mole ratio) is 4 mol NO_2/3 mol O_2.
   The number of moles NO_2 produced = $(4NO_2/3O_2) \times 1.313$ mol O_2.
                                    = 1.75 mol NO_2.

Convert back to mass:
   1.75 mole of nitrogen dioxide = 1.75 mol × 46.0 g mol^{-1} = 81 grams.

15. Actual yield = 75 grams
    Theoretical yield = 81 grams
    Percent yield = (75 grams / 81 grams) × 100% = 93% yield of nitrogen dioxide.

16. Actual yield = 3.0 grams
    Theoretical yield = 3.8 grams
    Percent yield = (3.0 grams / 3.8 grams) × 100% = 79% yield.

Challenge 2

Nitric acid is produced by the three step process:

$$4NH_3(g) + 5O_2(g) \rightarrow 4NO(g) + 6H_2O(g)$$

$$2NO(g) + O_2(g) \rightarrow 2NO_2(g)$$

$$3NO_2(g) + H_2O(l) \rightarrow 2HNO_3(aq) + NO(g)$$

If each step is achieved with a 90% yield, what is the overall yield of nitric acid from 1.00 kg of ammonia?

CHAPTER 5

Introduction

Many reactions occur in solution, especially in water (aqueous) solution. In solution, the molecules and ions are free to move about and interact. So in this chapter we will examine aqueous solutions to understand the behavior of ions in solution, to understand why some salts dissolve and others don't, to classify the types of reactions that occur in solution, and to determine the quantitative mass relationships for these reactions.

Contents

5.1 Properties of Compounds in Aqueous Solution

A **solution** is a homogeneous mixture of two or more substances. The substance that determines the state of the solution is the **solvent**. The substance that dissolves in the solvent is called the **solute**. Solutions in which the solvent is water are called **aqueous solutions**.

When an ionic solid dissolves in water, the solid breaks up and the individual ions become surrounded by water molecules. Positive ions attract the negative (oxygen) end of the water molecule; negative ions attract the positive (hydrogen) end of the water molecule. This process is called **solvation**. The solvated ions are free to move about in solution—and carry an electrical current. Solutes whose solutions conduct electricity are called **electrolytes**.

Electrolytes are strong or weak. A **strong electrolyte** produces a high concentration of ions in solution and is a good conductor of electricity—the compound is essentially completely ionized in solution. A **weak electrolyte** produces relatively few ions in solution and conducts poorly. Most molecules of a weak electrolyte remain molecules—only a few molecules break up to form ions.

Not all ionic compounds dissolve in water; the interaction between the ions in the solid is too strong to break up. Some solubility rules offer guidelines for determining the solubility of a salt.

Common salt NaCl is an example of a strong electrolyte. It dissociates almost completely in solution producing a high concentration of ions.

There is however considerable association between ions in aqueous solution, particularly between ions with higher charges.

Acetic acid (in vinegar) is an example of a weak electrolyte. Most acetic acid exists as molecules; only a few break up to form ions.

You may be asked to memorize the guidelines for determining solubility—the solubility rules (Figure 5.4, page 184).

5.2 Precipitation Reactions

A **precipitation reaction** produces a product that precipitates from solution—it is insoluble. Two reactants may well be soluble, but the combination of the cation of one with the anion of the other is insoluble. An example is the precipitation of silver chloride when solutions of silver nitrate and sodium chloride are mixed:

Because carbonates, chromates, sulfides, and hydroxides are often insoluble, they are often involved in precipitation reactions.

$$AgNO_3(aq) + NaCl(aq) \rightarrow AgCl(s) + NaNO_3(aq)$$

It is useful to write **net-ionic equations** for reactions in aqueous solution because these equations present the essential process. Consider, for example, the precipitation reaction above. The silver nitrate, sodium chloride, and sodium nitrate are all salts; they exist in solution completely ionized:

The (aq) symbols are omitted to make the equation clearer; all ions are in solution.

$$Ag^+ + NO_3^- + Na^+ + Cl^- \rightarrow AgCl(s) + Na^+ + NO_3^-$$

When writing a detailed ionic equation, write all species in the form in which they predominantly exist in solution. Solids, gases, non and weak electrolytes should be written in the molecular form. Strong electrolytes should be written in the ionic form.

This type of equation is sometimes referred to as a **detailed ionic equation**. Notice that nothing happens to the sodium ion in this reaction; likewise, nothing happens to the nitrate ion. They are not involved in the reaction and are called **spectator ions**. If these spectator ions are cancelled from both sides, the net-ionic equation results:

The net-ionic equation is still balanced.

$$Ag^+ \qquad + Cl^- \rightarrow AgCl(s)$$

The net ionic equation more clearly illustrates what is happening in the reaction. In this case, silver ions (from any soluble salt) combine with chloride ions (from any soluble salt) to produce the insoluble silver chloride.

5.3 Acids and Bases

Acids and bases are two important classes of compounds. An **acid** is a substance that, when dissolved in water, increases the concentration of hydrogen ions $H^+(aq)$. For example, hydrochloric acid ionizes in water to produce hydrogen ions and chloride ions. This acid is a **strong acid** because it is completely ionized in solution:

Strong acids:

hydrochloric acid HCl
nitric acid HNO_3
perchloric acid $HClO_4$
sulfuric acid H_2SO_4

also HBr, HI, and $HClO_3$

$$HCl(aq) \rightarrow H^+(aq) + Cl^-(aq)$$

An acid may produce more than one hydrogen ion per molecule, in which case it is called a **polyprotic acid**. An example is sulfuric acid (a diprotic acid):

Polyprotic acids:

sulfuric acid H_2SO_4 (strong)
phosphoric acid H_3PO_4 (weak)
carbonic acid H_2CO_3 (weak)

$$H_2SO_4(aq) \rightarrow H^+(aq) + HSO_4^-(aq) \qquad \text{complete ionization}$$

$$HSO_4^-(aq) \rightarrow H^+(aq) + SO_4^{2-}(aq) \qquad \text{only partial ionization}$$

Loss of the first hydrogen ion is virtually complete. So sulfuric acid is classified as strong. The second hydrogen ion is more difficult to remove and the bisulfate ion is only partially ionized. The bisulfate ion is a **weak acid**.

Some weak acids:

bisulfate ion HSO_4^-
acetic acid CH_3CO_2H
carbonic acid H_2CO_3
phosphoric acid H_3PO_4
hydrocyanic acid HCN
lactic acid $CH_3CH(OH)CO_2H$
hypochlorous acid HOCl

A **base** is a substance that, when dissolved in water, increases the concentration of hydroxide ions $OH^-(aq)$. Hydroxides of the Group 1A and 2A metals are common **strong bases**. These bases dissociate completely in water to produce metal cations and the hydroxide anions:

$$NaOH(aq) \rightarrow Na^+(aq) + OH^-(aq)$$

Ammonia, NH_3, is a **weak base**. It produces a relatively low concentration of hydroxide ions in solution (most ammonia in solution remains as NH_3):

$$NH_3(aq) + H_2O(l) \rightarrow NH_4^+(aq) + OH^-(aq)$$

Some nonmetal oxides dissolve in water to form acidic solutions. For example:

$$CO_2(g) + H_2O(l) \rightarrow H_2CO_3(aq) \qquad [\rightarrow H_3O^+(aq) + HCO_3^-(aq)]$$

$$SO_2(g) + H_2O(l) \rightarrow H_2SO_3(aq) \qquad [\rightarrow H_3O^+(aq) + HSO_3^-(aq)]$$

$$SO_3(g) + H_2O(l) \rightarrow H_2SO_4(aq) \qquad [\rightarrow H_3O^+(aq) + HSO_4^-(aq)]$$

$$P_4O_{10}(s) + 6H_2O(l) \rightarrow 4H_3PO_4(aq) \qquad [\rightarrow H_3O^+(aq) + H_2PO_4^-(aq)]$$

Some metal oxides, if they dissolve in water, form basic solutions. For example:

$$CaO(s) + H_2O(l) \rightarrow Ca(OH)_2(s) \rightarrow Ca^+(aq) + 2OH^-(aq)$$

Notice that ammonia does not contain an OH^- ion as part of its structure.

Some common strong bases:

sodium hydroxide NaOH
potassium hydroxide KOH
calcium hydroxide $Ca(OH)_2$

The common weak base:

ammonia NH_3

The nonmetal oxides that produce acidic solutions are called acidic oxides.

The metal oxides that produce basic solutions are called basic oxides.

5.4 Reactions of Acids and Bases

Acids react with bases to produce a salt and water. The reaction is called neutralization:

$$HCl(aq) + NaOH(aq) \rightarrow NaCl(aq) + H_2O(l)$$

Even though the solution at the end of the reaction may not be neutral.

The anion (Cl^-) of the salt comes from the acid, and the cation (Na^+) of the salt comes from the base. Both hydrochloric acid and sodium hydroxide are strong electrolytes so the detailed ionic equation for the neutralization reaction is:

$$H^+(aq) + Cl^-(aq) + Na^+(aq) + OH^-(aq) \rightarrow Na^+(aq) + Cl^-(aq) + H_2O(l)$$

When the spectator ions are deleted, the net ionic equation becomes:

$$H^+(aq) + OH^-(aq) \rightarrow H_2O(l)$$

This is the net ionic equation for the neutralization of any strong acid by any strong base. The only change when the acid or base is different is the salt that is formed in the reaction. Weak acids and weak bases also undergo neutralization reactions. The difference is that, being weak electrolytes, the acids or bases exist predominantly in the molecular form. For example, consider the reaction of ammonia (a weak base) with nitric acid (a strong acid):

$$HNO_3(aq) + NH_3(aq) \rightarrow NH_4NO_3(aq) \qquad \text{overall}$$

$$H^+(aq) + NO_3^-(aq) + NH_3(aq) \rightarrow NH_4^+(aq) + NO_3^-(aq) \qquad \text{detailed}$$

$$H^+(aq) + NH_3(aq) \rightarrow NH_4^+(aq) \qquad \text{net ionic}$$

5.5 Gas-forming Reactions

A carbonate or bicarbonate salt reacts with acid to produce carbon dioxide gas. Similarly, sulfite salts react with acid to produce sulfur dioxide gas. Other salts react similarly. For example, nitric acid reacts with magnesium carbonate:

$$HNO_3(aq) + MgCO_3(s) \rightarrow Mg(NO_3)_2(aq) + H_2CO_3(aq)$$
$$\rightarrow Mg^+(aq) + NO_3^-(aq) + H_2O(l) + CO_2(g)$$

The reaction is driven by the formation of the weak electrolyte, carbonic acid, which decomposes rapidly to form carbon dioxide and water.

5.6 Organizing Reactions in Aqueous Solution

Check the Problem Solving Tips and Ideas 5.1 on page 203 of the text.

Chemical reactions in aqueous solution maybe described as exchange reactions. They can be summarized as follows:

Precipitation—the formation of an insoluble product
Acid-Base—the formation of water (and a salt)
Gas-forming—the formation of a gas
In each case a weak electrolyte or insoluble compound is formed.

A fourth reaction type is called **oxidation-reduction** or **redox** and involves electron transfer. These reactions are described in the next section.

5.7 Oxidation-Reduction Reactions

The production of metals from their ores was an art developed by ancient civilizations.

When iron rusts, it combines with oxygen to form iron(III) oxide. This is a process called **oxidation**. If a metal oxide or sulfide ore is heated with carbon, it is converted into the metal. In this case the metal is said to be **reduced**. These are examples of a huge class of reactions called oxidation and reduction reactions, or **redox reactions**.

The oxidizing agent is always reduced.
The reducing agent is always oxidized.

In every redox reaction, some substance is oxidized and another substance is reduced. One cannot occur without the other. The substance that is reduced is by definition the substance that oxidizes the other—it is called the **oxidizing agent**. Likewise, the substance that is oxidized must reduce the other substance—it is called the **reducing agent**. For example, consider the reaction of iron(III) oxide with carbon monoxide:

$$Fe_2O_3(s) + 3CO(g) \rightarrow 2Fe(s) + 3CO_2(g)$$

The carbon monoxide is the reducing agent: it removes oxygen from the iron(III) oxide. It therefore must be oxidized—it is, to carbon dioxide. The iron(III) oxide is the oxidizing agent—it is responsible for oxidizing the carbon monoxide to carbon dioxide.

Sometimes there is an obvious transfer of electrons; other times not so obvious.

However, not all redox reactions involve oxygen. What is common to all redox reactions is a transfer of electrons between substances:

Reduction—the addition of electron(s)
Oxidation—the removal or loss of electron(s)
Reducing agent—supplies electrons
Oxidizing agent—removes electrons

When a substance loses electrons (when it is oxidized), the positive charge on an atom of the substance increases. When a substance gains electrons (when it is reduced), the positive charge on an atom of the substance decreases (or the negative charge increases).

Consider the oxidation of magnesium by oxygen:

$$2Mg(s) \ + \ O_2(g) \ \rightarrow \ 2Mg^{2+} \ + \ 2O^{2-} \ (as \ MgO \ solid)$$

The magnesium is oxidized; it loses two electrons.
The oxygen is reduced; it gains two electrons per atom.
The magnesium is the reducing agent; it supplies the electrons.
The oxygen is the oxidizing agent; it gains the electrons.

Recognition of a redox reaction is easier if oxidation numbers are assigned to the elements in the reactants and products. If there is a change in the oxidation number of an element, then the reaction is a redox reaction. The assignment of oxidation numbers is a very useful bookkeeping device. They are assigned according to the rules on page 207. However, it is important to understand that the numbers do not represent (except for monatomic ions) the actual charges on the atoms. For example, the oxidation number of manganese in the permanganate ion MnO_4^- is +7—a number much higher than the actual charge on the manganese atom.

Common oxidizing agents are listed in Table 5.4 on page 209 of the text and include oxygen itself, the halogens, nitric acid, permanganate and dichromate, and other reactants in which an element is in a high oxidation state. Common reducing agents are hydrogen, the alkali and alkaline earth metals, and carbon.

Check the Guidelines for determining oxidation numbers at the top of page 207 in the text.

You should be able to distinguish between an acid-base reaction and an oxidation-reduction reaction.

Look at Table 5.5 on page 210 for ways to recognize oxidation-reduction reactions.

5.8 Measuring Concentrations of Compounds in Solution

Most reactions done in the lab are reactions in solution. In solution the reactant molecules are able to move about and interact with one another. When dealing with solutions, it is often more convenient to measure volumes of solution rather than masses or moles of reactants. The relationship between the amount of solute and the volume of solution is the **concentration**. If the amount of solute is expressed in moles, the concentration is the **molarity**.

$$Molarity = \frac{moles \ of \ solute}{liters \ of \ solution}$$

If the molarity of a solution is known, then the number of moles of solute can be determined from the volume of solution.

$$Number \ of \ moles \ of \ solute = molarity \times volume \ of \ solution$$

When a salt dissolves in water, the salt dissociates into its separate ions. The concentration of the individual ions depends upon the stoichiometry of the salt. For example, if one mole of magnesium nitrate $Mg(NO_3)_2$ is dissolved in water to make 1 liter of solution, the concentration of magnesium ions is one molar (1M) but the concentration of the nitrate ions is 2 molar (2M). This is

$$Concentration = \frac{amnt \ of \ solute}{vol \ of \ solution}$$

This is the second fundamental relationship in stoichiometry that you must know. The first, if you recall, was:

moles = mass/molar mass

This is the same equation rearranged.

Molar means moles per liter. The symbol M means moles per liter or mol/L.

because one magnesium nitrate formula unit produces one magnesium ion but two nitrate ions.

The extent to which salts actually dissociate varies considerably. In dilute solutions of the alkali metal salts the dissociation is virtually 100%. However, in more concentrated solutions of the salts of divalent alkaline earth metals (and other metals), the association between ions can be quite high. For example, there are ions such as MgCl⁻ in a solution of magnesium chloride.

To prepare a solution of a known concentration (molarity) the required quantity of solute is dissolved in sufficient water to make the required volume of solution. The two relationships are required:

$$\text{\# moles} = \text{molarity} \times \text{volume}$$
$$\text{mass} = \text{\# moles} \times \text{molar mass}$$

Another method used to prepare a solution of a known concentration is to dilute a more concentrated solution. The key to such dilution tasks is to realize that the number of moles of solute remains unchanged when water is added. Therefore the molarity of the dilute solution × its volume must equal the molarity of the concentrated solution × its volume:

This is illustrated in the Problem-Solving Tips and Ideas 5.3 on page 218.

dilute solution		concentrated solution
molarity × volume	=	molarity × volume

5.9 Stoichiometry of reactions in aqueous solution

The stoichiometry of a chemical reaction is described by the coefficients in the balanced equation. These coefficients represent the mole ratio of reactants required and products formed. Because the number of moles can be expressed as the molarity of a solution multiplied by the volume of solution, stoichiometric calculations can be done equally well using molarity and volume information. There is just one additional step added to the series of calculations described in Chapter 4.

The sequence of steps in these calculations is illustrated in the Problem-Solving Tips and Ideas 5.4 on page 221.

When a solution of an acid reacts with a solution of a base in a neutralization reaction the experiment is called a **titration**. An unknown quantity of an acid, or an unknown quantity of a base, can be determined in a titration provided that all other variables are known. For example:

solution of acid		solution of base
number of moles of acid	=	number of moles of base
molarity × volume of acid	=	molarity × volume of base

The actual relationship between the number of moles of acid and the number of moles of base is determined by the stoichiometry of the neutralization reaction. Often simply 1:1, it is, for example, 2:1 for a diprotic acid such as H_2SO_4 titrated against NaOH, or 1:2 for HCl titrated against $Mg(OH)_2$.

In a titration, when the number of hydrogen ions supplied by the acid exactly equals the number of hydroxide ions supplied by the base, the **equivalence point** has been reached. Experimentally this equivalence point is recognized by the use of an acid-base **indicator**. This indicator has different colors in acid solution and in basic solution. The precise concentration of the acid or base used in the titration is determined by a process called **standardization**. The standardization is usually another titration in which the acid or base is titrated against a **primary standard**—the amount of which is known very accurately.

The other major type of titration involves a redox reaction. These titrations are particularly convenient when one of the reactants changes color when it is oxidized or reduced. An example is the reaction of permanganate against oxalate in which the permanganate ion is reduced. The color of the permanganate solution is an intense purple and a very slight excess of permanganate at the equivalence point imparts a faint pink color to the solution.

The stoichiometry of redox reactions is often more complicated than acid-base reactions.

Review Questions

1. Define solute, solvent, and solution.

2. Classify solutes in terms of their ability to conduct electricity in aqueous solution.

3. Summarize the guidelines for determining the solubility of a salt in water.

4. Explain why some reactions lead to the formation of a precipitate.

5. What is the difference between a detailed ionic equation and a net ionic equation?

6. Why are net ionic equations useful?

7. Describe what substances must be written in an ionic form, and which must be written in the molecular form, when writing a detailed ionic equation.

8. What is a neutralization reaction? Does neutralization result in a neutral solution?

9. Summarize the different types of reactions that can occur in aqueous solution. What drives these reactions to form the products?

10. Define oxidation and reduction in terms of electron transfer.

11. What is a reducing agent; what is a oxidizing agent?

12. How could you predict whether or not a substance is a good oxidizing agent?

13. How can you recognize whether a reaction is a redox reaction or not?

14. What is an oxidation number?

15. What is the molarity of a solution and how is it calculated?

16. How are the concentrations of individual ions in a solution of a salt related to the stoichiometry of the salt?

17. Describe the process you would use to calculate the mass of solute required to make a 5.0 liters of a 3.0 M solution.

18. Describe how a solution of particular molarity can be prepared from a more concentrated solution.

19. Quantitative chemical analysis often involves volumetric titration. Describe two major kinds of titration.

20. What is the equivalence point in a titration?

21. Explain why calculations based upon an acid-base titration, or a redox titration, depend upon a knowledge of the stoichiometry of the reaction.

Answers to Review Questions

1. The solute is the substance that dissolves in the solvent to make the solution. The solvent is the component of the solution that determines the state of the solution. The solution is a homogeneous mixture of two or more substances.

2. A solute that dissolves to form ions in solution is called an electrolyte. The ions carry the electrical current through the solution. A high concentration of ions means that the solution can carry a high current. Solutes such as salts dissociate almost completely in solution and produce a high concentration of ions —they are strong electrolytes. Other solutes, like acetic acid, produce a relatively low concentration of ions in solution and conduct poorly —these solutes are weak electrolytes. Some solutes produce no ions at all when they dissolve. These solutions do not conduct an electrical current. The solutes are nonelectrolytes.

3. Check Figure 5.4 in the text. You may, or may not, be required to memorize these rules.

There is actually considerable association between ions in solution. We will assume that the dissociation is complete.

4. When a salt dissolves in water, it dissociates completely. Once the ions are released from the salt in the solution process, the origin of the ions is immaterial. Any combination of cation and anion that is insoluble will precipitate from solution. For example, silver nitrate and sodium chloride solutions lead to a precipitate of silver chloride when mixed. The net ionic reaction is the reaction between the silver ions and the chloride ions. The reaction of potassium chloride and silver acetate leads to exactly the same precipitate.

5. A detailed ionic equation is derived from the initial chemical equation by writing all the strong electrolytes in their ionic form. All strong acids, strong bases, and all salts are written as ions because that is the form in which they predominantly exist in solution. A net ionic equation is then derived from the detailed ionic equation by cancelling all the spectator ions from the two sides of the equation.

6. The net ionic equation is particularly useful because it represents exactly what happens in the solution. For example, the reaction of a strong acid with a strong base is always:

    $$H^+(aq) \ + \ OH^-(aq) \ \rightarrow \ H_2O(l)$$

7. When writing a detailed ionic equation, all strong electrolytes (in other words, all strong acids, all strong bases, and all salts) are written in the ionic form. Other solutes are weak electrolytes (weak acids and weak bases) and these are written in the molecular form because this is the form in which they predominantly exist.
    Other substances that are not in solution are also written in the molecular form (gases that bubble out of the solution) or as formula units (salts that precipitate out of the solution).

8. A neutralization reaction is the reaction of an acid with a base; the base neutralizes the acid and *vice versa*. However the process does not necessarily result in a neutral solution. For example the reaction of equimolar quantities of a weak acid and a strong base leads to a solution that is slightly basic.

9. These reactions are described in section 5.6 of the text:

    Precipitation—the formation of an insoluble product.
    Acid-Base—the formation of water (and a salt).
    Gas-forming—the formation of a gas.
    Redox—oxidation and reduction.

    These reactions are all driven to completion because the products are thermodynamically more stable than the reactants—the reactions are product-favored.   See Chapter 6.

10. Oxidation—the loss of electron(s).
    Reduction—the gain of electron(s).

11. The reducing agent reduces some other reactant; it supplies electrons; and therefore is itself oxidized.
    The oxidizing agent oxidizes some other reactant; it removes electrons; and therefore is itself reduced.

12. An oxidizing agent often contains an element in a high oxidation state (for example: $KMnO_4$ or $K_2Cr_2O_7$). It is a substance that is readily reduced; it readily accepts electrons (for example: F_2 or O_2).

13. Look for a change in the oxidation number of any one of the reactants.   Oxidation numbers can be assigned following a set of simple rules; see page 207 of the text.

14. The oxidation number is a number assigned to an atom in a compound in order to keep account of the movement of electrons in a chemical reaction.

15. The molarity of a solution is a measure of the concentration of the solute in the solution. Specifically, the amount of solute is expressed in moles and the amount of solution is expressed in liters. The molarity is the number of moles of solute per liter of solution.

16. When a salt dissolves it dissociates into ions. The concentrations of the ions produced depends upon how many ions are present in one formula unit of the salt. For example, when one mole of sodium chloride NaCl dissolves, it produces one mole of sodium ions and one mole of chloride ions. However, when one mole of magnesium nitrate $Mg(NO_3)_2$ dissolves, it produces one mole of magnesium ions and two moles of nitrate ions. The concentration of nitrate ions is twice the concentration of magnesium ions.

17. There are two relationships that you will use over and over again. The first relates the mass of a substance and the number of moles:

$$mass = \# \ moles \times molar \ mass$$

The second relates the number of moles and the molarity and volume:

$$\# \ moles = molarity \times volume$$

It is the number of moles that is common to both relationships and in all stoichiometric calculations it is the number of moles that is important. Always work in moles!
For example, the number of moles in 5.0 liters of a 3.0 M solution
= 5.0L × 3.0 moles/L = 15 moles.
The mass of substance required = number of moles (15 moles) × molar mass of the substance.

18. The key to all dilution problems like this is to realize that the number of moles of solute does not change in a dilution. Only additional solvent (water) is added to the solution.

dilute solution   concentrated solution
molarity × volume   =   molarity × volume

The volume of concentrated solution required equals the molarity x volume of the dilute solution divided by the molarity of the concentrated solution.

19. Quantitative chemical analysis can be achieved by the volumetric titration of two reactants when the equivalence point is characterized by an observable change in a physical property of the solution. This is very often a change in color measured directly or through the use of an indicator, but it could, for example, be a change in conductivity. Two major types of titration are acid-base and oxidation-reduction.

20. The equivalence point in an acid-base titration is reached when the number of moles of H^+ (from the acid) exactly equals the number of moles of OH^- (from the base). The equivalence point in a redox titration is reached when the reactants have been added in the quantities indicated by the redox equation—one reactant has been exactly consumed by the other.

21. All stoichiometric calculations depend upon the mole ratio of reactants and products in a balanced equation. In an acid-base reaction, the balanced equation indicates the number of moles of acid and the number of moles of base required to reach the equivalence point. The same is true for a redox reaction.

Study Questions and Problems

1. Classify each of the following solutes as a strong electrolyte, weak electrolyte, or nonelectrolyte:

   sugar
   common salt (NaCl)
   alcohol
   acetic acid

   sodium hydroxide
   hydrochloric acid
   copper sulfate
   carbonic acid

2. Predict the solubility of the following salts:

   sodium sulfate
   potassium chromate
   silver bromide
   nickel(II) hydroxide
   aluminum nitrate
   barium sulfide
   ammonium acetate
   strontium iodide

3. Write the ions that are produced when the following substances dissolve in water:
   $Mg(OH)_2$
   K_2SO_4
   $NaHCO_3$
   $(NH_4)_3PO_4$
   $NaClO$

4. Predict whether or not the following reactions will lead to a precipitate. Write detailed and net ionic equations for all the reactions.

   a. potassium chromate and lead acetate
   b. silver perchlorate and ammonium chloride
   c. potassium carbonate and copper acetate
   d. sodium fluoride and magnesium iodide
   e. barium nitrate and potassium sulfate

A life spent making mistakes is not only more honorable but more useful than a life spent doing nothing.

George Bernard Shaw
(1856-1950)

5. Write the overall chemical reaction, the detailed ionic equation, and the net ionic equation, for the following acid-base reactions:

   a. acetic acid and potassium hydroxide
   b. hydrocyanic acid and ammonia
   c. nitric acid and sodium hydroxide
   d. sulfuric acid and ammonia (1:1 mole ratio)
   e. carbonic acid and sodium hydroxide (1:2 mole ratio)

6. Write equations for, and classify, the following reactions:

   a. nitric acid and cobalt(II) carbonate
   b. sodium sulfate and barium nitrate
   c. ammonia and acetic acid
   d. hydrochloric acid and calcium carbonate
   e. sodium hydroxide and nickel carbonate
   f. lead acetate and hydrochloric acid
   g. iron(III) nitrate and sodium hydroxide

7. Assign oxidation numbers to the underlined atoms in the following compounds:

   a. $\underline{N}O_3^-$
   b. $\underline{Cl}F_3$
   c. $NaH_2\underline{P}O_4$
   d. $Na\underline{Cl}O_4$

   e. $\underline{S}O_2$
   f. $K_2\underline{Cr}_2O_7$
   g. $\underline{Cu}(NO_3)_2$
   h. $K_2\underline{S}_2O_3$

8. In the following reactions determine which element is oxidized, which element is reduced, which reactant is the reducing agent, and which reactant is the oxidizing agent:

   a. $N_2H_4(aq) + 2O_2(g) \rightarrow N_2(g) + 2H_2O(g)$

   b. $3Cl_2(g) + NaI(aq) + 3H_2O(l) \rightarrow 6HCl(aq) + NaIO_3(aq)$

9. Determine whether the following reactions are acid-base reactions or oxidation-reduction reactions:

   a. $2N_2O(g) + 3O_2(g) \rightarrow 4NO_2(g)$

   b. $2NO_2(g) + H_2O(l) \rightarrow HNO_2(aq) + HNO_3(aq)$

   c. $MgCO_3(s) + H_2SO_4(aq) \rightarrow MgSO_4(s) + H_2O(l) + CO_2(g)$

   d. $Cu(s) + CuCl_2(aq) + 2Cl^-(aq) \rightarrow 2[CuCl_2]^-(s)$

   e. $HNO_3(aq) + KOH(aq) \rightarrow KNO_3(aq) + H_2O(l)$

10. a. If 5.00 grams of sodium hydroxide is dissolved to make 600 mL of solution, what is its molarity?

   b. How much potassium chloride has to be dissolved in water to produce 2.0 liters of a 2.45 M solution.

   c. If 24.63 grams of magnesium chloride is dissolved to make exactly 3 liters of solution, calculate the concentrations of the ions in the solution (in moles/liter).

11. a. What is the molarity of the solution that results from adding 25 mL of a 0.15 M solution of sodium hydroxide to sufficient water to make 500 mL of solution?

   b. What volume of a 2.50 M solution of hydrochloric acid is required to prepare 2.0 liters of a 0.30 M solution?

12. When excess silver nitrate was added to a 25.0 mL sample of a solution of calcium chloride, 0.9256 gram of silver chloride precipitated. What is the concentration of the calcium chloride solution?

13. What volume of a 0.291 M solution of NaOH is required to reach the equivalence point in a titration against 25.0 mL of 0.350 M HCl?

14. KHP (potassium hydrogen phthalate $KHC_8H_4O_4$) is often used as a primary standard for sodium hydroxide standardization:

   The molar mass of KHP is 204.224 g/mol.

   $$HC_8H_4O_4^-(aq) + OH^-(aq) \rightarrow C_8H_4O_4^{2-}(aq) + H_2O(l)$$

   If a 0.4856 gram sample of KHP is dissolved in sufficient water to prepare 250 mL of solution, and 25 mL of the solution requires 18.76 mL of sodium hydroxide solution to reach the equivalence point, what is the molarity of the sodium hydroxide?

Answers to Study Questions and Problems

1.
sugar	nonelectrolyte
common salt (NaCl)	strong electrolyte
alcohol	nonelectrolyte
acetic acid	weak electrolyte
sodium hydroxide	strong electrolyte
hydrochloric acid	strong electrolyte
copper sulfate	strong electrolyte
carbonic acid	weak electrolyte

2. | | | |
    |---|---|---|
    | sodium sulfate | soluble | alkali metal salt |
    | potassium chromate | soluble | alkali metal salt |
    | silver bromide | insoluble | one of 3 exceptions for the halides |
    | nickel(II) hydroxide | insoluble | metal hydroxide |
    | aluminum nitrate | soluble | nitrate |
    | barium sulfide | insoluble | metal sulfide |
    | ammonium acetate | soluble | both NH_4^+ and $CH_3CO_2^-$ confer solubility |
    | strontium iodide | soluble | almost all iodides are soluble |

3. $Mg(OH)_2 \rightarrow$ One Mg^{2+} and two OH^-
    $K_2SO_4 \rightarrow$ Two K^+ and one SO_4^{2-}
    $NaHCO_3 \rightarrow$ Na^+ and HCO_3^-
    $(NH_4)_3PO_4 \rightarrow$ Three NH_4^+ and one PO_4^{3-}
    $NaClO \rightarrow$ Na^+ and ClO^-

The *(aq)* notation is omitted for clarity.

4. a. potassium chromate + lead acetate → insoluble lead chromate

    $$K_2CrO_4 + Pb(CH_3CO_2)_2 \rightarrow PbCrO_4(s) + 2KCH_3CO_2$$
    $$2K^+ + CrO_4^{2-} + Pb^{2+} + 2CH_3CO_2^- \rightarrow PbCrO_4(s) + 2K^+ + 2CH_3CO_2^-$$
    $$CrO_4^{2-} + Pb^{2+} \rightarrow PbCrO_4(s)$$

    b. silver perchlorate + ammonium chloride → insoluble silver chloride

    $$AgClO_4 + NH_4Cl \rightarrow AgCl(s) + NH_4ClO_4$$
    $$Ag^+ + ClO_4^- + NH_4^+ + Cl^- \rightarrow AgCl(s) + NH_4^+ + ClO_4^-$$
    $$Ag^+ + Cl^- \rightarrow AgCl(s)$$

    c. potassium carbonate + copper acetate → insoluble copper carbonate

    $$K_2CO_3 + Cu(CH_3CO_2)_2 \rightarrow CuCO_3(s) + 2KCH_3CO_2$$
    $$2K^+ + CO_3^{2-} + Cu^{2+} + 2CH_3CO_2^- \rightarrow CuCO_3(s) + 2K^+ + 2CH_3CO_2^-$$
    $$CO_3^{2-} + Cu^{2+} \rightarrow CuCO_3(s)$$

    d. sodium fluoride + magnesium iodide → insoluble magnesium fluoride

    $$2NaF + MgI_2 \rightarrow MgF_2(s) + 2NaI$$
    $$2Na^+ + 2F^- + Mg^{2+} + 2I^- \rightarrow MgF_2(s) + 2Na^+ + 2I^-$$
    $$2F^- + Mg^{2+} \rightarrow MgF_2(s)$$

    e. barium nitrate + potassium sulfate → insoluble barium sulfate

    $$Ba(NO_3)_2 + K_2SO_4 \rightarrow BaSO_4(s) + 2KNO_3$$
    $$Ba^{2+} + 2NO_3^- + 2K^+ + SO_4^{2-} \rightarrow BaSO_4(s) + 2K^+ + 2NO_3^-$$
    $$Ba^{2+} + SO_4^{2-} \rightarrow BaSO_4(s)$$

5.   a.   acetic acid and potassium hydroxide:

$CH_3CO_2H + KOH \rightarrow KCH_3CO_2 + H_2O$

$CH_3CO_2H + K^+ + OH^- \rightarrow K^+ + CH_3CO_2^- + H_2O$

$CH_3CO_2H + OH^- \rightarrow CH_3CO_2^- + H_2O$

The *(aq)* have been omitted to simplify the equations. Normally, all species dissolved in water would have the notation *(aq)*.

   b.   hydrocyanic acid and ammonia:

$HCN + NH_3 \rightarrow NH_4CN$

$HCN + NH_3 \rightarrow NH_4^+ + CN^-$ (detailed; no change for the net)

   c.   nitric acid and sodium hydroxide:

$HNO_3 + NaOH \rightarrow NaNO_3 + H_2O$

$H^+ + NO_3^- + Na^+ + OH^- \rightarrow Na^+ + NO_3^- + H_2O$

$H^+ + OH^- \rightarrow H_2O$ (same for any strong acid-strong base reaction)

   d.   sulfuric acid and ammonia (1:1 mole ratio):

$H_2SO_4 + NH_3 \rightarrow NH_4HSO_4$

$H^+ + HSO_4^- + NH_3 \rightarrow NH_4^+ + HSO_4^-$

$H^+ + NH_3 \rightarrow NH_4^+$

   e.   carbonic acid and sodium hydroxide (1:2 mole ratio):

$H_2CO_3 + 2NaOH \rightarrow Na_2CO_3 + 2H_2O$

$H_2CO_3 + 2Na^+ + 2OH^- \rightarrow 2Na^+ + CO_3^{2-} + 2H_2O$

$H_2CO_3 + 2OH^- \rightarrow CO_3^{2-} + 2H_2O$

6.   a.   nitric acid and cobalt(II) carbonate:

$2HNO_3 + CoCO_3 \rightarrow Co(NO_3)_2 + H_2O + CO_2(g)$

gas-forming reaction

   b.   sodium sulfate and barium nitrate:

$Na_2SO_4 + Ba(NO_3)_2 \rightarrow BaSO_4(s) + 2NaNO_3$

precipitate-forming reaction

   c.   ammonia and acetic acid:

$CH_3CO_2H + NH_3 \rightarrow NH_4CH_3CO_2$

acid-base reaction

   d.   hydrochloric acid and calcium carbonate:

$2HCl + CaCO_3 \rightarrow CaCl_2 + H_2O + CO_2(g)$

gas-forming reaction

e. sodium hydroxide and nickel carbonate:

$2NaOH + NiCO_3(s) \rightarrow Ni(OH)_2(s) + Na_2CO_3$

both nickel(II) carbonate and nickel(II) hydroxide are insoluble.

f. lead acetate and hydrochloric acid:

$Pb(CH_3CO_2)_2 + 2HCl \rightarrow 2CH_3CO_2H + PbCl_2(s)$

precipitation; $PbCl_2$ is one of three common insoluble chlorides

g. iron(III) nitrate and sodium hydroxide:

$Fe(NO_3)_3 + 3NaOH \rightarrow Fe(OH)_3(s) + 3NaNO_3$

precipitation; $Fe(OH)_3$ is insoluble.

7.
a.	$\underline{N}O_3^-$	+5	e.	$\underline{S}O_2$	+4
b.	$\underline{Cl}F_3$	+3	f.	$K_2\underline{Cr}_2O_7$	+6
c.	$NaH_2\underline{P}O_4$	+5	g.	$\underline{Cu}(NO_3)_2$	+2
d.	$Na\underline{Cl}O_4$	+7	h.	$K_2\underline{S}_2O_3$	+2 (average)

8. a. $N_2H_4(aq) + 2O_2(g) \rightarrow N_2(g) + 2H_2O(g)$

nitrogen: –2 in N_2H_4 to zero in N_2 loses electrons, is oxidized, is the reducing agent.

oxygen: zero in O_2 to –2 in H_2O gains electrons, is reduced, is the oxidizing agent.

b. $3Cl_2(g) + NaI(aq) + 3H_2O(l) \rightarrow 6HCl(aq) + NaIO_3(aq)$

chlorine: zero in Cl_2 to –1 in HCl   gains electrons, is reduced, is the oxidizing agent.

iodine: –1 in NaI to +5 in $NaIO_3$ loses electrons, is oxidized, is the reducing agent.

9. a. $2N_2O(g) + 3O_2(g) \rightarrow 4NO_2(g)$
redox; observe the change in oxidation state of the O for example: in O_2 it is zero and in NO_2 it is –2.

b. $2NO_2(g) + H_2O(l) \rightarrow HNO_2(aq) + HNO_3(aq)$
redox; a disproportionation reaction; observe the change in oxidation state of the nitrogen, it is both oxidized and reduced.

c. $MgCO_3(s) + H_2SO_4(aq) \rightarrow MgSO_4(s) + H_2O(l) + CO_2(g)$
acid-base; no change in oxidation state.

d. $Cu(s) + CuCl_2(aq) + 2Cl^-(aq) \rightarrow 2[CuCl_2]^-(s)$
redox; the copper metal is oxidized; the copper(II) is reduced.

e. $HNO_3(aq) + KOH(aq) \rightarrow KNO_3(aq) + H_2O(l)$
acid-base; no change in oxidation state.

10. a. The molar mass of sodium hydroxide is 40 g/mol.
So 5.00 grams is 5.0/40 = 0.125 mol NaOH.

> In general, always work in moles. Convert from mass as the first step and convert back to mass as the last step.

Molarity is moles/liter.
So the molarity of the solution is 0.125 mol/0.600 liter = 0.208 M.

> Remember to convert mL to L.

b. Molarity × volume = #moles.
2.45 M × 2.0 liters = 4.90 moles.
4.90 moles × 74.55 g/mol = 365 grams.

c. 24.63 grams of magnesium chloride = 24.63/95.21 = 0.259 mol.
The concentration of $MgCl_2$ = 0.259 mol/3 liters = 0.0862 mol/L.

> Molar mass of $MgCl_2$ = 95.21 g/mol.

The concentration of Mg^{2+} ions must be the same, since there is one Mg^{2+} ion in every $MgCl_2$ formula unit.
However, the concentration of Cl^- ions must be twice this concentration, since each $MgCl_2$ formula unit produces two Cl^- ions.
The concentration of Cl^- ions = 0.172 mol/L.

11. a. In this dilution, 0.15 M × 25 mL must equal the unknown molarity × 500 mL.
So the unknown molarity is 0.0075 M (the solution is more dilute).

> Remember that in a dilution problem, the number of moles of solute does not change.

b. This is the same kind of problem, in this case the unknown volume × 2.50 M hydrochloric acid must equal 2.0 liters × 0.30 M HCl.
So the unknown volume = 0.24 L or 240 mL.

12. 0.9256 gram of AgCl = 0.9256/143.32 = 0.006458 mol.
The number of moles of Cl^- ion is the same, since each AgCl formula unit contains one Cl^- ion.
However, one formula unit of calcium chloride $CaCl_2$ contains two Cl^- ions.
So the number of moles of calcium chloride $CaCl_2$ is one-half the number of moles of Cl^- ions = 0.003229 mol.

> The molar mass of AgCl = 143.32 g/mol.

This is the number of moles in 25 mL, or 0.025 L.
So the concentration of calcium chloride = 0.003229 mol/0.025 L = 0.129M (or 14.34 g/L).

> The molar mass of $CaCl_2$ = 110.98 g/mol.

13. This neutralization reaction has a 1:1 stoichiometry:

$$HCl(aq) + NaOH(aq) \rightarrow NaCl(aq) + H_2O(l)$$

The number of moles of NaOH must equal the number of moles of HCl at the equivalence point:
Volume of NaOH × 0.291 M NaOH = 25.0 mL HCl × 0.350 M HCl.

> Number of moles = volume × molarity.

So the volume of NaOH = (25.0 mL × 0.350 M)/0.291 M = 30.1 mL.

14. This neutralization reaction also has a 1:1 stoichiometry:

$$HC_8H_4O_4^-(aq) \ + \ OH^-(aq) \ \rightarrow \ C_8H_4O_4^{2-}(aq) \ + \ H_2O(l)$$

The molar mass of KHP is 204.224 g/mol.

0.4856 gram of KHP = 0.4856 g/ 204.224 g/mol = 0.002378 mol.
This is the number of moles of KHP in 250 mL.

Therefore, the number of moles in 25 mL = 0.0002378 mol KHP.
This must also be the number of moles of sodium hydroxide in 18.76 mL (this is the result of the titration).
Therefore the concentration of NaOH = 0.0002378 mol/ 0.01876 liter.
Molarity of NaOH = 0.01268 M.

CHAPTER 6

Introduction

Why do some substances react whereas others do not? How can we predict if a reaction will happen? If it does happen, does it happen quickly? We begin to study these questions in this chapter.

The rates of chemical reactions are examined in Chapter 15.

Almost invariably, a chemical reaction either liberates or absorbs energy in the form of heat, light or electricity. In this chapter we will look at the energy liberated or absorbed in the form of heat.

A chemical reaction is a rearrangement of atoms; in this rearrangement bonds between atoms are broken and new bonds between atoms are formed. Depending upon the relative strengths of the bonds, the difference in energy is required by, or released in, the reaction.

Contents

6.1 Types of Chemical Reactions and Thermodynamics

When a reaction proceeds from reactants to products **spontaneously**, it is said to be **product-favored**. An example is the combination of hydrogen and oxygen to form water.

A spontaneous reaction is defined specifically as a reaction that will happen by itself. Spontaneity does not in any way indicate how fast the reaction will occur. It may occur very slowly.

$$2H_2(g) \ + \ O_2(g) \ \rightarrow \ 2H_2O(l)$$

The reverse reaction does not happen without some outside intervention. An example of such an intervention could be an electrical current passed through the water to break up the molecules to form hydrogen and oxygen — a process called electrolysis:

$$2H_2O(l) \ \rightarrow \ 2H_2(g) \ + \ O_2(g)$$

The transfer of heat when chemical reactions occur, and the relationship between heat and work, is the subject of **thermodynamics**.

6.2 Energy: Its Forms and Units

Energy is the capacity to do work; work requires energy, and work and energy have the same units. Energy can be classified as kinetic or potential energy. **Kinetic energy** is the energy of movement; for example:

Thermal energy   — random motion of molecules at the particulate level.
Mechanical energy— movement of a macroscopic object.
Electrical energy   — movement of electrons through a conductor.
Sound energy   — organized compression and expansion of the spaces between molecules.

Potential energy is the energy due to the position of an object; for example:

Chemical energy   — the energy of electrons and their positions relative to the nuclei.
Gravitational energy   — due to position in a gravitational field.
Electrostatic energy   — relative positions of positive & negative charges.

Energy can neither be created nor destroyed; the energy of the universe is constant. Heat, work, and other forms of energy may be changed into one another but the total remains constant.

Temperature is a measure of how hot an object is. A knowledge of the temperatures of two objects in contact allows a prediction of the direction of energy transfer between them—energy always transfers spontaneously from the hotter object to the colder object. Heat transfer will occur until both objects are at the same temperature.

The kinetic energy of molecules (or other particulate matter) is called thermal energy. The thermal energy of an object is the sum of the kinetic energies of the individual particles (atoms or molecules) in the object.

The unit of energy or work most commonly used by chemists is the **joule** J (or kilojoule kJ). This is the SI unit. An older unit is the **calorie** which is defined as the quantity of heat required to raise the temperature of 1 gram of water from 14.5 to 15.5°C.

Power is the rate at which work is done. A more powerful person can do a certain amount of work more quickly than a weaker person. The units of power then are Js^{-1} (defined as a watt W). Therefore work or energy can be expressed as power × time (as in kilowatt-hours kWh).

6.3 Specific Heat Capacity and Thermal Energy Transfer

The relationship between the heat gained by an object, and the temperature change that results, is called the **heat capacity** of the object. The higher the heat capacity, the more heat is required for a particular temperature change:

Heat (J) = heat capacity (JK^{-1}) × temperature change (K)

Heat capacity is an extensive property; it depends upon the size of the object. It is often more useful to use the **specific heat capacity** which is defined as the heat capacity per gram of material. The specific heat capacity is therefore an intensive property. The units of specific heat capacity are $JK^{-1}g^{-1}$.

This is known as the First Law of Thermodynamics.

Heating a system does not always increase the temperature of the system. An example is heating a block of ice, which remains at O°C until the ice has all melted—the heat (energy) is used in breaking the bonds between the water molecules.

An object may also possess kinetic energy by virtue of the macroscopic motion of the object as a whole. This energy can do work and the difference in the two kinetic energies is a way to distinguish between work and heat.

One calorie = 4.184 joules.

See the Problem Solving Tips and Ideas 6.1 on page 251.

The specific heat capacity is often called simply the specific heat.

Heat (J) = specific heat capacity ($JK^{-1}g^{-1}$) × mass (g) × temp change (K)

The specific heat capacity can also be expressed per mole of material, in which case it is called the **molar heat capacity**.

The specific heat capacity of water is much larger than for most substances. The reason for this is related to the unusually strong bonds between water molecules in the liquid state which are gradually broken as more and more heat is added.

6.4 Energy and Changes of State

When a solid melts, the atoms, molecules, or ions that make up the solid have sufficient energy to loosen and break the bonds between themselves. A liquid is formed. When a liquid vaporizes, any attraction between the structural units becomes ineffective, the intermolecular bonds break completely, and a gas is formed. Both processes involve breaking bonds and both therefore require energy. The energy required to melt a substance is called the **heat of fusion**. The energy required to vaporize a liquid is called the **heat of vaporization**.

These **changes of state**, from solid to liquid, and from liquid to gas, occur at constant temperatures. For example, when ice melts at one atmosphere pressure, the temperature is 0°C and remains 0°C until all the ice has melted. Both heats of fusion and vaporization expressed in J are extensive properties.

6.5 Enthalpy

The entire **universe** can be divided into a system and its surroundings. The **system** is that part of the universe that is under examination; for example: a beaker on a hot plate, a balloon full of gas, an automobile engine, a house, or the planet earth. The **surroundings** are the rest of the universe.

Thermodynamics is concerned with how **work** and **heat** are transferred between the system and its surroundings.

By definition, any work (w) done on, or heat (q) transferred into, the system (resulting in an increase in the energy of the system) is positive. Work done by the system or heat leaving the system is negative. The change in energy of the system is therefore:

$$\Delta E = q + w$$

If there is no mechanical or electrical connection between the system and its surroundings, then no work can be exchanged between them. In this case, the change in the energy of the system equals the quantity of heat put into the system from the surroundings.

If the system does not change in volume and no work is done, w = zero and

$$\Delta E = q_v \text{ where the subscript v indicates constant volume}$$

Mechanical work can be done, for example, if the system expands at constant pressure. In this case work is done by the system on the surroundings. The work done is equal to the pressure multiplied by the change in volume:

$$\text{work done by the system} = -w = P\Delta V$$

Therefore, heat capacity = specific heat capacity × mass

Note that specific heat capacity refers to a particular homogeneous material – steel, wood, polystyrene, copper, etc. The heat capacity refers to an object that may be heterogeneous.

The bonds between water molecules are called hydrogen bonds. See page 188 (Chapter 13) in this study guide.

You should be able to use the expressions for heat capacity and specific heat to calculate heat from a temperature change and a temperature change from the heat.

Fusion means melting.

Note that these heats of fusion and vaporization can be expressed in units of Jg^{-1} or $Jmol^{-1}$ which are intensive quantities.

This is another expression of the 1st law of thermodynamics (the law of conservation of energy).

The sign is negative because the system loses energy when it does work.

So $\Delta E = q + w = q - P\Delta V$

or $q_p = \Delta E + P\Delta V$, where the subscript p indicates constant pressure.

Most reactions occur under constant atmospheric pressure, so most heats of reaction measured in the laboratory are q_p rather than q_v. Because the quantity q_p (or $\Delta E + P\Delta V$) is such a fundamental quantity (it is a state function), it is given its own name and symbol. It is called the **enthalpy change** ΔH.

Typically, a system is heated (and therefore gains energy q), and does some work on its surroundings (losing some energy –w), resulting in an increase in energy of the system equal to q + w (where in this case q is positive and w is negative).

If a process requires energy (as heat) then the system will absorb heat as the process occurs. The enthalpy of the system will increase and ΔH will be positive. Such as process is called **endothermic**. An example is the vaporization of water—energy is required to break the attractions between the water molecules. If, on the other hand, a process releases energy, then ΔH is negative and the process is called **exothermic**. An example is water freezing—the formation of the bonds between the water molecules releases energy.

6.6 Enthalpy Changes for Chemical Reactions

When reactants in a chemical reaction form products, the atoms rearrange, bonds are broken and new ones formed. The electrons are redistributed. The chemical potential energy of these electrons changes as the reactants form products. The difference in the energies of the reactants and products results in energy being absorbed (endothermic) or released (exothermic). For example, when hydrogen and oxygen combine to form water, a considerable amount of energy is released:

$$H_2(g) \ + \ 1/2O_2(g) \ \rightarrow \ H_2O(l) \ + \ 241.8 \text{ kJ}$$

Energy is a product of the reaction. The amount of energy released is 241.8 kJ for every mole of water that is produced. The enthalpy change for the reaction is –241.8 kJ (exothermic) and the reaction is usually represented as:

$$H_2(g) \ + \ 1/2O_2(g) \ \rightarrow \ H_2O(l) \quad \Delta H = -241.8 \text{ kJ}$$

The negative sign for ΔH indicates that the heat is released. The reverse reaction, breaking up water to form hydrogen and oxygen, requires exactly the same quantity of energy. This process is endothermic:

$$H_2O(l) \ \rightarrow \ H_2(g) \ + \ 1/2O_2(g) \quad \Delta H = +241.8 \text{ kJ}$$

Where does the energy come from in an endothermic process? And where does it go? To answer the second question first, the process is endothermic because the potential energies of the electrons and nuclei of the product molecules are higher than their potential energies in the reactants. Energy is required to raise the potential energies of these particles. There are two sources for this energy. The energy may come from the surroundings (the system could be heated for example). Or the energy could come from the system itself if the

---

An example of a system that can do work on its surroundings without any change in the volume of the system is an electrical cell or battery.

State functions are described in section 6.8 on page 81 of this study guide.

Using ΔH rather than ΔE avoids the necessity of figuring out what the change in volume is—which could be very difficult.

Remember that if heat is transferred into the system, q is positive. If work is done on the system, w is positive.

It is sometimes difficult to imagine the freezing of ice to be a process that releases heat (why doesn't the heat melt the ice?). Heat must be removed to cause the water to freeze – this is what a freezer does.

If ΔH is positive for a process in one direction, then ΔH is opposite in sign, but equal in magnitude, for the same process in the reverse direction.

Note that ΔH is an extensive property. The energy released or absorbed in a chemical reaction or physical process depends upon the quantities of matter involved.

The positive sign is often omitted.

kinetic energy of the atoms and molecules in the system is reduced. In this case the temperature of the system decreases.

6.7 Hess's law

If a reaction is the sum of two other reactions, then ΔH for the overall reaction is the sum of the two ΔH for the two constituent reactions. This is **Hess's law of constant heat summation**. In other words, ΔH is same regardless of the route from the starting point (the reactants) to the finishing point (the products).

Hess's law is useful because it allows you to calculate the enthalpy change for a reaction that might be difficult to determine experimentally.

There are two approaches to take in solving Hess's law problems. The first is the construction of an energy cycle, where two routes between the initial and final states are shown. The enthalpy change for each route must be the same, and setting them equal allows you to calculate the unknown enthalpy change. The second approach involves listing all the given equations so that when they are added together, the desired equation is produced. Adding the enthalpy changes therefore produces the desired ΔH. See the solution to problem #11.

It is important to remember that ΔH is an extensive property: If the stoichiometric coefficients are all doubled, the ΔH must also be doubled. If the reaction is reversed, the sign of ΔH must be reversed.

6.8 State Functions

A **state function** is a property of a system that does not depend upon how the state was arrived at. For example, temperature is a state function. If a system is at 25°C, it could have cooled down from 100°C or it could have warmed up from 0°C. Regardless of what the temperature has been, it is now 25°C.

Enthalpy is a state function. Because enthalpy is a state function, the enthalpy change in going from an initial state to a final state is always the same, regardless of the route taken.

6.9 Standard Enthalpies of Formation

The **standard state** of a substance is the most stable state in which that substance exists at 25°C (usually) and a pressure of 1 bar. When a reaction occurs such that all reactants and products are in their standard states, the heat of reaction ΔH is called the **standard enthalpy change of reaction $\Delta H°$**, where the superscript ° indicates the standard conditions.

When one mole of a compound in its standard state is formed from its constituent elements all in their standard states, the heat of reaction is called the **standard molar enthalpy of formation $\Delta H_f°$**. Since forming an element in its standard state from the same element in its standard state involves no change, the standard enthalpies of formation of all elements are zero.

Most $\Delta H_f°$ are negative; in other words, most compounds are formed in product-favored exothermic reactions. Some compounds however, such as acety-

> This is why water evaporating from your skin makes your skin feel cold. Evaporation is an endothermic process.

> This is just another statement of the first law of thermodynamics (energy is conserved).

> Try problems 6.11 involving Hess's law.

> This is Hess's law.

> 1 bar is almost the same as 1 atm pressure.
>
> 1 bar = 0.98692 atm and 1 bar = 10^5 Pa exactly.
>
> The enthalpy change for a reaction is commonly called the heat of reaction (at constant pressure).
>
> $\Delta H_f°$ of an element = zero.

lene C_2H_2 and hydrazine N_2H_2, have positive ΔH_f° values. Such compounds store considerable chemical potential energy, which is released when the compounds are broken down, and the compounds are therefore good fuels. Hydrazine, for example, is used as a rocket fuel.

See Table 6.2 (p 270) and Appendix L.
See also www.nist.gov. for current values.

Standard enthalpies of formation are known for many compounds. They are useful because the enthalpy change for any reaction can be calculated if the standard enthalpies of formation of the reactants and products are known.

$$\Delta H^\circ_{reaction} = \Sigma(\Delta H_f^\circ \text{(products)} - \Delta H_f^\circ \text{(reactants)})$$

This equation is simply another way of expressing the 1st law of thermodynamics. It can be derived using Hess's law (see page 272 in the text and review question 18 in this study guide).

6.10   Determining Enthalpies of Reaction

The heat evolved or absorbed in a chemical reaction or physical process is determined by measuring a temperature change. The technique is called **calorimetry**, literally heat-measurement. The device in which the reaction takes place is called a calorimeter. If the heat capacity of the calorimeter and its contents is known, then the heat of reaction can be calculated from the temperature change:

Heat (J) = heat capacity (JK^{-1}) × temperature change (K)

Look at problems 14 and 15 for examples of the use of a bomb calorimetry and a constant pressure calorimeter.

If the calorimeter is a constant volume calorimeter (a **bomb calorimeter**), then the heat equals ΔE, the change in the energy of the system. If the calorimeter and contents are under constant external pressure, then the heat equals ΔH, the change in enthalpy of the system.

6.11   Applications of Thermodynamics

Thermodynamics is the science of the transfer of energy as heat and work. As such, it is relevant in the study of how energy is used in our economy.

Sources of energy include wood (biomass), fossil fuels (petroleum, coal, natural gas), hydroelectric, solar, geothermal, wind, ocean currents, and nuclear. Wood and fossil fuels contain chemical potential energy; a chemical reaction (combustion) releases this energy for use. This chemical potential energy derived originally from solar energy and was captured through photosynthesis. Wind, ocean currents, and hydroelectric sources all owe their potential and kinetic energies to the action of the sun.

Energy is not *used up* (the energy of the universe is constant), but it is *used*. It is most *useful* when it does work.

Solar energy can be transformed directly into electrical energy by photovoltaic cells.

Hydrogen gas is a clean fuel; the product of its combustion is water. A hydrogen economy, where the hydrogen is produced in artificial photosynthesis and used in a fuel cell, is very attractive.

Review Questions

1.   Define a spontaneous reaction. How can you tell whether a reaction is spontaneous?

2.   Which reactions studied in Chapter 5 are "product-favored"?

3.   List and describe various forms of energy.

4.   What is the relation between power, energy, and time?

5.   Which is the smaller unit, the calorie or the joule? How are they related?

6.   Define heat capacity, specific heat capacity, and molar heat capacity, indicating clearly the differences between them.

7.   Why does water have a high specific heat capacity? What does this mean?

8.   Write different expressions for the 1st Law of Thermodynamics. How are they related?

9.   Define the heat of fusion and the heat of vaporization. What do these two quantities depend upon?

10.   Define or describe heat and work.

11.   What is the enthalpy change for a process? How is it related to the change in the energy of a system? When does the enthalpy change equal the energy change?

12.   Define an exothermic and an endothermic process.

13.   Where does the energy come from in an endothermic process? And where does it go?

14.   What happens to ΔH for a reaction if the coefficients in the equation are all tripled.

15.   What does standard state mean?

16.   Why is the standard molar enthalpy of formation of an element equal to zero?

17.   Define the standard enthalpy change for a reaction $\Delta H°$. Define standard molar enthalpy of formation $\Delta H_f°$.

18. Derive diagrammatically, or by using Hess's law, the expression for the standard enthalpy change for a reaction in terms of the standard molar enthalpies of formation of the reactants and products.

Answers to Review Questions

1. A spontaneous reaction is a reaction that happens by itself. It may happen quickly or slowly but it does happen. Calculations in thermodynamics can be done to determine whether or not a reaction is spontaneous. However, if a process or reaction does happen by itself, you can be sure that it is spontaneous.

Note, however, that a nonspontaneous reaction will happen when it is coupled with a spontaneous reaction—see Chapter 20.

2. Reactions discussed in Chapter 5 that are product-favored are:
precipitate-forming reactions.
gas-forming reactions.
reactions in which a weak electrolyte is formed.
reduction-oxidation reactions.

3. Some examples of kinetic energy are: thermal, mechanical, electrical, and sound. Examples of potential energy include chemical, gravitational, and electrostatic.

4. Power is the rate at which work is done. Power = work/time. The SI unit for power is the watt W which is defined as Js^{-1}.

5. The joule is smaller than the calorie: one calorie = 4.184 joules.

6. The heat capacity is the relationship between the heat gained by an object and the temperature change that results. The higher the heat capacity, the more heat is required for a particular temperature change. The units are JK^{-1}.
The specific heat capacity is defined as the heat capacity per gram of material. The units of specific heat capacity are $JK^{-1}g^{-1}$.
The molar heat capacity is defined as the heat capacity expressed per mole of material. The units are $JK^{-1}mol^{-1}$.

7. The specific heat capacity of water is much larger than for most substances because of the unusually strong bonds between the water molecules. These intermolecular bonds are progressively broken as more and more heat is added. What this means is that a considerable quantity of heat is required to heat water and a considerable amount of heat must be transferred out of the water before it cools down appreciably.

8. Some expressions for the 1st law are:
$\Delta E = q + w$ where ΔE refers to the system

$q_{in} = q_{out}$ heat gained = heat lost

$\Delta E = $ zero   where ΔE refers in this case to the entire universe

$\Delta H^\circ_{reaction} = \Sigma(\Delta H^\circ_f \text{(products)} - \Delta H^\circ_f \text{(reactants)})$

All these expressions represent an energy balance, reflecting the fact that energy can neither be created nor destroyed.

9. The heat of fusion is the quantity of heat required to melt a certain quantity of a substance. The heat of vaporization is similarly the heat required to vaporize a certain quantity of a substance. The heat required is extensive and therefore heats are normally expressed as quantities per gram or per mole. Both quantities depend upon the strength of the bonds between the structural units in the solid or liquid.

10. Heat, or thermal energy, is a chaotic or incoherent motion of particles at the atomic and molecular level. The hotter a substance, the more vigorous is the motion of the particles. Work is the result of a concerted or coherent movement of particles—for example, a wind blowing can do work by rotating a windmill regardless of the temperature (or thermal energy) of the air.

11. Enthalpy can be interpreted literally as the "heat within". If the enthalpy of a system decreases, then it loses heat. If the enthalpy increases, then it gains heat. In these instances, the heat refers to the heat released or absorbed at constant external pressure q_p. The energy of a system may change if either heat leaves or enters the system or work is done by or on the system. Enthalpy is a useful concept because it incorporates any work done when the system changes volume at constant pressure. The relation between the energy change ΔE and the enthalpy change ΔH is $\Delta H = \Delta E + P\Delta V$.

12. An exothermic process is a process that liberates energy. Invariably the system gets hot because the kinetic energy of the particles in the system increases. An endothermic process is one which requires or absorbs energy. In this case the temperature of the system falls because the energy required in the reaction is taken from the kinetic energy of the particles in the system. In both cases there is probably movement of energy between the system and its surroundings—only in an isolated (well-insulated) system is this prevented.

13. A process is endothermic when the chemical potential energy of the product molecules is higher than the chemical potential energy of the reactant molecules. Energy is required to raise the potential energy in the reaction. There are two sources for this energy. The energy may come from the surroundings if the system is heated. Or the energy could come from the system itself if the kinetic energy of the atoms and molecules in the system is reduced. In this case the temperature of the system decreases.

14.   The enthalpy change for a particular reaction is a characteristic of that reaction, and as such, is constant under the prescribed conditions. However, the reaction is represented by a chemical equation, and the ΔH is listed for the equation as it is written, where the stoichiometric coefficients represent moles of reactants and products. If these coefficients are tripled, then the value listed for ΔH must also be tripled. For example, if one mole of propane releases 2220 kJ when burned, then three moles will release three times as much, or 6660 kJ.

15.   The standard state of a substance is the most stable form of that substance at a pressure of 1 bar and a specified temperature (usually 25°C).

16.   The standard molar enthalpy of formation of a substance is the enthalpy change for a reaction in which one mole of the substance in its standard state is made from its constituent elements in their standard states. For a substance that is an element, such a reaction represents no change, and therefore the enthalpy change must be zero.

17.   The standard molar enthalpy of reaction is the enthalpy change for the reaction in which all reactants and products are in their standard states. For the standard molar enthalpy of formation of a substance see the answer to question 16 (preceding).

18.   For the constituent elements of the reactants $\rightarrow$ the reactants
$\Delta H° = \Delta H_f° \text{(reactants)}$
For the same set of constituent elements $\rightarrow$ the products
$\Delta H° = \Delta H_f° \text{(products)}$
According to Hess's law, the $\Delta H°$ for the reaction is independent of the route:

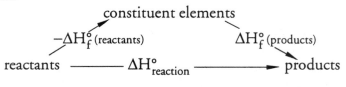

Therefore, $\Delta H°_{\text{reaction}} = \Delta H_f° \text{(products)} - \Delta H_f° \text{(reactants)}$

Note the negative sign for $-\Delta H_f° \text{(reactants)}$ **—the reaction is reversed.**

Study Questions and Problems

If you can measure that of which you speak, and you can express it by a number, you know something about your subject; but if you cannot measure it, your knowledge is meager and unsatisfactory.

Lord Kelvin
(1824-1907)

1.   What transfer of energy takes place when

   a.   A saucepan of water is heated to boiling
   b.   A tennis ball bounces to rest on the floor
   c.   Two automobiles collide head on
   d.   A flashlight is left on until the battery runs down
   e.   A fire burns
   f.   A wind turbine generates electricity
   g.   A chemical reaction liberates heat

2.   a.   Convert 800 kWh into J
      b.   Convert 377 kcal into J

3.   If the temperature of a 50.0 gram block of aluminum increases by 10.9 K
      when heated by 500 joules, calculate the
      a.   heat capacity of the aluminum block
      b.   molar heat capacity of aluminum
      c.   specific heat of aluminum

4.   The specific heat of gold is 0.128 $JK^{-1}g^{-1}$ and the specific heat of iron is
      0.451 $JK^{-1}g^{-1}$. Calculate the molar heat capacities of these two metals
      and compare to the value for aluminum calculated in question 3.

5.   Calculate the heat necessary to change the temperature of one kg of iron
      from 25°C to 1000°C. The specific heat of iron is 0.451 $JK^{-1}g^{-1}$.

6.   If a 40 gram block of copper at 100°C is added to 100 grams of water at
      25°C, calculate the final temperature assuming no heat is lost to the
      surroundings. The specific heat of copper is 0.385 $JK^{-1}g^{-1}$ and the spe-
      cific heat of water is 4.184 $JK^{-1}g^{-1}$.

7.   Calculate the amount of heat necessary to melt 27.0 grams of ice if the
      heat of fusion of ice is 6.009 kJ/mol.

8.   If 27.0 grams of ice at 0°C is added to 123 grams of water at 100°C in an
      insulated container, calculate the final temperature. Assume that the spe-
      cific heat of water is 4.184 $JK^{-1}g^{-1}$.

9.   A 50 gram block of an unknown metal alloy at 100°C is dropped into an
      insulated flask containing approximately 200 grams of ice. It was deter-
      mined that 10.5 grams of the ice melted. What is the specific heat capac-
      ity of the unknown alloy?

10.   If the enthalpy change for the combustion of propane is –2220 kJ/mol
      propane, what quantity of heat is released when 1 kg of propane is burned?

      $C_3H_8(g) + 5O_2(g) \rightarrow 3CO_2(g) + 4H_2O(l)$ $\Delta H = -2220$ kJ

      How much heat is released when 2 kg of propane is burned?

11.   Using the following thermochemical data, calculate the molar heat of
      combustion ΔH_f° of methane CH_4:

      $CH_4(g) + 2O_2(g) \rightarrow CO_2(g) + 2H_2O(l)$

      $2CH_4(g) + 3O_2(g) \rightarrow 2CO(g) + 4H_2O(l)$ $\Delta H^\circ = -1215$ kJ

      $2C(s) + O_2(g) \rightarrow 2CO(g)$ $\Delta H^\circ = -221$ kJ

      $C(s) + O_2(g) \rightarrow CO_2(g)$ $\Delta H^\circ = -394$ kJ

12. Calculate the standard molar enthalpy of formation of methane from the data given in question 11, your answer to question 11, and the following:

$$\Delta H^\circ_f {}_{(H_2O(l))} = -286 \text{ kJ/mol}$$

13. When ammonia is oxidized to nitrogen dioxide and water, the quantity of heat released equals 349 kJ per mol of ammonia:

$$2NH_3(g) + 7/2O_2(g) \rightarrow 2NO_2(g) + 3H_2O(l) \quad \Delta H^\circ = -698 \text{ kJ}$$

Calculate the standard molar enthalpy of formation of ammonia if

$$\Delta H^\circ_f {}_{(H_2O(l))} = -286 \text{ kJ/mol}$$

$$\Delta H^\circ_f {}_{(NO_2(g))} = +33 \text{ kJ/mol}$$

14. A 0.915 gram sample of sugar ($C_{12}H_{22}O_{11}$; molar mass 342 g/mol) was ignited in a bomb calorimeter in the presence of excess oxygen. Combustion was complete. The temperature of the calorimeter and its contents rose by 3.53°C. If the heat capacity of the calorimeter and its contents is 4250 J K^{-1}, calculate the heat released per mole of sugar.

15. When 40 grams of ammonium nitrate is dissolved in 100 grams of water in a constant-pressure coffee-cup calorimeter, the temperature of the solution drops by 22.4°C. If the specific heat capacity of the solution is 4.18 J K^{-1}g^{-1}, calculate the enthalpy of solution of ammonium nitrate.

The enthalpy of solution is the heat released or absorbed when one mole of the solute dissolves in a great excess of water.

Answers to Study Questions and Problems

1. a. When a saucepan of water is heated to boiling, the heat derived from the burning gas or the electric hot plate causes the kinetic energy of the water molecules to increase. Therefore the temperature of the water increases. Eventually the water molecules have sufficient energy to break the intermolecular bonds and exert a pressure equal to that of the external atmospheric pressure. The water boils. The initial heat is derived from chemical energy either directly in a gas burner or indirectly in an electrical power generating plant.

   b. The tennis ball starts with a high gravitational potential energy. When dropped, the potential energy is converted to kinetic energy. With each bounce on the floor some of the macroscopic kinetic energy of the ball is converted into microscopic thermal kinetic energy in the floor and the ball (both the floor and the ball get warmer).

   c. The kinetic energy of the two automobiles is converted into thermal energy of the materials of which the cars are made. Some parts of the cars may temporarily gain some gravitational potential energy before returning to the ground.

d. The chemical energy of the flashlight battery is converted to electrical energy. In the light bulb, the electrical energy is converted to light and heat. The light and heat energy is converted in turn to other forms of energy.

e. The chemical potential energy of the fuel is converted to light and heat as the fire burns. Originally the fuel gained its chemical potential energy from the sun through photosynthesis.

f. The macroscopic movement in the air is a concerted movement of all the molecules (not the random movement of thermal energy). This concerted movement can do work as in the driving of a windmill. The mechanical energy of the windmill can be converted into electrical energy or it can pump water increasing the gravitational potential energy of the water.

g. Most chemical reactions liberate or absorb heat. Reactions that liberate heat are called exothermic; those that absorb heat are called endothermic. In exothermic reactions, the chemical potential energy of the system is converted into thermal energy of the system and its surroundings.

2. a. 800 kWh × 1000 W/kW × 60 min/1 h × 60 sec/1 min
   $= 2.88 \times 10^6$ J

   b. 377 kcal × 1000 cal/1 kcal × 4.184J/1 cal
   $= 1.58 \times 10^6$ J

   Use dimensional analysis for problems like these.

3. a. heat capacity of the aluminum block = 500 J /10.9 K = 45.9 JK^{-1}
   b. molar heat capacity of aluminum
   $= 45.9 \ JK^{-1} \times 26.98 \ g \ mol^{-1}/50.0 \ g = 24.8 \ JK^{-1}mol^{-1}$
   c. specific heat of aluminum
   $= 500 \ J/ (10.9K \times 50.0 \ grams) = 0.917 \ JK^{-1}g^{-1}$

4. The molar heat capacity of gold = specific heat of gold x molar mass
   $= 0.128 \ JK^{-1}g^{-1} \times 196.97 \ g/mol$
   $= 25.2 \ JK^{-1}mol^{-1}$
   The molar heat capacity of iron = specific heat of iron x molar mass
   $= 0.451 \ JK^{-1}g^{-1} \times 55.85 \ g/mol$
   $= 25.2 \ JK^{-1}mol^{-1}$
   The molar heat capacities of all metals at room temperature are approximately the same (theoretically = 3R = 24.9 $JK^{-1}mol^{-1}$).

   This is known as the Law of Dulong and Petit.

   R is the gas constant = 8.314 $JK^{-1}mol^{-1}$.

5. Heat and temperature are related by the heat capacity, where the heat capacity equals the specific heat capacity × mass:

   Heat (J) = heat capacity (JK^{-1}) × temperature change (K)

   Heat (J) = specific heat capacity ($JK^{-1}g^{-1}$) × mass (g) × temp change (K)

   Heat (J) = 0.451 $JK^{-1}g^{-1}$ × 1000 g × (1000-25) K = 440,000 J = 440 kJ

   The temperature is a temperature change where a Celsius degree has the same magnitude as a degree Kelvin.

6. Heat lost by the copper block:
Heat (J) = specific heat capacity ($JK^{-1}g^{-1}$) × mass (g) × temp change (K)

Heat (J) = 0.385 $JK^{-1}g^{-1}$ × 40 g × (100–T) °C where T is the final temp.

Heat gained by the water:
Heat (J) = specific heat capacity ($JK^{-1}g^{-1}$) × mass (g) × temp change (K)

Heat (J) = 4.184 $JK^{-1}g^{-1}$ × 100 g × (T–25) °C where T is the final temp.

The heat lost by the copper must be the same as the heat gained by the water:

T, the final temperature = 27.7°C.

Notice that the temperature of the water does not increase very much; water has a very high specific heat capacity.

7. 27.0 grams of ice = 27.0/18.016 moles of ice = 1.50 moles
Heat required = 6.009 kJ mol^{-1} × 1.50 mol = 9.01 kJ.

1000 J = 1 kJ

Be very careful with the signs in problems like these!

If you state each heat as a positive quantity, then because the heat lost = heat gained, set the two equal and solve for T.

Alternatively, let the heat lost have its negative sign, the heat gained its positive sign, and then state that the sum of the two equals zero (energy must be conserved), and solve for T.

8. Heat is required to melt the ice at 0°C, *and* to heat the water produced from 0°C to the final temperature T.

The heat is provided by the 123 grams of water at 100°C which cools down to the same final temperature T.

The heat lost by the hot water must equal the heat needed to melt the ice and heat the water produced:

Heat required to melt the ice
= 6.009 kJ mol^{-1} × 1.50 mol = 9014 J (*cf.* question 7)

Heat required to heat the water produced
= 4.184 $JK^{-1}g^{-1}$ × 27.0 g × (T–0) °C.

Heat lost by the hot water
= 4.184 $JK^{-1}g^{-1}$ × 123 g × (100–T) °C

Therefore,

9014 + 113T = 51463 – 515T
T = 67.6°C

The metal block must be at a temperature of 0°C at the end of the experiment—not all the ice melted.

9. The quantity of heat required to melt 10.5 grams of ice
= 10.5 g × 1 mol/18.016 g × 6.009 kJ mol^{-1}
= 3.50 kJ or 3500 J

For the metal alloy:

Heat (J) = specific heat capacity ($JK^{-1}g^{-1}$) × mass (g) × temp change (K)

3500 J = specific heat capacity ($JK^{-1}g^{-1}$) × 50 g × 100°C

Therefore,

Specific heat capacity of the metal = 3500/(50 × 100) $JK^{-1}g^{-1}$
= 0.70 $JK^{-1}g^{-1}$

10. The enthalpy change for a chemical reaction is an extensive property—it depends upon the quantities of substances involved. A large fire generates more heat than a small fire. If the molar enthalpy of combustion of propane is −2220 kJ, this means that 2220 kJ of heat is generated for every mole (44 grams) of propane burned.

The quantity of heat released when 1 kg propane is burned
= 2220 kJ × 1000/44 = 50,450 kJ

The quantity of heat released when 2 kg propane is burned is twice as much = 50,450 kJ × 2 = 100,900 kJ.

11. Manipulate the data provided so that when the equations are added together, the desired equation is produced:

$$CH_4(g) + 3/2 O_2(g) \rightarrow CO(g) + 2H_2O(l) \quad \Delta H° = -1215 \text{ kJ} \quad /2$$
$$= -607.5 \text{ kJ}$$

$$CO(g) \rightarrow C(s) + 1/2 O_2(g) \quad \Delta H° = -221 \text{ kJ} \times -1 \quad /2$$
$$= +110.5 \text{ kJ}$$

$$C(s) + O_2(g) \rightarrow CO_2(g) \quad \Delta H° = -394 \text{ kJ}$$

_____

$$CH_4(g) + 2O_2(g) \rightarrow CO_2(g) + 2H_2O(l) \quad \Delta H° = -891 \text{ kJ}$$

Divide equation 1 by 2 since only one mole of CH_4 is required on the left side of the equation.

We need a CO on the left to cancel the CO on the right in the first equation, but only one, so reverse equation 2 and divide by 2.

We need a C on the left to cancel the C on the right in the second equation, equation 3.

12. The enthalpy of reaction can be calculated from the molar enthalpies of formation of the participants:

$$\Delta H°_{reaction} = \Sigma(\Delta H°_f \text{(products)} - \Delta H°_f \text{(reactants)})$$

$$CH_4(g) + 2O_2(g) \rightarrow CO_2(g) + 2H_2O(l) \quad \Delta H° = -891 \text{ kJ}$$

$\Delta H°_f (H_2O(l)) = -286$ kJ/mol
$\Delta H°_f (CO_2(g)) = -394$ kJ/mol (from the third equation in question 11)

$\Delta H°_{reaction} = \Sigma(\Delta H°_f \text{(products)} - \Delta H°_f \text{(reactants)})$

$-891 = [(-394 + (2 \times -286)) - (\Delta H°_f \text{(methane)})]$

$\Delta H°_f \text{(methane)} = +891\ -394\ +(2 \times -286)$

$\qquad\qquad = -75$ kJ/mol

$\Delta H°$ is extensive−whatever you do to the equation you must do to $\Delta H°$.

Be careful with the signs!

Remember that "heat released" means that the process is exothermic, and that therefore ΔH is negative.

13. $2NH_3(g) + 7/2 O_2(g) \rightarrow 2NO_2(g) + 3H_2O(l) \quad \Delta H° = -698$ kJ

$\Delta H°_f (H_2O(l)) = -286$ kJ/mol
$\Delta H°_f (NO_2(g)) = +33$ kJ/mol

$\Delta H°_{reaction} = \Sigma(\Delta H°_f \text{(products)} - \Delta H°_f \text{(reactants)})$

$-698 = [((2 \times +33) + (3 \times -286)) - (2 \times \Delta H°_f \text{(ammonia)})]$

$\Delta H°_f \text{(ammonia)} = [+698\ +(2 \times +33)\ +(3 \times -286)]/2 \quad = -47$ kJ/mol

Oxygen is an element, and has a $\Delta H°_f$ of zero.

14.   Heat (J) = heat capacity (JK^{-1}) × temperature change (K)

Heat (J) = 4250 JK^{-1} × 3.53°C = 15,000 J

This is the heat released when 0.91 gram of sugar is burned.
For one mole:

Heat (J) = 15,000 J × (342 g/0.915 g) × (1 kJ/1000 J)= 5600 kJ.

15.   The mass of the solution = 100 g water + 40 grams ammonium nitrate.

Heat (J) = specific heat capacity $(JK^{-1}g^{-1})$ × mass (g) × temp change (K)

= 4.18 $JK^{-1}g^{-1}$ × 140 g × 22.4°C
= 13,108 J (for 40 grams)

For one mole:

Heat (J) = 13,108 × 80.04/40 = 26,230 J = 26 kJ/mol

$\Delta H^{\circ}_{solution}$= +26 kJ/mol (an endothermic process)

Remember that for a
temperature *difference*, K = °C.

EXAMINATION 1

Introduction

This examination tests your knowledge and understanding of the chemistry in the first six chapters of Kotz & Treichel. The questions are formatted as true–false questions and multiple choice questions—the sort you might encounter on your own examinations. Try the exam before looking at the answers provided at the end of this study guide.

True–false questions

1. All atoms of an element are identical.

2. Iodine is a noble gas.

3. The number of atoms on both sides of a chemical equation is the same.

4. Heat capacity = specific heat capacity × mass.

5. A Celsius degree interval is equal in magnitude to a Kelvin degree interval.

6. The base SI unit for mass is the gram.

7. If a reaction is exothermic in one direction, then it will be endothermic in the reverse direction.

8. The prefix milli– means $\times 10^{-3}$.

9. A compound must consist of two or more different elements.

10. Neon–22 has 12 neutrons in its nucleus.

11. An empirical formula for a compound indicates the simplest possible integer ratio of the elements in the compound.

12. Energy (in joules J) can be expressed as power (in watts W) × time (in seconds).

13. Volume, pressure, and temperature are all state functions.

14. Perchloric acid $HClO_4$ is a strong acid but chlorous acid $HClO_2$ is weak.

15. All molecules are compounds.

16. Solutions are always neutral at the end of a neutralization reaction.

17. $\Delta H_f^\circ (Br_2(g)) = \Delta H_{vaporization}^\circ (Br_2(l))$

18. Mass × number of moles = molar mass.

19. Melting is always an endothermic process.

20. The oxidizing agent in a redox reaction is always reduced.

21. An exothermic reaction is always product-favored.

22. Elements in the Periodic Table are arranged in order of their atomic masses.

23. An allotrope is a naturally occurring mixture of all the isotopes of an element.

24. Lead sulfate is an insoluble salt.

25. Non-metal oxides produce acidic solutions when dissolved in water.

Multiple choice questions

1. The international (SI) unit for force is the newton (symbol N). Force is mass × acceleration. What are the units of a newton in base SI units?

   a. ms^{-2} c. $m\ s^{-1}$ e. $kg\ m\ s^{-2}$
   b. $kg\ s^{-1}$ d. $m\ s^{-2}$ f. $kg^2\ m\ s^{-1}$

2. What are the units of the answer to the following problem?

   55 mile/hr × 4 weeks × 24 hr/day × 7 days/week × 5280 feet/mile

   a. weeks   d. feet/hour   f. feet/day
   b. feet   e. mile/week   g. 1/feet
   c. there are no units; all cancel

3. In a neutral atom,
   a. the number of neutrons always equals the number of protons.
   b. the number of protons plus neutrons equals the number of electrons.
   c. the atomic number equals the number of neutrons.
   d. the mass number equals the number of neutrons.
   e. the number of electrons equals the number of protons.
   f. the mass number equals the number of protons.

4. An ion that contains 27 protons and 24 electrons must be

   a. Cr^{3+} c. Fe^{3+} e. Al^{3+} g. Co^{3+}
   b. Co^{2+} d. N^{3+} f. V^{3+} h. B^{3+}

5.   If there are two possible polyatomic oxyanions of an element, the anion having the fewer oxygens has a name ending in

  a. –ite   c. –ate   e. –ide
  b. –ium   d. –ous   f. –ic

6.   Rutherford's gold foil experiment indicated that

  a. electrons and protons are of approximately the same mass.
  b. most of the mass of the atom is concentrated at a nucleus.
  c. electrons are moving around in the atom at great speeds.
  d. electrons and protons are strongly attracted to one another.
  e. alpha particles have a charge opposite to that of an electron.

7.   If 6 moles of nitrous oxide N_2O and 8 moles of oxygen O_2 react to produce as much nitrogen dioxide NO_2 as possible, how many moles of which reactant will remain at the end of the reaction?

$$2N_2O + 3O_2 \rightarrow 4NO_2$$

  a. 1/3 mol N_2O c. 1.5 mol O_2 e. 2/3 mol N_2O g. 2/3 mol O_2
  b. 2 mol O_2 d. 1 mol N_2O f. 1 mol O_2 h. 1/3 mol N_2O

8.   Which one of the following formulas is incorrect?

  a. K_2SO_4 potassium sulfate
  b. NH_4NO_3 ammonium nitrate
  c. $Ca(CO_3)_2$ calcium carbonate
  d. N_2O_5 nitrogen(V) oxide
  e. $NaOH$ sodium hydroxide
  f. Cr_2O_3 chromium(III) oxide

9.   How many moles of sulfur atoms are there in a 256 gram sample?

  a. 1   c. 3   e. 6   g. 12
  b. 2   d. 4   f. 8   h. 16

10.   A mole is
  a. the mass of exactly 12 grams of the carbon-12 isotope.
  b. the volume occupied by 22.4 liters of any gas.
  c. Avogadro's number of anything.
  d. the number of molecules in 1 gram of any substance.
  e. the number of hydrogen atoms in 2 grams of H_2 molecules.
  f. the mass of Avogadro's number of atoms of any element.

11.   How many moles of nitrogen atoms are there in three moles of ammonium nitrate?

  a. 3   c. 6   e. 12
  b. 4   d. 9   f. 18

12. A flask contains 8 moles of sulfur dioxide gas. What is the mass of the gas in grams?

    a. 2   c. 4   e. 8   g. 16
    b. 64   d. 128   f. 256   h. 512

13. If 75 mL of 0.10 M sodium sulfate is added to 25 mL of 0.20 M barium nitrate, and as much barium sulfate precipitates as possible, what are the concentrations of the ions remaining in solution?

[barium ions]	[sodium ions]	[sulfate ions]	[nitrate ions]
a. 0.050 M	0.15 M	0.075 M	0.10 M
b. 0.10 M	0.10 M	0.050 M	0.20 M
c. 0.20 M	0.10 M	0.10 M	0.20 M
d. 0.0 M	0.15 M	0.025 M	0.10 M
e. 0.0 M	0.075 M	0.0 M	0.050 M
f. 0.0 M	0.075 M	0.025 M	0.050 M
g. 0.050 M	0.15 M	0.0 M	0.10 M

14. What is the name of the $CO_3{}^{2-}$ ion?

    a. bicarbonate   c. sulfide   e. chloride   g. nitrate
    b. sulfate   d. carbonate   f. chlorate   h. chromate

15. What set of coefficients is required to balance the following equation?

    $$\_Be_3N_2 \; + \; \_H_2O \; \rightarrow \; \_Be(OH)_2 \; + \; \_NH_3$$

a. 1	2	1	2
b. 1	3	3	2
c. 1	3	1	1
d. 1	4	3	2
e. 1	6	2	2
f. 1	6	3	2
g. 1	6	3	3

16. Which sample contains the greatest amount of oxygen?

    a. 0.10 mol $KMnO_4$ d. 0.30 mol $Ba(OCl)_2$
    b. 0.15 mol Na_2SO_4 e. 0.50 mol Fe_2O_3
    c. 1.0 mol NaOH   f. 0.20 mol $Ca_3(PO_4)_2$

17. What is the percent by mass of sodium hydroxide NaOH (molar mass 40 g/mol) in an aqueous solution containing 2.5 moles in 1600 g of water?

    a. 0.42   c. 6.25   e. 8.21
    b. 5.88   d. 12.3   f. 16.5

18. What is the oxidation state of phosphorus in Na_2HPO_4?

    a. 1+   c. 2+   e. 3+   g. 4+   e. 6+
    b. −1   d. −2   f. −3   h. 5+   f. 7+

19. A compound of nitrogen and sulfur contains 30% nitrogen by mass. What is the mole ratio of nitrogen to sulfur in this compound?

    a. 1:1   c. 1:2   e. 1:3   g. 1:4   i. 2:3
    b. 2:1   d. 3:1   f. 4:1   h. 5:1   j. 3:7

20. The difference between a strong acid and a weak acid is that
    a. a strong acid is always more concentrated.
    b. weak electrolytes are less soluble than strong electrolytes.
    c. it's impossible to prepare a dilute solution of a strong electrolyte.
    d. the degree of ionization of a strong electrolyte is complete whereas for a weak electrolyte it is not.
    e. solutions of weak acids cannot conduct an electrical current except when they are concentrated.

21. How many moles of sodium hydroxide are there in a mixture of 4.0 grams of sodium hydroxide and 12.0 grams of sodium chloride?

    a. 0.25   c. 2.5   e. 10.0   g. 4.0
    b. 0.10   d. 1.0   f. 1.6   h. 0.16

22. The element chlorine (Cl) is

    a. a metal   c. an alkali metal   e. a noble gas
    b. a transition element   d. a lanthanide   f. a halogen

23. In which of the following ions, molecules, or compounds does oxygen have an oxidation number of −2?

    a. O_2 c. O_2^{2-} e. OF_2
    b. NO   d. O_2^- f. O_3

24. The net ionic equation for the reaction between aqueous ammonia solution and hydrochloric acid solution is

    a. NH_3 + HCl → NH_4Cl

    b. NH_4OH + HCl → NH_4Cl + H_2O

    c. NH_3 + H^+ + Cl^- → NH_4^+ + Cl^-

    d. NH_3 + H^+ → NH_4^+

    e. NH_4OH + Cl^- → NH_4Cl + OH^-

25. A sample of hydrated copper(II) sulfate, $CuSO_4 \cdot xH_2O$ weighs 5.29 g. When the water is driven off, the anhydrous $CuSO_4$ salt (molar mass 159.6) weighs 3.38 g. What is the value of x in the formula of the hydrated salt?

    a. 1   c. 2   e. 3
    b. 4   d. 5   f. 6

26. Using the data given, calculate the enthalpy change for the reaction in which 56 grams of ethylene (C_2H_4) is hydrogenated to form ethane (C_2H_6).

$$C_2H_4(g) \ + \ H_2(g) \ \rightarrow \ C_2H_6(g)$$

$2C(s) \ + \ 3H_2(g) \ \rightarrow \ C_2H_6(g)$	$\Delta H° \ = -84.7 \ kJ$
$2C(s) \ + \ 2H_2(g) \ \rightarrow \ C_2H_4(g)$	$\Delta H° \ = +52.3 \ kJ$

    a. –137 kJ   c. –32.4 kJ   e. –274 kJ
    b. +137 kJ   d. +64.8 kJ   f. –64.8 kJ

27. A block of lead (mass 30 grams; specific heat capacity 0.128 $JK^{-1}g^{-1}$) at 90.0°C is dropped into water at 20.0°C in a well insulated coffee-cup calorimeter. The final temperature of the water and the lead block is 21.6°C. How much water was in the calorimeter? Assume the heat capacity of the calorimeter is zero JK^{-1}. The specific heat capacity of water = 4.184 $JK^{-1}g^{-1}$.

    a. 19 g   c. 29 g   e. 39 g
    b. 24 g   d. 34 g   f. 44 g

28. A system is heated by its surroundings and, as a result, absorbs 100 kJ of heat. The system, in turn, does 60 kJ of work on the surroundings. What is the change in the energy of the system?

    a. –160kJ   c. –60 kJ   e. –40 kJ
    b. +40 kJ   d. +60 kJ   f. +160kJ

29. Assume that the salt aluminum nitrate is fully dissociated in aqueous solution. How many moles of nitrate ions are there in 200 mL of a 0.020 M solution of aluminum nitrate?

    a. 0.12   c. 0.060   e. 0.012
    b. 0.0040   d. 0.30   f. 0.020

30. Malonic acid has the composition 34.6% carbon, 3.90% hydrogen, and 61.5 % oxygen. Calculate the empirical formula of malonic acid.

    a. CHO_2 c. C_2HO_2 e. CH_2O_2 g. $C_2H_4O_3$
    b. $C_3H_4O_4$ d. $C_2H_2O_3$ f. $C_3H_3O_4$ h. $C_2H_3O_3$

31. The molar heat of combustion ($\Delta H_{combustion}$) of methane (CH_4; natural gas) is 890 kJ mol^{-1}. Sufficient methane was burned to produce 9.0 grams of water. How much heat was released?

    a. 55.6 kJ   c. 222 kJ   e. 890 kJ   g. 3560 kJ
    b. 111 kJ   d. 445 kJ   f. 1780 kJ   h. 7120 kJ

CHAPTER 7

Introduction

Chemical properties and reactions depend upon the behavior of the electrons in atoms and molecules. In order to make some sense out of the thousands of reactions and structures of chemical compounds, it is essential to know something about the behavior of the electrons in atoms and molecules. In this chapter the electronic structure of atoms is examined.

Contents

7.1 Electromagnetic Radiation

Radiation, such as microwaves, visible light, and x-rays, is called **electromagnetic radiation**. This radiation is composed of two mutually perpendicular oscillating electric and magnetic fields traveling through space. These waves, like other traveling waves, are characterized by their frequency, wavelength, and velocity. The **frequency** is the rate at which complete waves pass a particular point. The **wavelength**, as its name suggests, is the length of a complete wave—the distance between two closest identical points on the wave. The frequency multiplied by the wavelength equals the **velocity** at which the wave is moving. The height of the wave above the axis of propagation is called the **amplitude**.

The electromagnetic spectrum is divided into different regions, some of which may be familiar to you. It is convenient to organize the regions by their wavelengths, each region differing by an approximate factor of 10^3:

Radiowaves	1 m
Microwaves	10^{-3} m
Visible	10^{-6} m
x-rays	10^{-9} m
γ-rays	10^{-12} m

The **visible spectrum** consists of the colors red, orange, yellow, green, blue, indigo, and violet, from low frequency to high frequency, or approximately 700 nm to 400 nm in wavelength. The infrared region is just beyond the long wavelength (low frequency) end of the visible region and the ultraviolet region is just beyond the short wavelength (high frequency) end of the visible region.

The symbols are:
frequency ν
wavelength λ
velocity   v

Do not confuse the greek ν and the v for velocity.

The velocity of electromagnetic radiation in a vacuum is 2.998×10^8 m s^{-1}.

These wavelengths are only approximate. For example, radio waves extend from 1 m all the way to 10^8 m. Microwaves from 10^{-3} m to 1 m and IR from 10^{-6} to 10^{-3} m.

You can remember these colors by their initials ROY G BIV.

Be able to calculate the frequency from the wavelength of electromagnetic radiation and vice versa.

Sound waves, and electromagnetic waves, are **traveling waves**. Sound waves propagate through some medium, the denser the better. Electromagnetic waves propagate best through a vacuum. Both travel from their source to some other point.

Another type of wave motion is **stationary**. This is the type of wave motion seen in a vibrating violin string or on the surface of a drum. For stationary waves, only certain wavelengths are possible. The wavelength of a vibrating string, for example, has to satisfy the condition that a (the distance between the points at which the string is fastened) equals $n(\lambda/2)$, where n must be an integer (1, 2, 3, 4...). The waves that satisfy this condition have points at which there is zero motion—these points are called **nodes**. A vibrating string is an example of a **quantized system**—only certain solutions, or energies, are allowed.

The emission from a hot object depends upon the temperature of the object. As the temperature is increased, the color moves from the red end of the visible spectrum to the blue end. An object that is "white hot" is hotter than an object that is merely "red hot". A human body, at only 98.6°F, emits radiation only in the infrared region. Explanation of the radiation emitted by hot objects was a problem puzzling physicists at the end of the 19th century. The problem was solved by Max Planck in 1900. He knew that the atoms of the heated object vibrated and that it was this vibration that caused the emission. He suggested that the vibrations of the atoms were quantized—only vibrations of specific energies were allowed. **Planck's equation**, one of the fundamental equations of **quantum mechanics**, relates the energy and frequency:

$$E = h\nu \qquad \text{where h is Planck's constant } 6.626 \times 10^{34} \text{ J.s}$$

When light strikes the surface of a metal, electrons may be emitted from the surface. These electrons are called photoelectrons and the phenomenon is called the **photoelectric effect**. The reason for this effect was not understood and could not be explained at all using classical physics. Einstein, in 1905, applied Planck's idea of the quantization of energy to explain the effect.

Einstein suggested that light had particle-like properties. He said that these particles (light quanta, now called **photons**) possessed an energy proportional to the frequency. The energy of the light is therefore proportional to the frequency—not the amplitude as classical physics would suggest. If light has a sufficiently high frequency, then the photons have a sufficiently high energy to cause the emission of an electron—one electron for each photon. Increasing the frequency of the light increases the kinetic energy of the photoelectrons. Increasing the intensity of the light increases the number of photons, and therefore increases the number of electrons (the current). This is how a photocell measures the intensity of light—by measuring the current.

7.3 Atomic Line Spectra and Niels Bohr

After atoms of an element in the gas state are excited by a high voltage, the atoms return to their ground (lowest energy) state by emission of radiation. If the emission is in the visible region, the emission is colored. When the radiation is dispersed into individual wavelengths, it is seen to consist of lines at particular

---

A stationary wave is often referred to as a standing wave. See Figure 7.4 on page 297.

The two ends, where the string is fixed, are obvious points of zero motion. Sometimes these are included in the total number of nodes, sometimes not. In the text they are included.

A white color indicates the presence of all visible wavelengths with intensities symmetrical about the center.

Some atoms vibrate at high frequency and emit high frequency radiation; other vibrations are low frequency and emit low frequency radiation; most vibrations are somewhere in the middle. The result is the spectral envelope illustrated in Figure 7.5 on page 298.

Planck suggested the idea that the vibrations of atoms in a heated object were quantized. Einstein suggested that the electromagnetic radiation itself was quantized.

Dispersion can be accomplished through a prism or by diffraction.

wavelengths, rather than a continuous (rainbow) spectrum. Every element has its own characteristic **line spectrum**.

Explanation of the existence of line spectra, rather than continuous spectra, puzzled physicists. An empirical mathematical expression was discovered by Balmer and Rydberg that could predict the wavelengths of the lines in the hydrogen spectrum, but why the lines existed was not understood.

Niels Bohr solved the problem by incorporating Planck and Einstein's ideas of the quantization of energy in the Rutherford model of the atom. Bohr suggested that the electron in the hydrogen atom could only occupy certain energy levels (or orbits)—the system was quantized. According to Bohr, the allowed energies of the electron in the hydrogen atom were restricted to $-Rhc/n^2$, where R is the Rydberg constant and n is a quantum number with integer values from 1 to infinity.

The fact that each element has a unique spectrum is the basis for spectroscopic analysis. Some line spectra are illustrated in Figure 7.9 on page 305.

The Rydberg equation is shown on page 304 of the text.

$R = 1.097 \times 10^7 \text{ m}^{-1}$
$c = 2.998 \times 10^8 \text{ ms}^{-1}$.

7.4 The Wave Properties of the Electron

Louis deBroglie, considering Einstein's assertion that light (electromagnetic radiation) could have particle-like properties, wondered whether particles could have wavelike properties. He derived his equation:

$$\text{wavelength } (\lambda) \text{ x momentum } (mv) = h \text{ (Planck's constant)}$$

Wavelength (λ) is a wave property and momentum (mv) is a particle property. The equation relates the two and is another fundamental relationship of quantum mechanics. However, because h is so small, the wavelength is insignificant unless the particle (mass) is very small. In other words the wavelike properties of particles are unobservable unless the particle is very small—like an electron or an atom.

Experimental proof of the wavelike properties of electrons was established by Davisson and Germer, and G. P. Thompson, through the diffraction of an electron beam by a thin metal sheet.

7.5 The Wave Mechanical View of the Atom

A basic problem with Bohr's model for the behavior of an electron in an atom was a fundamental uncertainty in fixing both the position and momentum of the electron simultaneously. This is another principle determining the physics of small particles known as **Heisenberg's Uncertainty Principle**. In Bohr's model, the position (the orbit) and the momentum (angular momentum) of the electron are fixed and this is impossible. Electrons do not travel in fixed orbits.

The idea that small particles exhibited wavelike behavior prompted Schrödinger to write a **wave equation** to describe the behavior of the electron in a hydrogen atom. This wave equation was a 3-dimensional standing wave equation analogous to the equations for a vibrating violin string or a vibrating drum surface.

The solution of Schrödinger's wave equation yielded exactly the same energies as Bohr had obtained. The energy of the electron is quantized, just like the vibrations of a violin string are quantized. In addition, however, the solution yielded a **wavefunction** (φ) for each allowed energy level that described that energy level. Although the wavefunction (φ) itself has no physical significance,

This fundamental uncertainty comes about because in order to observe an object you have to use radiation with a wavelength comparable to the size of the object (that's why you cannot see objects smaller than about one μm using your eyes—no matter how good the microscope). So to observe an electron, the radiation wavelength would have to be very small, therefore the frequency would be very high, therefore the photon energy would be enormous, and the very act of observation would change the momentum of the electron. So you could use low frequency radiation and know the momentum but not the position, or use high frequency radiation and know the position but not the momentum. You *cannot* know both. The product of the two uncertainties is a function only of h.

What is a wavefunction? It is a mathematical function that describes a wave. For example y = sin x is a wavefunction that describes a sine wave.

An *orbit* is a fixed path. An *orbital* is a region of space—an electron domain.

If the three p orbitals are super-imposed on each other, or the five d orbitals are superimposed on each other, a spherically symmetric electron density distribution (just like the single s orbital) results.

Look at Figure 7.15 on page 320 for the shapes of the orbitals.

If you are interested in what the electron wavefunctions actually look like, examine the Closer Look on page 322 of the text.

You may one day, in a higher level chemistry course, solve Schrödinger's wave equation yourself. When you do, you will understand why the d_{z^2} orbital is drawn as it is.

the square of the wavefunction (φ^2) reflects the probability of finding the electron in any particular region of space around the nucleus—referred to as the **electron density**.

These patterns of electron density for each energy level are called **orbitals**. Just as the solutions to a one-dimensional standing wave equation for a violin string require one quantum number n, the solutions to Schrödinger's three dimensional wave equation require three quantum numbers n, ℓ, and m_ℓ.

The **principal quantum number** n indicates the energy of the orbital. As n increases, the electron is found on average further from the nucleus. The principal quantum n can have any integer value from 1, 2, 3, 4…to infinity.

The **secondary (or angular momentum) quantum number** ℓ indicates the shape of the orbital. It can have values from 0, 1, 2, 3, to a maximum of n-1. To avoid confusion, the value for ℓ is usually denoted by a letter, so that, if $\ell = 0$, the letter s is used, and the orbital is referred to as an s orbital. If $\ell = 1$, the letter p is used, and the orbital is referred to as a p orbital. If $\ell = 2$, the letter d is used, and the orbital is referred to as a d orbital. If $\ell = 3$, the letter f is used, and the orbital is referred to as an f orbital.

The **magnetic quantum number** m_ℓ indicates the orientation of the orbital in the space around the nucleus. It can have integer values from $-\ell$ through 0 to $+\ell$. If $\ell = 0$, then the only possible value for m_ℓ is 0, and there is only one orientation. Therefore s orbitals come in sets of one only. If $\ell = 1$, then there are three possible values for m_ℓ: -1, 0, and $+1$. There are three possible orientations for p orbital (p_x, p_y, and p_z), and p orbitals always come in sets of three. Likewise d orbitals always come in sets of 5, f orbitals always in sets of 7, and so on.

7.6 The Shapes of Atomic Orbitals

The chemical properties of a substance are determined by its electrons—particularly those electrons on the outside of the atoms, highest in energy, or "at the frontier". These electrons in the highest shell (highest value of n) are called **valence electrons**. The orbitals containing the valence electrons are called **valence orbitals** and the shapes and orientations of these orbitals influence the properties.

All s orbitals are spherically symmetric. The electron density (the likelihood of finding the electron—as a particle—per unit volume) is a maximum at the nucleus and decreases further and further from the nucleus.

All p orbitals come in sets of three and all p orbitals have the same shape. The shape is like a sphere that has been cut into two halves. The plane running between the two halves is a plane on which the probability of finding the electron is zero; it is called a **nodal plane**. Because the nodal plane passes through the nucleus, the probability of an electron in a p orbital existing at the nucleus is zero.

All d orbitals come in sets of five, and all d orbitals in a set are theoretically the same except in orientation. You will notice however, that one d orbital, the d_{z^2}, is drawn with a different shape—the reason why is something to be left for a higher level chemistry course.

Review Questions

1. Describe the electromagnetic spectrum. Describe the approximate wavelength range of the visible spectrum and list the colors of the spectrum in order of decreasing wavelength.

2. Define wavelength, frequency, velocity, and amplitude.

3. What is the relationship between frequency, wavelength, and velocity?

4. What is a node?

5. Two developments in physics took place in 1900 and 1905 that revolutionized our understanding of the interaction of matter and radiation. Describe these two developments.

6. When ν is the frequency of electromagnetic radiation, what is E in Planck's equation $E = h\nu$?

7. Why does an emission spectrum exist as lines rather than as a continuous rainbow?

8. Describe Niels Bohr's ideas about the behavior of electrons in atoms.

9. Derive the Rydberg-Balmer equation from Bohr's equation for the allowed energies for the electron in the hydrogen atom.

10. Why is Bohr's model for the behavior of an electron in an atom incorrect?

11. What information is derived from the solution of Schrödinger's wave equation apart from the allowed energies of the electron?

12. Describe the three quantum numbers that are used to characterize the solutions to Schrödinger's wave equation.

13. What letters are used to signify the various values of the secondary (angular momentum) quantum number?

14. Describe the physical significance of the three quantum numbers. In other words, what do the quantum numbers tell you about the orbitals?

15. How do the values allowed for the three quantum numbers depend upon each other?

Answers to Review Questions

1. The electromagnetic spectrum extends from 10^{-16} to 10^8 m in wavelength (corresponding to 10^{24} to 1 Hz in frequency). Decreasing in wavelength the ranges are called: radio waves, microwaves, IR, visible, UV X-rays, and γ rays. The wavelength range for the visible region is 700–400 nm (from red, orange, yellow, green, blue, indigo, to violet).

2. The wavelength is the length of a complete wave—the distance between two closest identical points on the wave. The frequency is the rate at which complete waves pass a particular point. The frequency multiplied by the wavelength equals the velocity at which the wave is moving. The height of the wave above the axis of propagation is called the amplitude.

3. Wavelength (λ) × frequency (ν) = velocity (v).

4. A node on a standing (stationary) wave is a point at which the amplitude is zero. In the vibration of a violin string it's where the string doesn't move up and down. In the wavefunction for an electron in a hydrogen atom, it's where the wavefunction changes sign and there is zero probability of finding the electron.

5. The two developments were black-body radiation and the photoelectric effect. The first describes the emission of radiation from a hot object. As the temperature is increased, the color changes from red to orange, to white, and then to blue. Explanation of the emission was accomplished by Max Planck in 1900. He suggested that the vibrations of the atoms responsible for the emission were quantized—only vibrations of specific energies were allowed. Planck's equation relates the energy and frequency: $E = h\nu$.

    When light strikes the surface of a metal, electrons may be emitted from the surface. This phenomenon is called the photoelectric effect. Einstein, in 1905, applied Planck's idea of the quantization of energy to explain the effect. Einstein suggested that light had particle-like properties. He said that these particles (photons) possessed an energy proportional to the frequency.

6. E in Planck's equation $E = h\nu$ is the energy of one photon of the radiation having a frequency ν. Multiply by Avogadro's number to obtain an energy in J/mol.

7. Only certain energies are allowed for the electron. As the electron moves from one energy level to another, it emits or absorbs radiation corresponding to the energy difference between the two levels: $\Delta E = h\nu$. Each energy difference therefore has a characteristic frequency—a line in the spectrum.

8.   Niels Bohr was the first person to incorporate the ideas of Planck and Einstein concerning the quantization of energy and radiation in a model for the behavior of electrons in atoms. He proposed that the electron could only exist in certain orbits at fixed energies which he established by imposing quantization on the angular momentum of the electron: $mvr = n(h/2\pi)$. He was able to derive Rydberg's equation for the line spectrum of hydrogen.

9.   The allowed energies for an electron in a hydrogen atom according to Niels Bohr were given by the equation: $E = -Rhc/n^2$ where R is the Rydberg constant equal to 1.097×10^7 m^{-1}. The difference between any two energy levels n_2 and n_1 is therefore:

$$\Delta E = h\nu = -Rhc/n_2^2 - Rhc/n_1^2 = Rhc(1/n_1^2 - 1/n_2^2)$$

10.   According to Heisenberg's uncertainty principle, the position and momentum of a small particle like an electron cannot both be known at the same time. Although an electron in the lowest orbital of a hydrogen atom is most likely to be found at the radius calculated by Bohr, it can also be found at other distances from the nucleus. The electron does not travel on a fixed path like a planet around the sun. All you can say is that there is a certain region of space somewhere around the nucleus in which it is probable you will find the electron. These regions of space are called orbitals.

11.   The solution of Schrödinger's wave equation provides the allowed energies of the electron and wavefunctions which describe the behavior of the electrons at these energy levels. These wavefunctions are squared to obtain the probabilities of finding the electron at particular points or at certain distances from the nucleus.

12.   The three quantum numbers that are used to characterize the solutions to Schrödinger's wave equation are the principal quantum number n, the secondary or angular momentum quantum number ℓ, and the magnetic quantum number m_ℓ.

13.   If $\ell = 0$, the letter s is used; if $\ell = 1$, the letter p is used; if $\ell = 2$, the letter d is used; and if $\ell = 3$, the letter f is used.

14.   The principal quantum number n indicates the energy of the orbital. As n increases, the electron is found on average further from the nucleus. The secondary (or angular momentum) quantum number ℓ indicates the shape of the orbital. The magnetic quantum number m_ℓ indicates the orientation of the orbital in the space around the nucleus.

The secondary quantum number ℓ can be related to the number of planar nodes in the wavefunction.

15. The values allowed for the three quantum numbers depend upon each other. The principal quantum n can have any integer value from 1, 2, 3, 4...to infinity. The secondary (or angular momentum) quantum number ℓ can have values from 0, 1, 2, 3, to a maximum of $n-1$. The magnetic quantum number m_ℓ can have integer values from $-\ell$ through 0 to $+\ell$. The number of different values m_ℓ can have equals $2\ell +1$. This equals the number of orbitals within a set. For example, s orbitals come in sets of one only, p orbitals always come in sets of three, d orbitals always come in sets of 5, and so on.

Natural science does not simply describe and explain nature, it is a part of the interplay between nature and ourselves.

Werner Heisenberg
(1901-1976)

Study Questions and Problems

1.   a. Calculate the wavelength of electromagnetic radiation that has a frequency of 5.56 MHz.
     b. Calculate the frequency of electromagnetic radiation that has a wavelength equal to 667 nm.

2.   Draw the first few stationary waves possible for a violin string fixed at ends 1 meter apart.

     a. What is the wavelength of the lowest frequency vibration?
     b. How many nodes (not including the ends) are there in the $n=5$ vibration?
     c. How many wavelengths fit within the boundaries in the $n=4$ vibration?
     d. At what points are the nodes in the $n=4$ vibration?
     e. What is the relationship between the frequency of the $n=2$ vibration and the $n=4$ vibration?

3.   Electromagnetic radiation at the blue end of the visible spectrum has a wavelength of 400nm.

     a. Calculate the frequency of the radiation.
     b. Calculate the energy of one photon of this radiation.
     c. Calculate the energy of one mole of photons of this radiation.

4.   Examine the emission spectrum of hydrogen (Figure 7.9 page 304, Figure 7.10 page 305, or Figure 7.12 page 309).

     a. How many lines appear in the visible region (Balmer series)?
     b. Why so few?

     The Lyman series occurs in the UV region.
     c. How many lines would you expect in the Lyman series?
     d. How many other series would you expect beyond the blue end of the visible spectrum?
     e. How many different series would you expect beyond the red end of the visible spectrum?

5. Calculate the frequency of the line in the hydrogen spectrum corresponding to the electron transition from $n=9$ to $n=8$. Whereabouts in the electromagnetic spectrum does this line occur?

6. The ionization energy of an element is the energy required to remove the most loosely held electron from atoms of the element in the gaseous state (*cf.* page 357). It is usually expressed in units of kJ mol^{-1}.
   Given that
   R, the Rydberg constant, is 1.097×10^7 m^{-1}
   h, Planck's constant, is 6.626×10^{-34} J s
   c, the speed of light, is 2.998×10^8 m s^{-1}
   calculate from these data the ionization energy of hydrogen in kJ mol^{-1}. Compare the results of your calculation with the value in your text (Appendix F).

7. Describe the electron density pattern for a
   a. 2s orbital
   b. $3p_x$ orbital
   c. $3d_{xy}$ orbital

8. Which statements are incorrect? Explain why.

   a. If $m_\ell = 1$, the orbital must be a p orbital.
   b. If $n = 2$, only two orbitals are allowed, one s and one p.
   c. If $\ell = 3$, there are 3 possible values for the quantum number m_ℓ.
   d. If $m_\ell = 0$, the value of ℓ must equal 0.

Answers to Study Questions and Problems

1. a. Wavelength (λ) × frequency (ν) = velocity (v).
      The velocity of electromagnetic radiation is 2.998×10^8 m s^{-1}.
      So the wavelength (λ) is 2.998×10^8 m s^{-1} / 5.56×10^6 Hz.
         = 53.9 m.
   b. Wavelength (λ) × frequency (ν) = velocity (v).
      The frequency of this radiation = 2.998×10^8 m s^{-1} / 6.67×10^{-7} m.   1 nm = 10^{-9} m
         = 4.50×10^{14} Hz.   Hz = s^{-1}

2. a. Lowest frequency vibration; $n = 1$, where 1 meter = n(λ/2).
      Wavelength = 2 meters
   b. Number of nodes = $n-1$ = 4 (not including the ends).   If the fixed ends were included,
   c. Two complete wavelengths—three nodes.   there would be 6 nodes.
   d. Equally spaced at 25 cm, 50 cm, and 75 cm.
   e. The frequency of the $n=4$ vibration is twice the frequency of the $n=2$ vibration (wavelength is inversely proportional to frequency).

3. 400nm = 4.00×10^{-7} m
   a. frequency = 2.998×10^8 m s^{-1} / 4.00×10^{-7} m = 7.5×10^{14} s^{-1}

b. energy = $h\nu$ = 6.626×10^{-34} J s $\times 7.5 \times 10^{14}$ s^{-1} = 5.0×10^{-19} J

c. energy of one mole of photons
    = 5.0×10^{-19} J $\times 6.022 \times 10^{23}$ photons/mol = 300 kJ/mol.

4. a. Four lines are visible in the Balmer series: red green, blue, and violet.

    b. There are more lines—an infinite number in fact. The fifth line occurs at 397 nm which is just beyond the end of the visible region.

    c. An infinite number again, but the lines get closer and closer together toward the high energy end, becoming indistinguishable from one another.

    d. No others, only the Lyman series—corresponding to transitions down to the n =1 energy level.

    e. An infinite number, but as the lines get closer, the series overlap and coalesce.

The five established series are:

Lyman (to n=1)   UV
Balmer (to n=2)   visible
Paschen (to n=3)   IR
Brackett (to n=4)   IR
Pfund (to n=5)   IR

5. Using the Rydberg equation:

    $\Delta E = h\nu$ = Rhc$(1/n_1^2 - 1/n_2^2)$ where n_1 = 8 and n_2 = 9.

    $\Delta E = h\nu$ = 2.179×10^{-18} (1/64 − 1/81) = 7.146×10^{-21}

Rhc = 2.179×10^{-18}.

    Frequency $\nu = \Delta E/h$ = 1.08×10^{13} s^{-1}.
    This corresponds to a wavelength of 2.78×10^{-5} m; in the IR region.

6. The ionization energy of an element is the energy required to remove the most loosely held electron from atoms of the element in the gaseous state. It is usually expressed in units of kJ mol^{-1}.
    R, the Rydberg constant, is 1.097×10^7 m^{-1}
    h, Planck's constant, is 6.626×10^{-34} J s
    c, the speed of light, is 2.998×10^8 m s^{-1}

    Use the Rydberg equation, where n_1 = 1, but n_2 = infinity (the electron is free).

Rhc = 2.179×10^{-18}.

1/infinity → zero.

    ΔE = Rhc$(1/n_1^2 - 1/n_2^2)$ where n_1 = 1 and n_2 = infinity.

    ΔE = Rhc(1 − 0)

    ΔE = Rhc

    ΔE = 2.179×10^{-18} J/electron.

    ΔE = 2.179×10^{-18} J/electron $\times$ 1kJ/1000J $\times 6.022 \times 10^{23}$ mol^{-1}.

    ΔE = 1312 kJ mol^{-1}.

7. a. 2s orbital: spherically symmetric; no planar nodes

See Figure 7.16a on page 321 of the text.

    b. 3p$_x$ orbital: an orbital with a planar node in the yz plane; because n = 3, there is also a spherical node near the nucleus; the orbital lies along the x axis.

See Figure 7.16b on page 321 of the text.

    c. 3d$_{xy}$ orbital: two planar nodes splitting the distribution into four lobes (a cloverleaf shape); the orbital lies in the xy plane.

8.   All statements are incorrect.

a.   If $m_\ell = 1$, the orbital cannot be an s orbital but it could be a p, d, or f orbital. It is the value of ℓ that determines the type of orbital.

b.   If $n = 2$, two types of orbitals are allowed, s and p. But p orbitals always come in sets of three, so for $n = 2$, there are 4 orbitals, one s and three p.

c.   If $\ell = 3$, there are 7 possible values for the quantum number m_ℓ. The number of values for m_ℓ is given by the expression $2\ell + 1$.

d.   There is an $m_\ell = 0$ orbital in each set, s, p, d, f ...

Challenge 3

Calculate the wavelength of alpha (α) particles (emitted from a radioactive isotope) with a kinetic energy of 6.0×10^{11} J mol^{-1}.

An α particle is a helium nucleus.

People

—a crossword puzzle

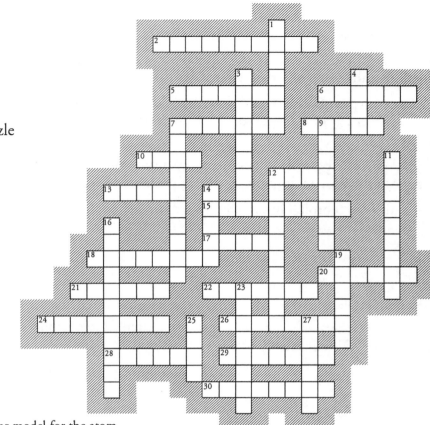

Across:

2. Proposed the nuclear model for the atom.
5. Described radiation in terms of electric and magnetic waves.
6. Developed the quantum theory in 1900.
7. Wrote the laws of electrolysis in 1833.
8. Proved the existence of electromagnetic radiation.
10. Incorporated ideas of quantization of energy into a model for the electronic structure of hydrogen.
12. Developed the law of constant heat summation.
13. A man of energy, a student of 27 down.
15. Proposed the law of conservation of matter.
17. Discovered polonium, named after her native land.
18. Discovered the neutron in 1932.
20. Proposed a temperature scale in 1848 for which the zero point was −273°C.
21. Proposed the law of constant composition.
22. Developed an equation used to predict the lines in the emission spectrum of hydrogen.
24. Measured the charge on a single electron.
26. An Italian whose ideas were a long time in being accepted.
28. See 23 down.
29. Introduced the term "mole" for an amount of substance.
30. Suggested the presence of electrons in all matter.

Down:

1. Determined that elements in the Periodic Table should be arranged by atomic number, not mass.
3. Discovered periodicity in the properties of elements and designed the first Periodic Table.
4. What was his name, the unit for power? A student of 14 down.
7. His degree is smaller than that of Celsius.
9. He used Planck's new quantum theory to explain the photoelectric effect in 1905.
11. Proposed wave–particle duality.
12. A man of some uncertainty.
14. Distinguished heat, temperature, and heat capacity.
16. Used 11 down's assertion to write a wave equation for the electron in a hydrogen atom.
19. The SI unit for force is named after him.
23. With 28 across established the wave–particle duality of an electron beam.
25. Suggested that φ^2 should be interpreted as the probability of finding the electron at any point about the nucleus.
27. Forcefully revived the idea of atoms in 1803.

CHAPTER 8

Introduction

The atomic orbitals described in the last chapter are really the orbitals for hydrogen—a simple case with only one proton in the nucleus and only one electron outside the nucleus. In this chapter, the idea of atomic orbitals and how electrons are configured in these orbitals is extended to the other elements in the Periodic Table. The dependence of the physical and chemical properties of atoms and ions on their electron configurations is described.

Contents

8.1 Electron Spin

The three quantum numbers described in Chapter 7 (n, ℓ, and m_ℓ) define an orbital in which an electron can exist. However, to characterize an electron in an atom completely, a fourth quantum number is required. This is the **electron spin magnetic quantum number**, m_s.

There are different types of magnetism: **Diamagnetic** substances are repelled by a magnetic field. **Paramagnetic** substances are drawn into a magnetic field. **Ferromagnetic** substances, such as iron, magnetite (Fe_3O_4), and alnico, possess a strong intrinsic magnetism. The different types of magnetism exhibited by different substances arises from their electron spins. A diamagnetic substance has no unpaired electrons; a paramagnetic substance has some unpaired electrons; and a ferromagnetic substance has all its unpaired electron spins lined up in the same direction.

Electron spin is quantized. In a magnetic field, the electron can line up in only two ways corresponding to only two possible values for the spin quantum number $m_s = \pm \frac{1}{2}$. When an electron is assigned to an orbital, it can take either value of m_s. However, if a second electron is assigned to the same orbital, it must take a value of m_s different from the first. The electrons become **paired**.

Two electrons can occupy the same orbital provided that their spin quantum numbers are different.

8.2 The Pauli Exclusion Principle

It is convenient to represent an orbital as a small box or circle, and the electrons as arrows, pointing up for an electron spin = +½ and pointing down for an electron spin = −½.

$\boxed{\uparrow\downarrow}$

For example, for n = 3:

There are three types of orbitals: s, p, and d.

There are 9 orbitals in total: One s, three p, and five d.

In 1925 Wolfgang Pauli devised his **Exclusion Principle**: No two electrons in a single atom can have the same set of values for the four quantum numbers. Since one orbital has the same set of three quantum numbers n, ℓ, and m_ℓ and there are only two possible values for m_s, this means that one orbital can accommodate only two electrons.

The number of subshells (types of orbitals) in a principal quantum level n is equal to n. The number of orbitals in a principal quantum level n is equal to n^2. Therefore the number of electrons that can be accommodated in a principal quantum level n is equal to $2n^2$—two in each orbital. For example, in the $n = 1$ level, there is only an s orbital, so the $n = 1$ level can accommodate two electrons. In the $n = 2$ level, there are two subshells: s and p orbitals. But p orbitals always come in sets of 3, so the total number of orbitals in the $n = 2$ level equals 4 (equal to 2^2). The number of electrons that can be accommodated is 8 (equal to $2n^2$ where $n = 2$).

8.3 Atomic Subshell Energies and Electron Assignments

The filling in order of increasing energy is sometimes referred to as the auf–bau principle.

See Figure 8.5 on page 339.

Electrons in a multi-electron atom are placed in orbitals in order of increasing energy. The electrons are assigned to **shells**, each shell characterized by the principal quantum number n. Within each shell there are **subshells** characterized by the secondary quantum number ℓ. The subshells are denoted by the letters: s, p, d, and f.

Experimental evidence for the existence of these shells and subshells can be obtained from the ionization energies for the elements. The ionization energy is the energy required to remove the most loosely held electron from an isolated atom or ion (*cf.* section 8.6).

All subshells within a principal quantum level n for hydrogen have the same energy. However, for any atom with more than 1 electron, the subshells are no longer equal in energy. This is because an electron in a multi-electron atom repels other electrons and interferes with the attraction between the nucleus and other electrons. The subshells are filled in the sequence of increasing energy. Within a principal quantum level, the order of filling is always s, p, d, and then f.

See Figure 8.6 on page 342.

For 4s, the value of $n + \ell$ is 4.
For 3d, the value of $n + \ell$ is 5.
So the 4s is filled before the 3d.

For 4p, the value of $n + \ell$ is 5.
For 3d, the value of $n + \ell$ is 5.
But 3 is lower than 4,
So the 3d is filled before the 4p.

The problem is that the principal quantum levels start to overlap, and therefore the s orbital of the nth level might be lower in energy than a d orbital of the $(n-1)^{th}$ level or an f orbital in the $(n-2)^{th}$ level. The rule is that electrons are assigned in order of increasing $n+\ell$, and if two orbitals have the same value for $n+\ell$, the one with the lower n is filled first.

The **effective nuclear charge** is the nuclear charge experienced by a particular electron in a multi-electron atom. This electron may be **shielded** from the full nuclear charge by intervening electrons. In general, s electrons are nearer to the nucleus than p electrons, therefore s electrons shield p electrons from the full nuclear charge. The s electrons are said to **penetrate** nearer to the nucleus. The shielding capability, and the degree to which electrons penetrate toward the nucleus decreases in the order s > p > d > f.

8.4 Atomic Electron Configurations

Hydrogen, element number 1, has one electron in the lowest possible orbital (1s). Its configuration is written $1s^1$, where the superscript 1 indicates one electron in the 1s orbital. Helium, element number 2, has the configuration $1s^2$ which fills this orbital and completes the $n = 1$ level. The next element, lithium has the configuration $1s^2\,2s^1$ and so on...

Be $1s^2\,2s^2$
B $1s^2\,2s^2\,2p^1$
C $1s^2\,2s^2\,2p^2$
N $1s^2\,2s^2\,2p^3$
O $1s^2\,2s^2\,2p^4$
F $1s^2\,2s^2\,2p^5$
Ne $1s^2\,2s^2\,2p^6$ —Ne completes the $n=2$ level
Na $1s^2\,2s^2\,2p^6\,3s^1$ —or, more simply, $[Ne]\,3s^1$

> You should be able to write the ground state electron configuration for any element in the Periodic Table—certainly the elements in the first 5 periods.

When electrons are assigned to a p, d, or f subshell, each successive electron is assigned to its own orbital until the set is half-full. Then the electrons are paired up in the orbitals. This minimizes the interelectronic repulsion leading a lower energy configuration. For example, the lowest energy configuration for nitrogen is

> This is known as Hund's rule of maximum multiplicity.

N $1s^2\,2s^2\,2p_x^1\,2p_y^1\,2p_z^1$ rather than $1s^2\,2s^2\,2p_x^2\,2p_y^1$

Electrons in the outermost orbitals that are involved in chemical reactions are called **valence electrons**. The inner **core** electrons are not involved in chemical reactions and are often represented in brackets as above. [Ne], for example, represents the core electrons $1s^2\,2s^2\,2p^6$.

Elements in the $n=4$ to $n=7$ shells also use d and f orbital subshells. The elements for which the d subshells are being filled are called the transition elements. These are the elements in the center block of 10 columns in the Periodic Table. The filling of the f orbital subshells occurs in the lanthanide and actinide series. These are the elements written in two rows at the bottom of the table.

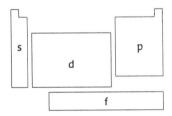

> The blocks of elements in the Periodic Table are commonly referred to as the s–block, the p–block, the d–block, and the f–block. See Figure 8.9 on page 347.

There are a few anomalies in what you might expect the lowest energy electron configuration to be. A half-filled or completely filled d or f subshell is particularly stable. Atoms often adopt such a configuration at the expense of the expected order. For example chromium has the valence electron configuration $3d^5\,4s^1$ rather than the expected $3d^4\,4s^2$. Copper is $3d^{10}\,4s^1$ rather than $3d^9\,4s^2$. Gadolinium is $4f^7\,5d^1\,6s^2$ rather than $4f^8\,6s^2$. Palladium is simply $4d^{10}$ rather than $3d^8\,4s^2$. These exceptions are not particularly important.

> Electron configurations can be written in order of increasing n or in the filling order, both are OK. The former is perhaps more common.

8.5 Electron Configurations of Ions

A cation is formed by the removal of electron(s) from an atom. The general rule is that the electron removed is the one with the highest n value. If there is more than one subshell available for the highest value of n, then the electron with the highest value of ℓ is removed. Some examples are:

Transition metals are characterized by their incompletely filled d orbital subshells. This feature greatly influences their chemical properties.

Ni $[Ar]3d^8 4s^2$ $\rightarrow$ Ni^{2+} $[Ar]3d^8$ removal of the two $4s^2$ electrons.

K $[Ar]4s^1$ $\rightarrow$ K^+ $[Ar]$ removal of $4s^1$; note the remaining noble gas configuration.

For the formation of anions, add electrons to the next available orbitals.

8.6 Atomic Properties and Periodic Trends

The similarity in the chemical and physical properties of elements in the same group results from their similar valence shell electron configurations. For example, all the alkali metals have the valence electron configuration ns^1. Not only do elements in the same group have similar properties, but there is a consistent and predictable variation in properties across a period.

Atomic size increases down a group as more and more shells of electrons are added. Atomic size decreases across a period as the nuclear charge increases and pulls the valence electrons closer to the nucleus.

Atomic sizes can be obtained from bond lengths for atoms in molecules. For metals, atomic sizes can be derived from the metallic structure.

Gallium (below aluminum) with 31 electrons, is smaller than aluminum with only 13 electrons (*cf.* Figure 8.10, page 355). The reason for this, and all the other anomalous properties of the elements gallium through to bromine, is the preceding block of 3d elements. For this 3d block, the nuclear charge increases by 10 from scandium to zinc without any compensating increase in the shielding—the electrons go into the poorly shielding d orbital subshell. The increase in the nuclear charge pulls in the valence s and p electrons and makes gallium smaller than it would otherwise have been.

It's interesting that a fluorine atom (72 pm), with 9 electrons, is smaller than a lithium atom (152 pm), with only 3 electrons. The electrons are much closer together in fluorine!

See Figure 8.12 on page 358.

Ionization energy is another property that varies systematically through the Periodic Table. The ionization energy is a measure of how strongly the outermost valence electron is held. It increases from left to right across a period, and decreases down a group, for exactly the same reasons as the atomic size changes. Small dips or peaks in the trends across a period are predictable. In general, the formation or breaking up of a completely filled or half-filled subshell of orbitals will influence the ionization energy. For example, removal of an electron to produce a half-filled p orbital subshell will be slightly favored, thus oxygen has a lower ionization energy than nitrogen, although the general trend would predict a higher ionization energy.

See Figure 8.13 on page 359.

Electron affinity is similar to the ionization energy; it is the energy required to remove an electron from a negative ion. To remove an electron from F^- (to form F) requires 328 kJ mol^{-1} and is a measure of how much the fluorine wants the electron. The 328 kJ mol^{-1} is called the electron affinity. Trends in electron affinity parallel those in ionization energy (for the same reasons).

If an electron is added to F to produce F^-, 328 kJ of energy is released. For that reason, electron affinities have in the past been represented as negative numbers.

See Figure 8.14 on page 361.

Isoelectronic means having the same number of electrons.

When an electron is added to an atom, there is always an increase in size. Similarly, if an electron is removed, there is always a decrease in size. Cations are always smaller than the atom; anions are always larger. The decrease in size is especially great when the last electron in a particular shell is removed. In a series of isoelectronic ions, there is a consistent decrease in size as the nuclear charge is increased.

8.7 Periodic Trends and Chemical Properties

Main group metals form cations with an electron configuration equivalent to the preceding noble gas. Likewise, nonmetals form anions with an electron configuration equivalent to the following noble gas. For example, all the alkali metals form 1+ ions, and all alkali metals have similar properties and form similar compounds. Periodicity is a function of electron configuration.

Products formed in chemical reactions, especially simple inorganic reactions, are generally the most (thermodynamically) stable products possible. Sodium and chlorine for example form NaCl, not Na_2Cl, or $NaCl_2$ because Na^+ is the stable sodium cation and Cl^- is the stable chlorine anion. Likewise, magnesium and oxygen form MgO where magnesium has lost 2 electrons and oxygen has gained 2 electrons.

Corresponding transition elements in the 2nd and 3rd series are very similar to each other. The reason for this is the block of 4f elements following lanthanum. For this 4f block, the nuclear charge increases by 14 across the series, without any compensating increase in the shielding—the electrons go into the very poorly shielding f orbital subshell. The increase in the nuclear charge pulls in the valence s and p electrons and makes the third series transition metal almost exactly the same size as the corresponding 2nd series metal. For example, platinum and palladium, tantalum and niobium, gold and silver. All pairs are difficult to separate from each other.

Niobe (Z=41) was the daughter of Tantalus (Z=73).

Review Questions

1. Describe the three types of magnetism. How is each related to the electron spins?

2. Describe the Pauli Exclusion Principle. Why does this principle limit the number of electrons in an orbital?

3. What are the expressions for:
   The number of subshells (types of orbitals) in a principal quantum level.
   The number of orbitals in a principal quantum level.
   The number of orbitals in a subshell (s, p, d, f, etc.)
   The number of electrons that can be accommodated within a subshell.
   The number of electrons that can be accommodated in a principal quantum level.

4. How does the ionization energy for an atom or an ion provide information about the existence of shells and subshells (*cf.* Figure 8.5 on page 339)?

5. Describe effective nuclear charge, the shielding effect, and the penetration effect.

6. What is the $n+\ell$ rule? How does it work?

7. Why does the 4s orbital lie lower in energy than the 3d orbital?

8. What is a core electron? How are the core electrons best represented in an electron configuration. What is a valence electron?

9. Why are the lanthanides and the actinides drawn in two rows at the bottom of the Periodic Table? Where do they really belong? If you draw a Periodic Table with the lanthanides and the actinides in their "proper" place, what pattern, if any, do you notice?

10. What are the general rules that determine which electron(s) to remove when a cation is formed from a neutral atom?

11. What are the general trends in atomic size, ionization energy, and electron affinity through the Periodic Table. Why are all three related?

12. Why are tantalum and niobium so difficult to separate from each other?

Answers to Review Questions

1. There are three types of magnetism: Diamagnetic substances are repelled by a magnetic field; they have no unpaired electrons. Paramagnetic substances are drawn into a magnetic field; they have at least one unpaired electron. Ferromagnetic substances possess a strong intrinsic magnetism; these substances have all the unpaired electron spins lined up in the same direction.

2. Pauli Exclusion Principle: No two electrons in a single atom can have the same set of values for the four quantum numbers n, ℓ, and m_ℓ and m_s. Since one orbital has the same set of three quantum numbers n, ℓ, and m_ℓ and there are only two possible values for m_s, this means that one orbital can accommodate only two electrons: each with a different value for m_s.

It's not necessary to memorize all these expressions; especially if you know that each orbital can accommodate 2 electrons.

3. The number of subshells in a principal quantum level = n.
The number of orbitals in a principal quantum level = n^2.
The number of orbitals in a subshell = $2\ell + 1$.
The number of electrons in a subshell = $2(2\ell + 1)$.
The number of electrons in a principal quantum level = $2n^2$.

4. The ionization energy for an atom or an ion indicates directly the amount of energy required to remove an electron from a particular orbital. It indicates therefore the energy of that orbital. As the ionization energies vary in a systematic way through the Periodic Table, so therefore do the energies of the orbitals from which the electron is being removed.

5. The effective nuclear charge is the nuclear charge experienced by a particular electron in a multi-electron atom. The shielding effect describes how an electron in one orbital shields an electron in another orbital from the nuclear charge. Electrons in s orbitals approach nearest to the nucleus and are the best shielders. In fact, the shielding capability decreases in the order s > p > d > f. The penetration effect describes how close to the nucleus a particular electron can get. The degree to which electrons penetrate toward the nucleus also decreases in the order s > p > d > f.

6. The $n + \ell$ rule is a rule that determines the order of the filling of orbitals in building up the electron configuration of a multi-electron atom. The orbitals are filled in the order of increasing $n + \ell$. For two equal values of $n + \ell$, the orbital with the lower value of n is filled first.
    The rule works because it takes into account (empirically) the splitting in energy of the subshells within a principal quantum level due to the shielding and penetration effects.

    > Use the Periodic Table, left to right, row by row, to determine the order of the filling of the orbitals.

7. The 4s orbital lies lower in energy than the 3d orbital because the 4s orbital penetrates nearer to the nucleus than the 3d orbital. Electrons in a 4s orbital are pulled in more by the nucleus and are lower in energy.

8. A core electron is an electron in an orbital of the period in the Periodic Table above the period of the valence shell. A core electron is not involved in chemical reactions. The core electrons are best represented in an electron configuration by a noble gas in square brackets. A valence electron is an electron in the outermost subshells of an atom. These are the electrons that are involved in chemical reactions. For example, the electron configuration of cobalt is written [Ar] $3d^7\ 4s^2$. The [Ar] represents the core electrons—the 18 electrons in the 1s, 2s, 2p, 3s, and 3p orbitals (Periods 1, 2, and 3). The $3d^7\ 4s^2$ is the valence shell configuration; these are the valence electrons.

9. The lanthanides and the actinides are drawn in two rows at the bottom of the Periodic Table so that the table fits conveniently on one 8.5 × 11 page. They really belong in horizontal rows between La and Hf and between Ac and Rf. When placed in these positions, the steps in the table are more obvious: The principal quantum numbers of the d orbital subshells are off by 1 compared to the s and p subshells of the same period. The principal quantum numbers of the f orbital subshells are off by 2 compared to the s and p subshells of the same period. For example, Period 6 is composed of the 6s and 6p, but the 5d, and the 4f subshells.

10. Electrons are removed from the highest energy orbitals—in other words, the electrons that are easiest to remove. The general rule is that the electron removed is the one with the highest n value. If there is more than one subshell available for the highest value of n, then the electron with the highest value of ℓ is removed.

11. All general trends in atomic size, ionization energy, and electron affinity depend upon the effective nuclear charge experienced by the valence (outermost) s and p electrons. The greater this charge, the smaller the size, the higher the ionization energy, and the greater the electron affinity. All three trends are therefore related in a systematic way.

12. Corresponding transition elements in the 2nd and 3rd series, such as niobium and tantalum, are very similar to each other. The reason for this is the block of 4f elements that occurs between the 2nd and 3rd series. Across the 4f block, the nuclear charge increases by 14+, without any compensating increase in the shielding—the electrons go into the poorly shielding f orbital subshell. The increase in the nuclear charge pulls in the valence s and p electrons and makes tantalum almost exactly the same size as niobium. The two metals behave very similarly and are difficult to separate.

No human investigation can be called real science if it cannot be demonstrated mathematically.

Leonardo da Vinci
(1452-1519)

Study Questions and Problems

1. Write the electron configurations of the following elements using the shorthand notation for the noble gas cores.
   a. phosphorus
   b. nickel
   c. osmium
   d. californium
   e. titanium

2. Which orbital is filled following these orbitals?
   a. 3d
   b. 4s
   c. 5p
   d. 5f

3. How many electrons can be accommodated in
   a. a d subshell
   b. a set of f orbitals
   c. the $n = 4$ shell
   d. the 7s orbital
   e. a p_x orbital?

4. What is wrong with the following ground state electron configurations?

5.   How many unpaired electrons are there in
      a. a nitrogen atom   c. a nickel(II) cation
      b. an iodine atom   d. an oxide ion?

6.   Which of the following sets of quantum numbers describe an impossible situation? Explain why.

      a. $n = 2$, $\ell = 1$, $m_\ell = 2$, $m_s = +\frac{1}{2}$
      b. $n = 5$, $\ell = 2$, $m_\ell = 1$, $m_s = -\frac{1}{2}$
      c. $n = 6$, $\ell = 5$, $m_\ell = 0$, $m_s = 0$
      d. $n = 3$, $\ell = 3$, $m_\ell = 1$, $m_s = -\frac{1}{2}$
      e. $n = 4$, $\ell = 2$, $m_\ell = 1$, $m_s = +\frac{1}{2}$

7.   Arrange the elements S, Ge, P, and Si in order of increasing atomic size.

8.   Arrange the ions Na^+, K^+, Cl^-, and Br^- in order of increasing size.

9.   Arrange the elements Be, Ca, N, and P in order of increasing ionization energy.

10.   Which one of each of the following pairs would you expect to have the highest electron affinity?
      a. Cl or Cl^-
      b. Na or K
      c. Br or I

11.   Which ions would you expect to exist, and which wouldn't you expect to exist?
      a. K^{2+} e. P^{3-}
      b. Cl^- f. Mn^{7+}
      c. Al^{2+} g. Fe^{2+}
      d. Ar^+ h. Na^-

12.   Which elements fit the following descriptions:
      a. the smallest alkaline earth metal
      b. has a valence shell configuration $4f^{14}\,5d^{10}\,6s^1$
      c. the halogen with the lowest ionization energy
      d. has 13 more electrons than argon
      e. the smallest non metal
      f. the Group 4A element with the largest ionization energy
      g. its 3+ ion has the electron configuration $[Kr]\,4d^{10}$

13.   Given the series of ionic radii:

      C^{4-} 260 pm; N^{3-} 171 pm; O^{2-} 126 pm; F^- 119 pm;
      Na^+ 116 pm; Mg^{2+} 86 pm; Al^{3+} 68 pm,

      estimate the atomic radius of neon. Do you think this is a fair estimate?

Answers to Study Questions and Problems

1. a. phosphorus   [Ne] $3s^2\,3p^3$
    b. nickel   [Ar] $3d^8\,4s^2$
    c. osmium   [Xe] $4f^{14}\,5d^6\,6s^2$
    d. californium   [Rn] $5f^{10}\,7s^2$
    e. titanium   [Ar] $3d^2\,4s^2$

2. The easiest way to determine the order in which the orbitals are filled is the follow the Periodic Table row by row left to right.
    a. 3d is followed by 4p
    b. 4s is followed by 3d
    c. 5p is followed by 6s
    d. 5f is followed 6d

3. a. a d subshell   10 electrons; two in each of 5 orbitals
    b. a set of f orbitals   14 electrons; two in each of 7 orbitals
    c. the n = 4 shell $2n^2 = 32$ (or 2 in s; 6 in p; 10 in d, 14 in f)
    d. the 7s orbital   only 2
    e. a p_x orbital   only 2

4. a. OK, 4s is filled before 3d.
    b. 4s should be filled before 4p subshell.
    c. 4s cannot have 2 electrons with same spin.
    d. OK, stability of half-filled 3d subshell.
    e. 4p should first fill one orbital at a time.

5. a. a nitrogen atom   3 unpaired electrons   ($2s^2\,2p^3$)
    b. an iodine atom   1 unpaired electron   ($5s^2\,5p^5$)
    c. a nickel(II) cation   2 unpaired electrons   ($3d^8$)
    d. an oxide ion   no unpaired electrons   (closed shell [Ne])

6. a. $n = 2,\ \ell = 1,\ m_\ell = 2,\ m_s = +\frac{1}{2}$ maximum value for m_ℓ is $+\ell$
    b. $n = 5,\ \ell = 2,\ m_\ell = 1,\ m_s = -\frac{1}{2}$ OK
    c. $n = 6,\ \ell = 5,\ m_\ell = 0,\ m_s = 0$ m_s must be either $+\frac{1}{2}$ or $-\frac{1}{2}$
    d. $n = 3,\ \ell = 3,\ m_\ell = 1,\ m_s = -\frac{1}{2}$ maximum value for ℓ is $n - 1$
    e. $n = 4,\ \ell = 2,\ m_\ell = 1,\ m_s = +\frac{1}{2}$ OK

7. Increase from right to left and down from top to bottom:
    S < P < Si < Ge.

8. K^+ and Cl^- are isoelectronic; the negative ion is the larger of the two (the nuclear charge is greater for the positive ion). Br^- is larger than K^+ and Cl^-. Na^+ is smaller than K^+ and Cl^-. So, in order of increasing size:
    $Na^+ < K^+ < Cl^- < Br^-$

9. Increase from left to right and up from bottom to top:
    Ca < Be < P < N.
    Elements lying close to a diagonal like Be and P are difficult to predict.
    Notice that Si, to the left of P, has an ionization energy less than Be.

10. a. Cl: the Cl^- ion has a complete shell and zero electron affinity. Cl is
        just one electron short and has a very high electron affinity.
    b. Na: you would expect the electron affinity to increase up a group.
    c. Br: you would expect the electron affinity to increase up a group; an
        exception is a decrease from Cl to F—attributed to electron-electron
        repulsion in the very small fluorine ion.

11. a. K^{2+} doesn't exist; the K^+ has the stable configuration [Ar].
    b. Cl^- this is the stable anion formed by chlorine.
    c. Al^{2+} doesn't exist; Al forms Al^{3+} —the same configuration as [Ne].
    d. Ar^+ Ar already has a stable configuration; never forms Ar^+.
    e. P^{3-} this is the phosphide ion; same configuration as [Ar].
    f. Mn^{7+} doesn't exist; charge is far too high.
    g. Fe^{2+} one of the stable ions formed by iron.
    h. Na^- sodide ion; actually can be made; configuration $3s^2$.

12. a. the smallest alkaline earth metal — beryllium Be.
    b. has a valence shell configuration $4f^{14} 5d^{10} 6s^1$ — gold Au.
    c. the halogen with the lowest ionization energy — astatine At.
    d. has 13 more electrons than argon — gallium [Ar] $3d^{10} 4s^1 4p^2$.
    e. the smallest non metal — hydrogen H.
    f. Group 4A element with the largest ionization energy — carbon C.
    g. its 3+ ion has the electron configuration [Kr] $4d^{10}$ — indium In^{3+}.

13. C^{4-} 260 pm; N^{3-} 171 pm; O^{2-} 126 pm; F^- 119 pm;
    Na^+ 116 pm; Mg^{2+} 86 pm; Al^{3+} 68 pm.

    All these species are isoelectronic; all have the configuration $[He]2s^2 2p^6$.
    However, the nuclear charge increases along the series and the result is a
    decrease in the radius as the electrons are pulled closer to the nucleus.

    Somewhere between 116 and 119 pm? This is where Ne fits in the iso-
    electronic series. Ne does not form compounds like the nonmetals in the
    preceding groups and measuring its atomic radius is therefore difficult.
    Some data could be obtained from the crystalline structure of solid neon
    but the bonding in such a solid is quite different from the bonding in
    elements like carbon or sulfur, or in compounds like sodium fluoride.
    There is wide variation is stated values for the atomic radius of neon—
    anywhere from 70 pm to 112 pm. Putting together a set of self-consis-
    tent data is not easy. For example, Na^+ itself is listed as having an atomic
    radius anywhere between 95 and 116 pm. You would expect Ne to be
    larger than Na^+. F is listed usually at about 72 pm — you might have
    expected Ne to be smaller. However, see the note in the margin.

In neon the interelectronic repulsion is high; the electrons are very crowded. As a result neon might well be larger than fluorine or oxygen. A value between 100 and 110 pm seems reasonable.

Challenge 4

In a world far from here, things are quite different. Planck's constant is much larger and, as a result, Heisenberg's uncertainty principle is a constant problem. The quantum numbers for electrons are actually the same but they are allowed different values:

n, the principle quantum number can have any integer value from 0, 1, 2, 3...to infinity.

ℓ, the angular momentum quantum number can have any integer value from 0, 1, 2...to $n + 1$. Curiously, the values of ℓ are denoted by the same letters s, p, d, f, g....

m_ℓ the magnetic quantum number can have any integer value from $+2\ell$ through 0 to -2ℓ.

m_s, the electron spin quantum number can have three possible values: $+\frac{1}{2}$, 0, and $-\frac{1}{2}$. Only three electrons can be placed in one orbital, and each one must have a different value for m_s.

As a visitor, you are asked:

a. What is the electron configuration of element number 57? Hund's rule still applies; like students in a residence hall, the electrons like to be as untripled as possible.

b. What is a general expression for the number of orbitals within a subshell?

c. What is a general expression for the number of electrons that fill a principal quantum level n?

CHAPTER 9

Introduction

The goal of this chapter, and the two chapters that follow, is to explain how atoms are arranged in chemical compounds and how they are held together. The relationship of the structure of a compound and the properties of that compound are examined and explained. The bonding and structural characteristics of individual atoms are very similar from one compound to another—they depend upon the valence shell electron configuration of the atom. This consistency permits the development of the various principles of structure and bonding that will be examined in these chapters. A knowledge of a compound's structure and an understanding of the bonding between the atoms of the compound are necessary in order to explain its chemical properties.

Contents

9.1 Valence Electrons

Electrons in an atom can be divided into two groups: the **core** electrons and the **valence** electrons. The core electrons are not involved in chemical reactions and are often represented in an electronic configuration by a noble gas in brackets. The valence electrons are in the outermost orbitals and are the electrons that are involved in chemical reactions.

It is sometimes useful to draw symbols representing the valence electrons of an atom of a main group element. This symbol is called a **Lewis electron dot symbol** and is the symbol for the element surrounded by a number of dots equal to the number of valence electrons. Dots are placed singly up to 4, and then paired from 5 to 8—four orbitals are available in the valence shell of a main group element.

The valence shells of the main group elements, therefore, can accommodate an **octet** of electrons. An octet of electrons is regarded as a stable electron configuration. It is the configuration of the noble gases that are chemically very unreactive.

This was described in Chapter 8. [Ne], for example, represents the core electrons $1s^2 \, 2s^2 \, 2p^6$.

The number of valence electrons for a main group element equals the group number. For transition metals, the valence electrons include the $(n-1)d$ subshell. This was described in Chapter 8.

$\cdot \overset{\cdot \cdot}{N} \cdot$

The Lewis electron dot symbol for nitrogen. Other symbols are shown in Table 9.2 on page 374 of the text.

9.2 Chemical Bond Formation

A third type of bond is the metallic bond—to be discussed later in Section 10.4 (page 144 in this study guide).

Each type of chemical bond describes a different way in which the electrons are redistributed:

covalent: electrons shared
ionic: electrons transferred
metallic: electrons delocalized

sodium is a *metal*,
chlorine is a *covalent* molecule. Between them, they produce an *ionic* compound: NaCl.

All three are different types of matter with quite different properties.

This was discussed in Chapter 8 section 7.

When two atoms combine, their valence electrons reorganize so that they decrease in energy. There is a net attractive force between the two atoms. This force is called a **chemical bond**. There are two general types of chemical bond—covalent and ionic.

An **ionic bond** forms when one or more valence electrons are transferred from one atom to another. Positive and negative ions are formed. The bond is the electrostatic attractive force between the ions.

A **covalent bond** is formed when the two atoms share one, two, or three pairs of electrons between them. The sharing, when the two atoms are different, is unequal.

Both processes usually involve the formation of complete octets around the atoms:

$$Na^+ \quad :\!\ddot{\underset{\cdot\cdot}{Cl}}\!:^- \qquad\qquad :\!\ddot{\underset{\cdot\cdot}{Cl}}\!:\!\ddot{\underset{\cdot\cdot}{Cl}}\!:$$

ionic bond   covalent bond

9.3 Bonding in Ionic Compounds

The $\Delta H°_{ion–pair}$ is not measured directly but is calculated from Coulomb's law:

$E_{ion–pair} = (n^+e)(n^-e)/d$

n^+e = charge on cation
n^+e = charge on anion
d = internuclear distance

In the gas phase the ions are infinitely far apart.

Lattice energy is usually expressed as an enthalpy change ΔH.

Hess's law was discussed in Chapter 6. See the calculation of lattice energy in the Closer Look on page 380.

Elements on the left side of the Periodic Table tend to form positive ions through the loss of electrons. Elements on the right side of the Periodic Table tend to form negative ions through the gain of electrons. For example:

$$Na(g) \rightarrow Na^+(g) + e^- \qquad \Delta H° = +502 \text{ kJ mol}^{-1} \quad \text{(ionization energy)}$$

$$Cl(g) + e^- \rightarrow Cl^-(g) \qquad \Delta H° = -349 \text{ kJ mol}^{-1} \quad \text{(electron affinity)}$$

These two processes together are endothermic (energy is required), but energy is released when the two ions form a pair:

$$Na^+(g) + Cl^-(g) \rightarrow NaCl(g) \quad \Delta H° = -552 \text{ kJ mol}^{-1} \quad \text{(ion-pair formation)}$$

The overall process is exothermic. The attraction between the ions depends upon the magnitude of the charges and the sizes of the ions (the distance between them).

Lattice energy is a measure of the bonding in a crystalline solid. It is defined as the energy of formation of one mole of the crystalline lattice from the ions in the gas phase:

$$Na^+(g) + Cl^-(g) \rightarrow NaCl(s) \quad \Delta H° = -786 \text{ kJ mol}^{-1} \text{ (lattice energy)}$$

There are no ion pairs in the crystalline lattice; each sodium ion is surrounded by six chloride ions and each chloride ion is surrounded by six sodium ions. The lattice energy is calculated from experimentally determined thermodynamic properties using Hess's law. Like the energy of ion–pair formation, the lattice energy depends upon the charges and sizes of the ions. For example, the lattice energy of magnesium oxide MgO is almost four times more than the lattice energy of sodium fluoride NaF; the charges on the magnesium and oxide ions are 2+ and 2−, compared to 1+ and 1− for sodium fluoride.

The lattice energy also depends upon the symmetry of the crystal—how well the ions pack together.

9.4 Covalent Bonding

Covalent bonding involves the sharing of electrons between the atoms involved in the bond. Elements that bond covalently are commonly the nonmetals in the upper right hand corner of the Periodic Table.

In a simple description of covalent bonding, one or more electron pairs are shared between two atoms. These electron pairs, for example in the hydrogen molecule H_2, are represented by a line (H–H) or as a pair of dots (H:H). Pairs of electrons that are not involved in the bonding are called nonbonding or lone pairs and can also be represented as bars or as pairs of dots.

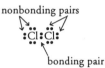

When two pairs of electrons are shared, the bond is called a **double bond** and is represented by two pairs of dots or by a double line. For example $H_2C::O$ or $H_2C=O$ for formaldehyde. Three pairs shared is a **triple bond**, as in dinitrogen $N:::N$, or acetylene $HC:::CH$. This bond is represented by three pairs of dots, or by three lines.

There is a tendency for atoms in simple covalent molecules to achieve an octet of electrons (four pairs) around themselves. Although there are many instances where the central atom in a molecule does not follow the rule, it is almost always followed by the terminal or outside atoms in a molecule. The major exception is hydrogen which can accommodate only two electrons. The procedure for determining the best Lewis electron dot structure is:

1. Determine the arrangement of atoms in the molecule.
2. Determine the total number of valence electrons.
3. Place one pair of electrons between each pair of bonded atoms.
4. Complete the octet of all terminal atoms (except hydrogen).
5. Place any electrons remaining on the central atom as lone pairs.
6. If the central atom has fewer than 8 electrons, consider moving nonbonding pairs of electrons from the terminal atoms into the bonds between the terminal atoms and the central atom.

The shapes of many molecules are related. Knowing the structure of one molecule will often allow you to predict the structure of another—when the pattern of bonding and nonbonding electrons is the same. For example, CCl_4 has the same structure as CH_4. The molecule NF_3 has the same structure as NH_3 or PCl_3. The nitrite ion NO_2^- has the same structure as sulfur dioxide SO_2.

In addition, in the formation of molecules with no charge, you will soon recognize that carbon always forms four bonds, nitrogen three bonds, oxygen two bonds, and fluorine just one bond. Because they follow the octet rule, this means that carbon has no lone pairs, nitrogen has one, oxygen has two, and fluorine has three pairs.

Isoelectronic molecules or ions have the same number of valence electrons. This is significant because it means the molecules will have the same structure.

Note that representative metals such as lithium, magnesium, beryllium, and aluminum can also form covalent bonds. There are also many covalent molecules or polyatomic ions of the transition metals.

Bond character is a continuum from 100% covalent (equal sharing) to more and more ionic as the sharing becomes more and more unequal.

single bond   :

double bond   ::

triple bond   :::

Try some of the problems at the end of this study guide.

Solutions using this procedure are provided.

Read the Problem Solving Tips and Ideas for Lewis Electron Dot Structures on page 391 in the text.

Patterns like these will become evident as you determine the structures of more and more simple molecules.

Note that there are some exceptions. For example, in a compound like CO, it is impossible for carbon to form 4 bonds; it forms three bonds and has a lone pair.

For example, the nitrate ion NO_3^- and the carbonate ion CO_3^{2-} are isoelectronic and have the same structure.

One Lewis structure for the nitrate ion (lone pairs on the O omitted).

The double headed arrows indicate resonance—the fact that all three structures contribute to the real structure.

Note that the double bond *does not move* from one pair of atoms to another; the fourth pair of electrons is shared equally by all three bonds.

See the Problem-Solving Tips and Ideas on page 393.

Elements in Period 3 down have d orbitals in their valence shells. The use of these d orbitals in expanding beyond the octet is the usual explanation of bonding in compounds like PCl_5 and SF_6.

When the Lewis electron dot structure for a molecule like ozone, or an ion like the nitrate ion, is drawn, some bonds are drawn as double bonds and some bonds are drawn as single bonds. In reality, all the bonds are the same. Lewis structures are often poor representations of the actual electronic structure. Linus Pauling proposed the concept of **resonance** to reconcile Lewis structures and the actual structures. The idea is to draw all the possible Lewis structures and then to indicate that the true structure is a composite (or resonance hybrid) of all the contributing Lewis structures. For example, in the nitrate ion, the double bond could have been drawn between the nitrogen atom and any one of the three oxygen atoms. So all three structures are drawn and the true structure said to be a composite of all three.

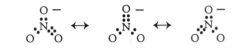

Although the majority of covalent compounds can be drawn so that the octet rule is obeyed, there are exceptions. These include compounds with fewer than eight electrons around the central atom (example BCl_3) and those with more than eight electrons around the central atom (examples SF_6, PF_5, ClF_3).

Some relatively common molecules have an odd number of valence electrons. This means that all the electrons cannot be paired. Examples are nitric oxide NO and nitrogen dioxide NO_2. These molecules are called **radicals**. Most radicals are more reactive than ordinary molecules—they participate in reactions leading to the pairing of the odd electron. For example both NO and NO_2 react with each other and themselves (dimerize) to form N_2O_2, N_2O_3, and N_2O_4.

9.5 Bond Properties

Bond order is the number of electron pairs shared between two atoms in a molecule. A single bond (one pair of electrons shared) has a bond order of 1. A double bond has a bond order of 2, and so on. Fractional bond orders are possible in molecules for which resonance is required. In the nitrate ion, for example, four pairs of electrons are shared by three pairs of atoms. Each bond has 1.33 pairs of electrons and therefore the bond order of each bond is 1.33.

Bond length is the distance between the nuclei of two atoms in a molecule.

See Table 9.8 on page 401 for typical bond lengths.

Bond lengths depend upon the sizes of the atoms, the number of electron pairs shared between the atoms (the bond order), and the positions of the atoms in the Periodic Table.

For example, H–F < H–Cl < H–Br < H–I because the size of the halogen increases. And C–O > C=O > C≡O because the bond order increases. A bond order between 1 and 2 results in a bond length between C–O and C=O. For example, the C–O bond in the carbonate ion, with a bond order of 1.33, has a bond length of 129 pm—between 143 pm for a single bond and 122 pm for a double bond.

Bond breaking is *always* endothermic.

Bond energy is the energy required to break the bond, or the energy released when the bond is made. The bond energy is related to the bond order; triple bonds are stronger than double bonds, and double bonds are stronger than single bonds. Although bond energies do vary from one compound to another, an estimate of the enthalpy of reaction $\Delta H°$ can be obtained from the difference between the bond energies of the bonds that need to be broken and the bond energies of the bonds that are formed in the reaction.

See Table 9.9 on page 403 for typical bond energies.

$$\Delta H°_{reaction} = \Sigma(\Delta H° \text{ (bonds broken)} - \Delta H°\text{(bonds made))}$$

9.6 Charge Distribution in Covalent Molecules

In a Lewis structure, it appears that the bonding electrons are shared equally between the bonded atoms. Most often they are not. Usually the bonding electrons are nearer to one atom than the other. This results in a **charge distribution** that is uneven.

The **formal charge** of an atom in a molecule is the charge calculated on the basis of equal sharing. If an atom contributes more electrons to the bonding than it gets back by sharing, it acquires a positive formal charge. Similarly, if an atom gets back more electrons than it gives, it acquires a negative formal charge. The sum of all the formal charges equals the charge on the molecule or ion (i.e. zero for a molecule).

The formal charge is calculated by adding up all the valence electrons an atom has in the molecule and comparing the number to what the atom has in a neutral isolated state. Lone pairs belong to the atom on which they reside. Bonding pairs are assumed to be divided equally between the two atoms participating in the bond. For example, in the carbonate ion:

The oxygen at the top has 2 lone pairs (4 electrons) and a half–share in 2 bonding pairs (2 electrons) for a total of 6 electrons. Oxygen is in Group 6A and should have 6 valence electrons. Therefore this oxygen atom has a zero formal charge.

The other two oxygen atoms are both the same; they have 3 lone pairs (6 electrons) and a half–share in one bonding pair (1 electron) for a total of 7 electrons. Both these oxygen atoms have one too many electrons and therefore a formal charge of -1.

The carbon atom has no lone pairs and a half–share in 4 bonding pairs (4 electrons). Carbon is in Group 4A and should have 4 valence electrons. It therefore has a zero formal charge. The sum of the formal charges is -2, the charge on the ion.

In reality, for the carbonate ion, all C–O bonds are the same, and the formal charge is delocalized over the entire ion, residing equally on each oxygen atom. Each oxygen atom has an average $-2/3$ formal charge.

Electronegativity (symbol χ) is a measure of the ability of an atom in a molecule to attract electrons to itself. If the two atoms participating in a bond have different electronegativities, then the bonding electrons are shared unequally, the electrons are pulled toward the more electronegative atom. In the extreme, an ionic bond results in which there is a transfer of the electrons to the more electronegative element. Thus, a bond between an element with a very low electronegativity (for example K) and an element with a very high electronegativity (for example F) is ionic—the sharing is very unequal. The closer the electronegativities of two nonmetals participating in a bond, the more equal the sharing of the electrons—the more covalent the bond.

Between the extremes there is a range of bonds with varying degrees of sharing; the more unequal the sharing, the more **polar** the bond.

Values of the electronegativities of the elements are listed in Figure 9.10 on page 410 of the text.

Bond polarities can be estimated from the electronegativity differences.

9.7 Molecular Shapes

The **valence shell electron pair repulsion theory** (VSEPR) was devised by N. V. Sidgewick and H. M. Powell soon after Lewis proposed his ideas of chemical bonding based upon electron pairs. The theory was developed and expanded by R. J. Gillespie and R. S. Nyholm in the 1950's. VSEPR theory is remarkably successful in predicting the shapes of the compounds of the main group elements.

The idea is that lone pairs and bonding pairs of electrons around an atom repel each other and get as far apart as possible. The direction in which the bonding pairs point determines the direction of the bond and therefore the shape of the molecule. The basic arrangements of electron pairs are:

Because the two pairs of electrons in a double bond, or the three pairs of electrons in a triple bond, share the same space, a more recent method for determining the shape of a molecule is called the electron domain (ED) model.
This reinforces the idea that it is the number of bonds and lone pairs that determines the shape, not simply the number of electron pairs.

These geometries are illustrated in Figures 9.11 and 9.12 on page 414.

2 electron pairs	linear arrangement
3 electron pairs	trigonal arrangement
4 electron pairs	tetrahedral arrangement
5 electron pairs	trigonal bipyramidal arrangement
6 electron pairs	octahedral arrangement

The actual shape of the molecule depends upon how many of the electron pairs are bonding and how many are nonbonding. The following table is divided into 5 sections corresponding to the total number of electron pairs:

Be sure to distinguish between the arrangement of electron pairs around the central atom and the shape of the molecule (which is the arrangement of atoms).

These shapes are illustrated in Figures 9.13 and 9.14 on pages 417 and 419 in the text.

2 bonding pairs		linear shape
3 bonding pairs		trigonal shape
2 bonding pairs	1 lone pair	V-shape (bent)
4 bonding pairs		tetrahedral shape
3 bonding pairs	1 lone pair	trigonal pyramidal shape
2 bonding pairs	2 lone pairs	V-shape (bent)
5 bonding pairs		trigonal bipyramidal shape
4 bonding pairs	1 lone pair	seesaw shape
3 bonding pairs	2 lone pairs	T-shape
2 bonding pairs	3 lone pairs	linear shape
6 bonding pairs		octahedral shape
5 bonding pairs	1 lone pair	square-pyramidal shape
4 bonding pairs	2 lone pairs	square-planar shape

The key to a successful use of the VSEPR theory is to determine the number electron pairs around the central atom in the molecule correctly. Note that when counting the number of electron pairs, double and triple bonds count the same as single bonds. This is because the two pairs of electrons in a double bond, or the three pairs of electrons in a triple bond, share the same space.

The procedure to follow:

1. Draw the Lewis electron dot structure of the molecule, or determine by counting the number of electron pairs around the central atom.
2. Determine the arrangement of the electron pairs around the central atom (one of the five basic arrangements).
3. Determine the shape of the molecule.

9.8 Molecular Polarity

Many molecules are polar—there is an asymmetry in the electron distribution in the molecule. This polarity of a molecule means that the molecule has a dipole moment. The dipole moment μ is defined as the product of the magnitude of the partial negative $\delta-$ and positive $\delta+$ charges at the ends of the molecule and the distance between them.

To determine whether a molecule is polar, you have to determine if any of the bonds in the molecule are polar, and if so, what is their position relative to one another. Bonds are often polar, but when positioned symmetrically, the bond polarities will cancel one another. For example, in carbon dioxide CO_2 the C=O bonds are polar, but they point symmetrically to either side of the molecule. So carbon dioxide is a nonpolar molecule even though it contains polar bonds.

The multiple bonds do take up more room and do distort the geometry.

You can determine the number of electron pairs around the central atom without drawing the Lewis structure as follows:

Add up the total number of valence electrons in the molecule.

Divide this total by 8; the answer is the number of bonding electron pairs; the remainder is the number of non-bonding electrons on the central atom (divide by 2 to obtain the number of pairs).

This will work provided that H is *not* one of the terminal atoms—in which case draw the Lewis structure.

Try some of the molecules in the problems at the end of this chapter.

Review Questions

1. What is a Lewis electron dot symbol?

2. Describe the two general types of chemical bond introduced in this chapter. How do they differ?

3. Why is the electron affinity expressed as a negative number, whereas the ionization energy is expressed as a positive number, in calculating the energy of ion–pair formation?

4. What is the difference between the energy of ion–pair formation and the lattice energy? Why is the lattice energy greater?

5. What is the difference between a bonding pair of electrons and a nonbonding pair of electrons?

6. What is the greatest number of valence electrons that can be shared between two main group atoms in the formation of a covalent bond?

7.   Is there such a thing as a 100% covalent bond? Is a bond ever 100% ionic?

8.   Summarize the procedure for drawing a Lewis electron dot structure and apply the procedure to the molecule BF_3.

9.   What does it mean when two molecules are described as isoelectronic? What significance is there in the two molecules being isoelectronic?

10.   What is resonance? How realistic is this concept?

11.   Do all atoms in all molecules obey the octet rule? What are the common exceptions?

12.   What is bond order?

13.   How is bond length related to bond order?

14.   How can you determine the enthalpy of reaction from the strengths of the bonds that are broken and the bonds that are formed in the reaction?

15.   What is the formal charge on the atom in a molecule? Does it differ from the oxidation number? Why?

16.   What is electronegativity? What is the difference between electronegativity and electron affinity?

17.   What is the basic idea behind VSEPR theory?

18.   How is the shape of a molecule or polyatomic ion related to the number of electron pairs surrounding the central atom?

19.   What is the difference between the "arrangement of electron pairs" and "the shape of the molecule"?

20.   How can a molecule contain polar bonds and yet be a nonpolar molecule?

21.   Is it true that all polyatomic ions, with charges anywhere between +1 and −3, must be polar?

22.   Are all trigonal pyramidal molecules polar? Are all tetrahedral molecules nonpolar?

23.   Are all molecules with an odd number of lone pairs around the central atom polar?

Answers to Review Questions

1. A Lewis electron dot symbol is the symbol for the element surrounded by a number of dots equal to the number of valence electrons. Dots are placed singly up to 4, and then paired from 5 to 8. The symbol represents in a simple way the number of valence electrons of an atom and its potential for forming bonds.

2. The attractive force between two atoms in a compound is called a chemical bond. There are two general types—covalent and ionic. An ionic bond forms when one or more valence electrons are transferred from one atom to another. A covalent bond is formed when the two atoms share one, two, or three pairs of electrons between them. The two types differ therefore in what happens to the valence electrons upon bond formation. The ionic bond involves a transfer of electrons; the covalent bond involves a sharing of electrons.

3. Electron affinity is defined as the energy required to remove an electron from a negative ion. In the calculation of the energy of ion–pair formation, an electron is added to the nonmetal to form the anion; energy is released and therefore $\Delta H°$ is negative (exothermic). The ionization energy is the energy required to remove an electron from the metal to form the cation—an endothermic process and $\Delta H°$ is positive.

    See chapter 8.

4. The energy of ion–pair formation is the energy released when an ion–pair is formed in the gas phase. The lattice energy is the energy released when the same gaseous ions form a solid crystal lattice. In the solid each cation is surrounded by several anions and vice versa; more bonds are formed and the lattice energy is always the greater of the two. The difference in energy equals the energy required to break up the solid lattice to form ion–pairs in the gaseous state.

5. A bonding pair of electrons is a pair of electrons shared between two atoms in a molecule. A nonbonding pairs of electrons is a pair of electrons that is not involved in the bonding; a nonbonding pair resides on just one atom in a molecule.

6. The greatest number of electrons that can be shared in a covalent bond is 6 (3 pairs in a triple bond). For example acetylene or nitrogen. Quadruple bonds are known but these occur between transition metal atoms.

7. A covalent bond is 100% covalent when the two atoms are identical (for example in O_2). When the two atoms differ in electronegativity, the bond is polar. However, a bond can never be 100% ionic—this would imply no sharing of electrons at all and an infinite distance between the two atoms.

8.   1.   Determine the arrangement of atoms in the molecule.
      2.   Determine the total number of valence electrons.
      3.   Place one pair of electrons between each pair of bonded atoms.
      4.   Complete the octet of all terminal atoms (except hydrogen).
      5.   Place any electrons remaining on the central atom as lone pairs.
      6.   If the central atom has fewer than 8 electrons, consider moving
           nonbonding pairs of electrons from the terminal atoms into the
           bonds between the terminal atoms and the central atom.

      Example BF_3

      1.   B at the center of a triangle of F atoms.
      2.   24 valence electrons (3 from B and 7 each from F).
      3.   One pair in each bond:
      4.   Complete octets of terminal atoms:
      5.   No lone pairs remaining for the B.
      6.   Possible double bonding between B and F
           to relieve the deficiency of electrons on the
           boron. One resonance structure shown:

9.   Isoelectronic molecules have the same number of valence electrons. This
      is significant because it means the molecules will have the same structure.

10.   The concept of resonance reconciles a Lewis structure and the actual
      structure. The idea is to draw all the possible Lewis structures and then to
      indicate that the true structure is a composite (or resonance hybrid) of all
      these contributing Lewis structures. It is often the Lewis structure that is
      unrealistic. For example a nitrate ion with the double bond fixed be-
      tween the nitrogen and only one oxygen atom does not exist. Resonance
      indicates that the true structure is a composite that cannot be represented
      by dots for electrons.

A better picture of the bonding
in cases where resonance is
required is provided by
molecular orbital theory in the
next chapter.

11.   There are exceptions to the octet rule:
      a.   hydrogen
      b.   compounds with fewer than eight electrons around the central atom.
      c.   compounds with more than eight electrons around the central atom.
      d.   compounds with an odd number of valence electrons.

12.   Bond order is the number of electron pairs shared between two atoms in
      a molecule. A single bond (one pair of electrons shared) has a bond order
      of 1. A double bond has a bond order of 2, and so on. Fractional bond
      orders occur when resonance is required.

13.   Bond length is the distance between the nuclei of two atoms in a mol-
      ecule. Bond lengths depend upon the bond order. For example, N–O
      (136 pm) > N=O (115 pm) > N≡O (108 pm). The bond length decreases
      as the bond order increases.

14. Although bond energies do vary from one compound to another, an estimate of the enthalpy of reaction $\Delta H°$ can be obtained from the difference between the bond energies of the bonds that need to be broken and the bond energies of the bonds that are formed in the reaction.

See Table 9.9 on page 403 for typical bond energies.

$$\Delta H°_{reaction} = \Sigma(\Delta H° \text{ (bonds broken)} - \Delta H°\text{(bonds made)})$$

15. The formal charge of an atom in a molecule is an indication of the electronic charge residing on the atom. It is, however, the charge calculated on the basis of equal sharing of bonding electrons. The formal charge is calculated by adding up all the valence electrons an atom has in the molecule and comparing the total to what the atom has in a neutral isolated state. If it has fewer electrons, then it has a formal positive charge; if it has more electrons, then it has a formal negative charge.

Except for a monatomic ion, the oxidation number has little relation to the charge of the atom. The use of oxidation numbers is an electron accounting system used in redox reactions. For example, the formal charge on Mn in the permanganate ion MnO_4^- is anywhere between 0 and +3 depending upon the Lewis structure drawn; the oxidation number is +7.

16. Electronegativity (χ) is a measure of the ability of an atom in a molecule to attract electrons to itself. Electron affinity is the energy required to remove an electron from a negative ion. Electronegativity refers to an atom in a molecule. Electron affinity refers to an isolated (gaseous) atom.

17. The basic idea behind VSEPR theory is that lone pairs and bonding pairs of electrons around an atom repel each other and get as far apart as possible. The direction in which the bonding pairs point determines the direction of the bond and therefore the shape of the molecule.

18. There are 5 basic arrangements of electron pairs: linear, trigonal, tetrahedral, trigonal bipyramidal, and octahedral. Depending upon how many lone pairs there are, subsets of various molecular shapes result.

19. The arrangement of electron pairs refers to the arrangement of both bonding and nonbonding pairs of electrons around the central atom. The shape of the molecule refers to the arrangement of atoms in the molecule (i.e. just the bonding pairs).

20. The polarities of the bonds must cancel each other. For example, in carbon dioxide CO_2 the C=O bonds are polar, but they point symmetrically to either side of the molecule. So carbon dioxide is a nonpolar molecule.

21. No, it is not true. For example, in NO_3^-, ClO_4^-, PO_4^{3-}, SO_4^{2-}, CO_3^{2-}, and NH_4^+, etc., the charges are symmetrically distributed over the entire ion and therefore these ions are nonpolar. Some ions are polar—it depends on their shape.

22. All trigonal pyramidal molecules are polar—they all have a lone pair of electrons at one end of a tetrahedral arrangement. Some tetrahedral molecules are nonpolar—for example: CH_4, CCl_4, NH_4^+, BF_4^-, SO_4^{2-}. But many are not—for example: $CHCl_3$, CH_2Cl_2, $S_2O_3^{2-}$.

23. No, not all. For example I_3^- and XeF_2 are not polar; they are linear.

In studies, whatsoever a man commandeth upon himself, let him set hours for it.

Francis Bacon
(1561-1626)

Study Questions and Problems

1. Classify the following substances as covalent molecules or ionic compounds:
   a. MgO
   b. NI_3
   c. CuS
   d. NO_2
   e. LiCl
   f. SF_4
   g. XeF_4
   h. CsF

2. How many valence electrons do the following atoms or ions have? Write their Lewis symbols.
   a. Ca
   b. S
   c. P
   d. O^{2-}
   e. Mg^{2+}
   f. C^{4-}
   g. Li
   h. Ne

3. Which of the following molecules do not obey the octet rule?
   a. $AlCl_3$
   b. PCl_3
   c. PCl_5
   d. $SiCl_4$
   e. SF_6
   f. $BeCl_2$
   g. NO_2
   h. XeF_4

4. Order the following salts in increasing lattice energy:
   CaS, MgO, KCl, CsI, NaF

5. Draw Lewis electron dot structures for the following molecules:
   a. NCl_3
   b. BCl_3
   c. ClO_2^-
   d. SF_4
   e. OCS
   f. SO_2

6. Assign formal charges to all the atoms in the following species:
   a. chlorite ion ClO_2^-
   b. hydroxylamine $HONH_2$
   c. phosphorous acid H_3PO_3
   d. ozone O_3
   e. nitrogen dioxide NO_2

7. Estimate, using the bond energies in Table 9.9 on page 403 of the text, the enthalpy change $\Delta H°$ for the conversion of propene to isopropanol:

$$CH_3-CH=CH_2 \ + \ H_2O \ \rightarrow \ CH_3-CH(OH)-CH_3$$

8. When ethanol burns in air, heat is released. Estimate the enthalpy of combustion of ethanol vapor $\Delta H°$ from the average bond energies listed in Table 9.9. Use thermochemical data in Chapter 6 to calculate the same thing. Compare the two values obtained.

9. What is the bond order of the listed bonds in the following molecules or ions?
   a. C–O in carbonate ion
   b. C–O in acetic acid
   c. N–O in nitrite ion
   d. C–C acetylene
   e. O–O in oxygen O_2
   f. C–N in hydrogen cyanide
   g. S–F in sulfur tetrafluoride
   h. C–O in carbon dioxide

10. Draw possible resonance structures for
    a. NO_2^-
    b. HCO_2^-
    c. NO_2Cl

11. What molecular shapes are associated with the following electron pairs around the central atom?
    a. 3 bonding pairs and 2 lone pairs
    b. 4 bonding pairs and 1 lone pair
    c. 3 bonding pairs and 1 lone pair
    d. 2 bonding pairs and 2 lone pairs
    e. 2 bonding pairs and 3 lone pairs
    f. 5 bonding pairs and 1 lone pair

12. What shapes are the following molecules or polyatomic ions?
    a. O_3
    b. GaH_3
    c. SO_2Cl_2
    d. XeO_4
    e. NO_2^+
    f. ClO_4^-
    g. IF_4^-
    h. ClF_2^-

13. Determine whether the following molecules are polar or nonpolar:
    a. CCl_4
    b. XeF_4
    c. PCl_5
    d. PCl_3
    e. BF_3
    f. $BeCl_2$
    g. SCl_2
    h. CS_2

Answers to Study Questions and Problems

An electronegativity difference of about 1.7 is often considered to be the dividing line between ionic and covalent bonding.

1. Look first at the relative positions of the two elements in the Periodic Table and then, if necessary, examine the electronegativity difference between the two elements. Remember that some ionic bonds have considerable covalent character.

   a. MgO   ionic   e. LiCl   ionic
   b. NI_3 covalent   f. SF_4 covalent
   c. CuS   ionic   g. XeF_4 covalent
   d. NO_2 covalent   h. CsF   ionic

2. a. Ca   2 valence electrons $\overset{\cdot}{Ca}\cdot$
    b. S   6 valence electrons $\cdot\overset{\cdot\cdot}{\underset{\cdot\cdot}{S}}\cdot$
    c. P   5 valence electrons $\cdot\overset{\cdot}{P}\cdot$
    d. O^{2-} 8 valence electrons $:\overset{\cdot\cdot}{\underset{\cdot\cdot}{O}}:^{2-}$
    e. Mg^{2+} 0 valence electrons Mg^{2+}
    f. C^{4-} 8 valence electrons $:\overset{\cdot\cdot}{\underset{\cdot\cdot}{C}}:^{4-}$
    g. Li   1 valence electron $Li\cdot$
    h. Ne   8 valence electrons $:\overset{\cdot\cdot}{\underset{\cdot\cdot}{Ne}}:$

Fewer than 8 electrons (an octet) usually results in dimerization or polymerization in a condensed phase. For example $AlCl_3$ exists in a nonpolar solvent as Al_2Cl_6 in which an octet is achieved. In the solid state there are 6 Cl around each Al.

3. Terminal atoms almost always obey the octet rule; examine the number of electrons around the central atom:

   a. $AlCl_3$ only 6 electrons   e. SF_6 12 electrons
   b. PCl_3 OK   f. $BeCl_2$ only 4 electrons
   c. PCl_5 10 electrons   g. NO_2 7 electrons (odd number)
   d. $SiCl_4$ OK   h. XeF_4 12 electrons

4. Lattice energy depends upon the charge/size ratio:
   MgO > CaS > NaF > KCl > CsI

5. a. NCl_3 d. SF_4

   b. BCl_3 e. OCS

   c. ClO_2^- f. SO_2

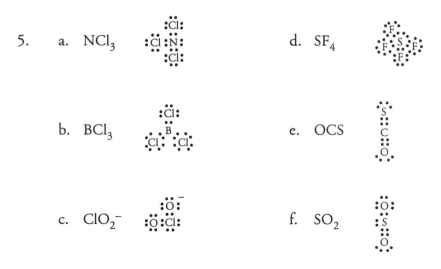

6. Formal charges:

    a. chlorite ion ClO_2^- (all single bonds)   Cl +1; both O −1
    b. hydroxylamine $HONH_2$ all zero
    c. phosphorous acid H_3PO_3 all zero
    d. ozone O_3 (one O=O; one O−O)   central O +1; singly
    bonded oxygen −1

    e. nitrogen dioxide NO_2 (odd electron on N)   nitrogen +1; singly
    bonded oxygen −1

7. $CH_3–CH=CH_2(g)$ + $H_2O(g)$ → $CH_3–CH(OH)–CH_3(g)$

Bonds broken:		Bonds made:	
C=C	611 kJ	C–C	347 kJ
O–H	464 kJ	C–O	351 kJ
		C–H	414 kJ
Total	1075 kJ		1112 kJ

    Difference = $\Delta H°$ = 1075 − 1112 kJ = −37 kJ mol^{-1}

8. $CH_3CH_2OH(g)$ + $3O_2(g)$ → $2CO_2(g)$ + $3H_2O(g)$

Bonds broken:		Bonds made:	
C–C	347 kJ	4 C=O	4 × 803 kJ
C–O	351 kJ	5 O–H	5 × 464 kJ
5 C–H	5 × 414 kJ		
3 O=O	3 × 498 kJ		
Total	4262 kJ		5532 kJ

    Difference = $\Delta H°$ = 4262 − 5532 kJ = −1270 kJ mol^{-1}

    From the heats of formation,
    $\Delta H°_{reaction}$ = $\Sigma(\Delta H°_f \text{(products)} − \Delta H°_f \text{(reactants)})$
    = [(2 × −393.5 + 3 × −241.8) − (−235.1)] = −1277 kJ mol^{-1} Often the agreement isn't quite
    so good as this.

9. Bond order:

a.	carbonate ion	1.33 (avg)	e.	oxygen O_2	2
b.	acetic acid	1.5 (avg)	f.	hydrogen cyanide	3
c.	nitrite ion	1.5 (avg)	g.	sulfur tetrafluoride	1
d.	acetylene	3	h.	carbon dioxide	2

10.

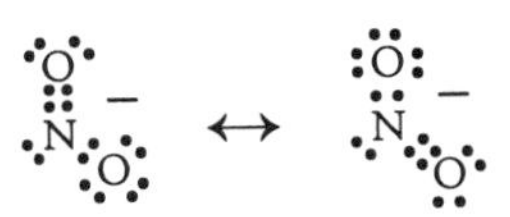

    a. nitrite ion
    NO_2^-

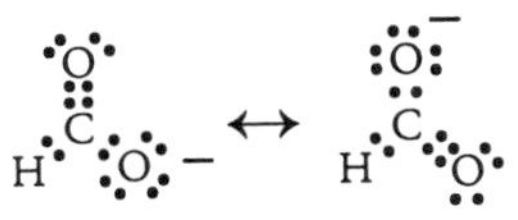

    b. formate ion
    HCO_2^-

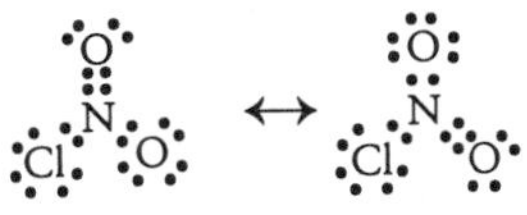

    c. nitryl chloride
    NO_2Cl

11. Determine the arrangement first; then the shape of the molecule:

		arrangement	shape
a.	3 bonding; 2 lone pairs	trigonal bipyramid	T–shape
b.	4 bonding; 1 lone pairs	trigonal bipyramid	Seesaw
c.	3 bonding; 1 lone pairs	tetrahedral	trig. pyramid
d.	2 bonding; 2 lone pairs	tetrahedral	V–shape
e.	2 bonding; 3 lone pairs	trigonal bipyramid	linear
f.	5 bonding; 1 lone pairs	octahedral	square pyramid

12.

a.	O_3	V-shape	e.	NO_2^+	linear
b.	GaH_3	trigonal pyramid	f.	ClO_4^-	tetrahedral
c.	SO_2Cl_2	tetrahedral	g.	IF_4^-	square planar
d.	XeO_4	tetrahedral	h.	ClF_2^-	linear

Look for some asymmetry in the molecular structure—specifically some asymmetry in the electron distribution in the molecule.

13.

a.	CCl_4	nonpolar tetrahedral	e.	BF_3	nonpolar trigonal planar
b.	XeF_4	nonpolar square planar	f.	$BeCl_2$	nonpolar linear
c.	PCl_5	nonpolar trigonal bipyramid	g.	SCl_2	polar V-shape
d.	PCl_3	polar trigonal pyramid	h.	CS_2	nonpolar linear

CHAPTER 10

Introduction

In this chapter we take a closer look at just how atoms are held together in molecules. Electrons in molecules exist in orbitals, just as they do in individual atoms. There are two approaches to deriving these orbitals. The first is the valence bond approach of Linus Pauling and the second is the molecular orbital approach of Robert Mullikan and others. We will examine both in this chapter.

Contents

10.1   Orbitals and Bonding Theories

Electrons are responsible for the bonds between atoms in a molecule. These electrons reside in orbitals belonging to more than just one atom. There are two common approaches used to derive these orbitals: valence bond (VB) theory and molecular orbital (MO) theory.

The **valence bond** approach is closely related to Lewis's idea of pairs of electrons shared between atoms in a molecule. The electron pairs are said to be in orbitals **localized** between adjacent atoms. In **molecular orbital theory**, on the other hand, orbitals are derived from *all* the atoms in the molecule and electrons in them are often **delocalized** (shared by more than just two atoms).

Valence bond theory is the simpler approach and gives an image of the bonding in a molecule in its ground state (lowest energy state) that's easy to picture. Molecular orbital theory is useful when excited states of a molecule need to be examined and for some molecules it is really the only theory to explain properties of the molecule satisfactorily.

10.2   Valence Bond Theory

When two atoms come together, the electrons of each come under the influence of the nuclei of both atoms and are pulled down in energy. The atoms do not approach too closely because the positive nuclei repel one another. The bond distance is the distance between the nuclei when the potential energy of the system is at a minimum.

See Figure 10.1 on page 440 in the text.

The orbitals of the individual atoms overlap as the atoms approach. The basic principle of valence bond theory is that bonds form when the orbitals overlap. An electron in an orbital on one atom pairs up in the shared space with another electron in an orbital on the other atom; this constitutes the chemical bond. Sometimes both electrons in the bond originate from the same atom.

When an s orbital on one atom overlaps with an s orbital on another atom, the region of overlap lies along the internuclear axis. The pair of bonding electrons has a high probability of being found along the internuclear axis. This type of bond is called a σ bond. Similar σ bonds can be formed by any atomic orbital directed along the internuclear axis. This could be an s, p, d, or hybrid atomic orbital.

Many molecules have shapes based upon the five geometries discussed in the last chapter (linear, trigonal, tetrahedral, trigonal bipyramid, and octahedral). Which orbitals on the central atom can be used to create the bonds in these molecules? The atomic orbitals point in the wrong directions!

To solve this problem, Pauling proposed the concept of **hybridization** as an integral part of valence bond theory. He proposed that a new set of **hybrid atomic orbitals** could be created that would accommodate the shape of the molecule. These orbitals are created by mixing together an appropriate set of atomic orbitals:

atomic orbitals	number of orbitals	hybrid orbitals	direction in which they point
s and p	2	sp	linear
s and two p	3	sp^2	trigonal
s and three p	4	sp^3	tetrahedral
s, three p, one d	5	dsp^3	trigonal bipyramid
s, three p, two d	6	d^2sp^3	octahedral

Hybridization creates a set of valence orbitals pointing in the correct direction for the formation of valence bonds. For example, in methane CH_4, the 2s and three 2p orbitals mix to form the four orbitals of the sp^3 hybrid set. These orbitals in the hybrid set are identical in everything but direction. Each points to one corner of a tetrahedron. The four valence electrons are distributed evenly among them. The hybrid orbitals then form σ bonds with the 1s orbitals on the four hydrogen atoms.

Hybrid orbitals have to accommodate the nonbonding pairs of electrons on the central atom as well. For example, in ammonia NH_3, a tetrahedral sp^3 set is still required. Three sp^3 hybrid orbitals are used for bonding, the fourth sp^3 hybrid orbital is used for the lone pair. Similarly, in the water molecule, two sp^3 hybrid orbitals are used for bonding and two sp^3 hybrid orbitals are used for the two lone pairs. The key to a successful use of VB theory is to make use of VSEPR theory:

1. Determine the number of electron pairs around the central atom in the molecule. Include any lone pairs.
2. Use VSEPR theory to determine the arrangement of those electron pairs.
3. Select the hybrid orbital set necessary to accommodate that arrangement according to the table above.
4. Superimpose any π bonding on the σ framework.

Bonds are named according to their symmetry about the internuclear axis. A σ bond, when viewed down the internuclear axis, looks circular, just like an s orbital. A π bond looks just like a p orbital, and so on. The Greek letters used correspond directly to the letters for the atomic orbitals:

Atomic	Molecular
s	σ
p	π
d	δ

See Figure 12 on page 450 of the text for pictures of the orbital sets.

Note that the number of hydrid orbitals created equals the number of atomic orbitals used.

It is usually unnecessary to hybridize the orbitals of the terminal atoms—just the central atom.

The participation of d orbitals in the bonding of main-group elements is now considered to be minimal (i.e hybrid sets like dsp^3 and d^2sp^3 are unrealistic). [See J.Chem.Educ **75** 910 (1998)]

In a situation like this the contribution of each atomic orbital to a particular hybrid orbital is not equal. The lone pair hybrid orbital has more s character than the bonding hybrids. This reconciles the slightly smaller than tetrahedral angles in the ammonia and water molecules.

Remember that a molecule is not tetrahedral because the hybridization is sp^3. The hybridization is sp^3 because the molecule is tetrahedral. For example, in molecular orbital theory, hybridization is not necessary, and the molecule is still tetrahedral!

A **multiple bond** is a bond in which more than one pair of electrons is shared. Although a single bond requires just one orbital in the region of overlap between the two atoms (for the single pair of bonding electrons), a double bond requires two orbitals, and a triple bond requires three orbitals (for the additional shared pairs of electrons).

When an atom like the carbon in ethylene, or the nitrogen in the nitrate ion, is sp^2 hybridized, there is a p orbital left over that is perpendicular to the plane of the molecule. This leftover p orbital can overlap sideways on with a similar p orbital on an adjacent atom. This overlap results in a π bond. Note that in a π bond the electron density lies to either side of the internuclear axis. Note also that the two p orbitals must be lined up in the same way in order to combine to form a π bond. The σ bond and the π bond together make up the double bond.

When atoms are sp hybridized, like carbon in acetylene, then two p orbitals are left over on each carbon atom, and these two p orbitals can form two π bonds with similar orbitals on an adjacent atom. A triple bond results.

A important feature of multiple bonds is that they restrict movement about the bond. Rotation about a single bond requires little energy; the bond can remain intact, with constant overlap of the atomic orbitals, throughout the rotation. When rotation of a double bond or a triple bond is attempted however, the π overlap of the p_π orbitals is destroyed in the process. Breaking the π bond requires energy and the rotation is impossible at normal temperatures.

A consequence of **restricted rotation** about multiple bonds is the existence of **structural isomers** called **stereoisomers**. An example is cis– and trans– 1,2–dichloroethylene. In the cis– isomer the chlorine atoms are on the same side of the molecule and in the trans– isomer the two chlorine atoms are on opposite sides.

The bonding in the six–membered **benzene ring** is particularly interesting. The σ bonding involves sp^2 hybridization at each carbon atom in the ring: two of the three sp^2 hybrid orbitals are used for bonding to the adjacent carbon atoms and the third sp^2 hybrid is used for bonding to the hydrogen atom. This leaves a p_π orbital on each carbon atom perpendicular to the molecular plane. The interesting aspect of the bonding is the interaction of the six p_π orbitals. Each p_π orbital overlaps with the p_π orbitals on its adjacent carbon atoms—the π interaction is unbroken around the six–membered ring. The π bond in benzene extends over all six carbon atoms. Two electrons in such a bond are delocalized over the entire benzene ring. In order to obtain this picture of the bonding in benzene using valence bond theory, resonance is required. Picturing the π bonding in terms of localized electron pairs between adjacent atoms is unsatisfactory.

For many commonly encountered molecules and polyatomic ions—those for which resonance was necessary in drawing a Lewis electron dot structure—the σ bonding is most easily described in terms of valence bond theory and the localized sharing of electron pairs. For the π bonding, however, a different approach is desirable—one which describes the **delocalization of electrons**. This is molecular orbital theory, the subject of the next section.

---

single bond $\vdots$

double bond $\vdots\vdots$

triple bond $\vdots\vdots\vdots$

one σ bond   = single bond

one σ bond and one π bond
    = double bond

one σ bond and two π bonds
    = triple bond

p orbitals can be designated p_σ or p_π:

A p_σ orbital is a p orbital lying along the internuclear axis able to form a σ bond.

A p_π orbital is an orbital perpendicular to the bond axis table to form a π bond.

See Figure 10.18 on Page 457 of the text.

See Figure 10.19 on Page 458 of the text.

10.3   Molecular Orbital Theory

In molecular orbital theory all the atomic orbitals on all the atoms are combined together to form a set of molecular orbitals. In reality, only the valence orbitals are of practical importance. The number of molecular orbitals constructed equals the number of atomic orbitals used (**1st principle**). This was not a criterion of valence bond formation.

When two atomic orbitals interact to form two molecular orbitals, the two molecular orbitals are different in energy: a **bonding** combination of atomic orbitals is lower in energy and an **antibonding** combination is higher in energy (**2nd principle**). Electrons in a bonding orbital are located between the nuclei and tend to hold the atoms together—hence bonding. Electrons in an antibonding orbital are largely absent from the space between the nuclei; the nuclei repel each other and destabilize the bond—hence antibonding.

When the molecular orbitals have been constructed, they are filled with the available electrons (**3rd principle**). Just as with atomic orbitals, the electrons are assigned in order of increasing energy (auf–bau principle), two per orbital with opposite spins (Pauli exclusion principle), and as unpaired as possible in a degenerate set (Hund's rule).

Three criteria are important in molecular orbital construction. The atomic orbitals must:

- overlap or share the same space; no interaction can occur if there is no overlap of the orbitals.
- be similar in energy; the further apart they are in energy, the weaker the interaction (**4th principle**).
- have the same symmetry with respect to each other; for example an orbital that wants to form a π bond cannot interact with one that can only σ bond.

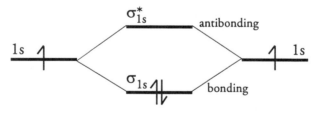

For dihydrogen H_2, the two 1s orbitals combine to produce two σ orbitals, one bonding σ_{1s} and one antibonding σ_{1s}^*. The two valence electrons go into the lower energy bonding orbital. So the electron configuration of H_2 is written $(\sigma_{1s})^2$.

For dihelium He_2, there are four electrons so the antibonding σ_{1s}^* is also filled. This more than cancels the bonding provided by the two electrons in the σ_{1s} and the dihelium molecule He_2 does not exist. Similar arguments support the existence of the dilithium molecule, but not the diberyllium molecule.

The **bond order** in a diatomic molecule is defined as the difference between the number of bonding and antibonding electrons divided by 2. If the molecule has a bond order of zero, then it doesn't exist. If the bond order is greater than zero then it may exist (a bond order of 1/2 indicates a weak bond). A bond order

---

Margin notes:

There is a node between the nuclei in the antibonding orbital.

When establishing a molecular electron configuration, the same three principles apply—just as they did for atomic orbitals:

- Bohr's auf–bau principle
- Pauli's exclusion principle
- Hund's rule of maximum spin multiplicity

Although dilithium exists in the gaseous state, dilithium crystals (in the solid state) do not exist. Lithium has a metallic structure in the solid state.

This is analogous to the comparison of the NaCl ion–pairs in the gaseous state and the NaCl crystal lattice in the solid state discussed in Chapter 9 (page 124 of this study guide).

of 1 is equivalent to the single bond of valence bond theory. A bond order of 2 is equivalent to the double bond of valence bond theory. Fractional bond orders are possible.

The molecules just discussed are called **homonuclear diatomic molecules** because the two atoms are the same. Molecular orbital energy diagrams can be drawn for all diatomic molecules Li_2 through F_2. Bond orders increase from 1 for B_2, to 2 for C_2, 3 for N_2, then down to 2 for O_2, 1 for F_2, and 0 for Ne_2 (doesn't exist). Bond energies and bond lengths correlate as expected with the bond orders and the electron configurations.

See Figure 10.27 on Page 465.

Increase bond order—
   increase bond energy
   decrease bond length.

The interactions between the atomic orbitals parallel those discussed already for valence bond theory: Orbitals lying on the internuclear axis are σ orbitals, and orbitals perpendicular to, and off, the internuclear axis are π orbitals. The difference is the existence of the antibonding orbitals in molecular orbital theory.

$$1s \;+\; 1s \;\rightarrow\; \sigma_{1s} \;+\; \sigma_{1s}^* \qquad \text{core electrons—negligible interaction}$$

$$2s \;+\; 2s \;\rightarrow\; \sigma_{2s} \;+\; \sigma_{2s}^* \qquad \sigma \text{ interaction}$$

$$2p_\sigma \;+\; 2p_\sigma \;\rightarrow\; \sigma_{2p} \;+\; \sigma_{2p}^* \qquad \sigma \text{ interaction}$$

$$2p_\pi \;+\; 2p_\pi \;\rightarrow\; \pi_{2p} \;+\; \pi_{2p}^* \qquad \pi \text{ interaction—two sets of these}$$

For example, the electronic configuration of dioxygen O_2 is written:

$$(\sigma_{1s})^2(\sigma_{1s}^*)^2(\sigma_{2s})^2(\sigma_{2s}^*)^2(\sigma_{2p})^2(\pi_{2p})^4(\pi_{2p}^*)^2$$

The π_{2p} bonding and antibonding orbitals exist in sets of two each and the two electrons in the π_{2p}^* orbitals of oxygen O_2 are in separate orbitals with parallel electron spins—accounting for the observed paramagnetism of the oxygen molecule.

One of the great successes of the molecular orbital approach was the explanation of the paramagnetism of dioxygen—something that could not be explained by valence bond theory.

Molecular orbital energy level diagrams can also be drawn for heteronuclear diatomic molecules like CO, NO, CN^-, OF^+, etc. The diagrams look similar to the diagrams for the homonuclear diatomics but are skewed because the initial atomic orbitals are no longer equal in energy.

See problem 6 at the end of this study guide.

Recall that resonance was proposed by Pauling to reconcile Lewis electron dot structures with the true structure of a molecule. The actual structure is described as a composite of all the Lewis structures. For example, the nitrate ion is described as a composite of the structures:

The existence of resonance merely highlights the inadequacy of the Lewis electron dot picture and is in fact unnecessary. A better approach is:

Use VSEPR and valence bond theory to establish the σ bonding framework of the nitrate ion—

- the three oxygen atoms are at the corners of a triangle with nitrogen in the middle.
- the three σ bonds are formed by the overlap of the three sp^2 hybrid orbitals on the nitrogen and σ–bonding valence orbitals on the oxygen atoms.

Molecular orbital theory now provides the finishing touch. On the nitrogen atom there is a p_π orbital perpendicular to the molecular plane. On the three oxygen atoms there are matching p_π orbitals. There are therefore a total of four orbitals perpendicular to the molecular plane with the correct orientation for π bonding. These four orbitals combine to produce four molecular orbitals:

Figuring out how many electrons to put in a diagram like this can sometimes be confusing. It is best to add up all the valence electrons and then distribute them logically.

There are 24 electrons in the nitrate ion; 6 are required for the σ bonding; each oxygen has three orbitals left—all with 2 electrons each which accounts for 18 more. Total = 24. Check the Lewis dot structure.

This diagram involves one N orbital with no electrons, and 3 oxygen orbitals with 2 electrons each, so 6 electrons must be placed in the diagram.

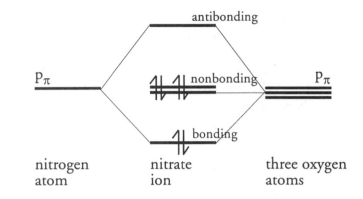

The orbital of lowest energy, the bonding orbital, is an orbital for which all the p_π orbitals are "in phase". The p_π orbitals constructively interfere to create the bonding orbital which extends across the entire molecule with no nodes. There are two electrons in this orbital. These two electrons correspond to the fourth pair of bonding electrons in the valence bond model that led to the necessity to invoke resonance. In the molecular orbital model, this pair of electrons is in an orbital that belongs to the entire molecule so that the electrons are delocalized over all four atoms—just as the valence bond resonance hybrid implied.

The next pair of orbitals in the diagram are nonbonding molecular orbitals. They contain two pairs of electrons. These electrons correspond to the nonbonding pairs of electrons on the oxygen atoms in the valence bond picture. The highest orbital is an empty antibonding π^* orbital.

10.4 Metals and Semiconductors

Molecular orbital theory can be used to explain the properties of metals. A metal sample can be viewed as a giant molecule in which all the valence orbitals of all the metal atoms in the sample participate in the construction of molecular orbitals extending over the entire metallic lattice. Electrons in these orbitals are **delocalized** over the entire lattice. And it is the movement of electrons in the orbitals that are singly occupied that is responsible for the electrical conductivity of metals. The number of molecular orbitals equals the number of atomic orbitals, and the number is enormous. They are all closely spaced together in a **band**—a virtual continuum of available energy levels. A **metal** is characterized by a partially filled valence band. An **insulator** has a valence band that is full, with a large energy gap before the next empty band. An **intrinsic semiconductor** also has a full valence band but the energy gap before the next available band (called a conduction band) is small. Semiconductors can be doped to improve their conductivity; such **extrinsic semiconductors** can be **n-type** or **p-type**.

Review Questions

1. What is the difference between valence bond theory and molecular orbital theory?

2. Explain the concept of hybridization. Why is it a necessary component of valence bond theory?

3. Describe a procedure for determining the hybridization of atomic orbitals around the central atom in a molecule.

4. What is a multiple bond? How is a multiple bond viewed in valence bond theory?

5. What is the difference between a σ bond and a π bond?

6. Is a single bond always a σ bond, or can it be a π bond?

7. Can an s orbital form a π bond or is a p orbital the only orbital that can form a π bond? Can a p orbital form a σ bond?

8. Explain how restricted rotation about a multiple bond leads to stereoisomerism.

9. Describe the bonding in an aromatic compound like benzene.

10. Describe the bonding in a molecule like sulfur dioxide.

11. Review the procedure for filling a set of (atomic or molecular) orbitals with electrons.

12. What are the four principles of molecular orbital theory discussed in the text book?

13. What is the difference between a bonding and an antibonding orbital?

14. Write the valence bond definition and the molecular orbital definition of bond order. Compare the two.

15. Describe the bonding in metals using molecular orbital theory.

16. What's the difference between an intrinsic semiconductor and an extrinsic semiconductor?

17. Describe the difference between an n-type extrinsic semiconductor and a p-type extrinsic semiconductor.

Answers to Review Questions

1. The valence bond approach is closely related to Lewis's idea of pairs of electrons shared between atoms in a molecule. The electron pairs are localized between adjacent atoms. In molecular orbital theory orbitals belong to the entire molecule and electrons in them are spread out or delocalized. Valence bond theory is the easier approach and gives a simple visual picture of the bonding in a molecule in its ground state. Molecular orbital theory does not require the use of resonance to reconcile fact and theory.

2. Linus Pauling proposed the concept of hybridization as an integral part of valence bond theory. He proposed that an appropriate set of atomic orbitals could be mixed together to create a set of hybrid atomic orbitals that would accommodate the shape of any molecule. The hybrid orbitals point in the correct direction for the formation of valence bonds with the terminal atoms. Hybridization is necessary because the s and p atomic orbitals do not point in the right direction and because they cannot generate the equivalent valence bonds observed in molecules like CH_4 or H_2O.

3. Make use of the VSEPR theory! Determine the number of electron pairs around the central atom in the molecule and remember to include the lone pairs. Use VSEPR theory to determine the arrangement of those electron pairs. Then select the hybrid orbital set necessary to accommodate that arrangement.

4. A multiple bond is a bond in which more than one pair of electrons is shared between the two atoms. A single bond requires one orbital in the region of overlap between the two atoms for the single pair of bonding electrons. A double bond requires two orbitals for the two pairs of electrons. A triple bond requires three orbitals for the three pairs of electrons. In valence bond theory, all electron pairs in a multiple bond are localized between the adjacent atoms—the first pair in a σ bond on the internuclear axis and the next two pairs in π bonds off the internuclear axis.

5. The electron density (the probability of finding the electron) in a σ bond is concentrated along the internuclear axis. When viewed down the internuclear axis a σ bond has a circular shape just like an s orbital. The electron density in a π bond is concentrated on either side of the internuclear axis. When viewed down the internuclear axis a π bond has a shape just like an p orbital.

6. A single bond is always a σ bond.

7.   No, an s orbital cannot form a π bond, it has the incorrect symmetry. There is no way to view an s orbital that will make it look like a p orbital! A p orbital can certainly form a π bond, but so can a d orbital. If a d orbital is viewed sideways on, it looks like a p orbital. And yes, a p orbital can form a σ bond; when it lies along the internuclear axis is has the correct symmetry to form a σ bond. In other words, when viewed lengthways on, it has the circular shape required for a σ bond.

Assuming the internuclear axis is the z axis:

s orbital – only σ

p_z orbital – σ
p_x orbital – π
p_y orbital – π

and, to complete the pattern, although you don't need to know these until later in your studies of chemistry :

d_{z^2} orbital – σ
d_{xz} orbital – π
d_{yz} orbital – π
d_{xy} orbital – δ
$d_{x^2-y^2}$ orbital – δ

8.   Multiple bonds restrict movement about the bond. When rotation of a double bond or a triple bond is attempted, the π overlap of the p_π orbitals and any π bonding is destroyed in the process. Breaking the π bond requires energy and the rotation is impossible at normal temperatures. A consequence of this restricted rotation is the existence of stereoisomers. An example is cis– and trans– 1,2–dichloroethylene. These two isomers cannot interconvert at normal temperatures:

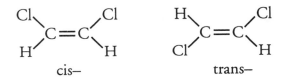

cis–　　　　　　　　　　trans–

9.   The σ bonding in the six–membered benzene ring involves sp² hybridization at each carbon atom in the ring: two sp² hybrid orbitals are used for bonding to the adjacent carbon atoms and the third sp² hybrid is used for bonding to the hydrogen atom. This leaves a p_π orbital on each carbon atom perpendicular to the molecular plane. The interaction of the six p_π orbitals is best described by molecular orbital theory. There are six p_π orbitals and therefore six molecular orbitals. There are six electrons and therefore three pairs:

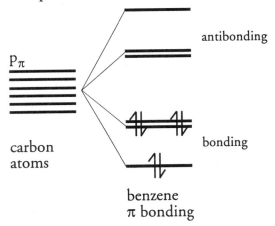

There are three bonding orbitals. The lowest energy orbital is the orbital that extends unbroken around the entire ring without any nodes. The next two bonding orbitals, slightly higher in energy, have one node each. The top three orbitals are antibonding and are empty.

The ozone molecule, which has a similar structure, is described in the text (pages 468-469).

10. In sulfur dioxide, as with most simple molecules, the σ bonding framework is best described using valence bond theory. The sulfur is sp^2 hybridized with two bonding pairs and one nonbonding pair. This leaves a p$_\pi$ orbital on the sulfur that can bond with an appropriate orbital on either oxygen—leading to resonance in valence bond theory. In molecular orbital theory, all three p$_\pi$ orbitals, the one on the sulfur and the two on the oxygens, combine to produce three molecular π orbitals. One is bonding and holds the bonding pair of electrons, one is nonbonding and is empty, and the third is antibonding and is also empty. The π bonding electron pair is delocalized over all three atoms.

11. When molecular orbitals have been constructed, they are filled with the available electrons according to the same rules that applied when establishing an atomic electron configuration: the electrons are assigned in order of increasing energy (auf–bau principle), only two per orbital with opposite spins (Pauli exclusion principle), and as unpaired as possible in a degenerate set (Hund's rule).

12. 1st principle:
The number of molecular orbitals constructed equals the number of atomic orbitals used

2nd principle:
When two atomic orbitals interact to form two molecular orbitals, a bonding combination of the atomic orbitals is lower in energy and an antibonding combination of the atomic orbitals is higher in energy.

3rd principle:
Molecular orbitals are filled with the available electrons according to the same rules that applied when establishing an atomic electron configuration.

4th principle:
The atomic orbitals must be similar in energy; the further apart they are in energy, the weaker the interaction.

13. The bonding combination of atomic orbitals is lower in energy—the electrons are located between the nuclei and tend to hold the atoms together. An antibonding combination of atomic orbitals is higher in energy—the electrons are largely absent from the space between the nuclei.

14. Valence bond definition: the number of electron pairs shared between two atoms; may be fractional due to resonance. Molecular orbital definition: the number of bonding electrons − the number of antibonding electrons divided by two. The bond orders should be the same using either method. For diatomic molecules and ions, the latter is easier to determine.

15. A metallic solid can be viewed as a giant molecule in which all the valence orbitals of all the metal atoms in the solid lattice participate in the construction of molecular orbitals. These orbitals extend over the entire lattice. Electrons in these orbitals are delocalized over the entire lattice. As usual, the number of molecular orbitals equals the number of atomic orbitals, and so the number of molecular orbitals is enormous. They are all closely spaced together within a finite energy range and form what is called a valence band—a virtual continuum of available energy levels for the electrons. In a metal this band of orbitals is only partially full. The highest filled orbital at 0K is called the Fermi level. At any temperature above 0K, there is a Boltzmann-like distribution of half-filled or half-empty orbitals about the Fermi energy.

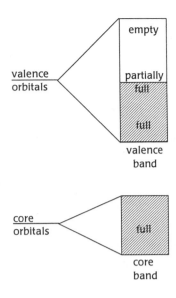

16. An intrinsic semiconductor is a pure semimetal such as germanium or silicon. An extrinsic semiconductor is a doped semiconductor. Dopants, such as gallium or arsenic, are added to enhance the conductivity of the semiconductor. (See also the answer to the next question.)

17. An n-type extrinsic semiconductor is doped with an element that has one additional electron (a Group 5A element). This extra electron is promoted relatively easily into the conduction band. The extra electrons in the conduction band increase the conductivity of the semiconductor.
     A p-type extrinsic semiconductor is doped with an element that has one less electron (a Group 3A element). The vacant orbital can relatively easily accept electrons from the valence band, providing more positive holes in the valence band. The extra holes in the valence band increase the conductivity of the semiconductor.

See Figure 10.34 on page 472 of the text.

Study Questions and Problems

1. What hybridization is required at the central atom of the following molecules or ions? (These are the same molecules examined in questions 3, 5, 12, and 13 in Chapter 9.)

    a. $AlCl_3$
    b. PCl_3
    c. PCl_5
    d. $SiCl_4$
    e. NCl_3
    f. BCl_3
    g. ClO_2^-
    h. O_3
    i. GaH_3
    j. SO_2Cl_2
    k. XeO_4
    l. CCl_4
    m. SCl_2

    n. SF_6
    o. $BeCl_2$
    p. NO_2
    q. XeF_4
    r. SF_4
    s. OCS
    t. SO_2
    u. NO_2^+
    v. ClO_4^-
    w. IF_4^-
    x. ClF_2^+
    y. BF_3
    z. CO_2

Nature has some sort of arithmetical-geometrical coordinate system, because nature has all kinds of models. What we experience of nature is in models, and all of nature's models are so beautiful.

R. Buckminster Fuller
(1895-1983)

2.   In the organic chemistry of carbon, three hybridizations are common. What are they, and why are they limited to three?

3.   Examine the bonding in the following molecules using valence bond theory. Describe each bond in terms of the orbitals on adjacent atoms that overlap to create the bond:

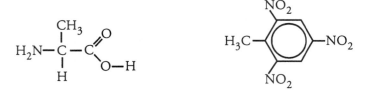

4.   Draw the Lewis electron dot structure of the nitrite ion. Describe the bonding in terms of valence bond theory. Then describe the π bonding in terms of molecular orbital theory.

5.   Describe the hybridization at each atom (other than H) in the following molecule (the anesthetic novocaine).

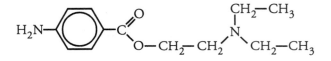

6.   Calculate the bond orders in He_2^+, Li_2, C_2, C_2^{2-}, N_2, O_2^{2-}, F_2. Which are paramagnetic?

7.   Draw the molecular orbital energy level diagrams for the heteronuclear diatomic ions NO^+ and CN^-. Write the electron configurations of both. What are the bond orders?

8.   Using valence bond theory describe the σ bonding and the π bonding in the molecule allene $CH_2=C=CH_2$.

9.   Use valence bond theory to describe the σ bonding and use molecular orbital theory to describe the π bonding in the molecule carbon dioxide CO_2.

10.   Carbon, silicon, and tin are all in Group 4A; they all have the same number of valence electrons. Yet carbon (diamond) is an insulator; silicon is an intrinsic semiconductor; and tin is a metal. Why?

11.   Do you think that the compound SiC would be a semiconductor, a metal, or an insulator? What about the compound GaAs (gallium arsenide)?

Answers to Study Questions and Problems

1.   First determine the arrangement of the electron pairs around the central atom; the hybridization must accommodate *all* electron pairs (not just the bonding pairs). So look at the arrangement of electron pairs, not the shape of the molecule. There are five arrangements and five corresponding hybridizations. The arrangements are listed in Chapter 9 (page 128 of this study guide).

linear	sp
trigonal	sp^2
tetrahedral	sp^3
trigonal bipyramid	dsp^3
octahedral	d^2sp^3

a. $AlCl_3$ sp^2
b. PCl_3 sp^3
c. PCl_5 dsp^3
d. $SiCl_4$ sp^3
e. NCl_3 sp^3
f. BCl_3 sp^2
g. ClO_2^- sp^3
h. O_3 sp^2
i. GaH_3 sp^2
j. SO_2Cl_2 sp^3
k. XeO_4 sp^3
l. CCl_4 sp^3
m. SCl_2 sp^2

n. SF_6 d^2sp^3
o. $BeCl_2$ sp
p. NO_2 sp^2
q. XeF_4 d^2sp^3
r. SF_4 dsp^3
s. OCS sp
t. SO_2 sp^2
u. NO_2^+ sp
v. ClO_4^- sp^3
w. IF_4^- d^2sp^3
x. ClF_2^- dsp^3
y. BF_3 sp^2
z. CO_2 sp

2.   The three hybridizations common in the organic chemistry of carbon are:

sp	linear arrangement of two σ bonds	two π bonds
sp2	trigonal arrangement of three σ bonds	one π bond
sp3	tetrahedral arrangement of four σ bonds	no π bonds

They are limited to these three because carbon only has s and p orbitals available; no d orbitals. Carbon forms 4 bonds.

3.

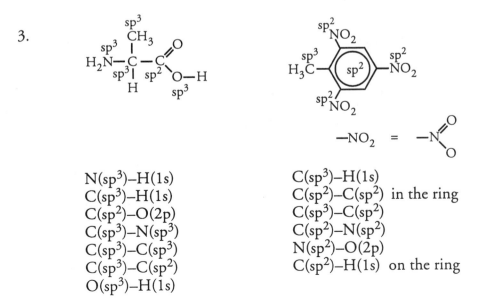

N(sp^3)–H(1s)
C(sp^3)–H(1s)
C(sp^2)–O(2p)
C(sp^3)–N(sp^3)
C(sp^3)–C(sp^3)
C(sp^3)–C(sp^2)
O(sp^3)–H(1s)

C(sp^3)–H(1s)
C(sp^2)–C(sp^2) in the ring
C(sp^3)–C(sp^2)
C(sp^2)–N(sp^2)
N(sp^2)–O(2p)
C(sp^2)–H(1s) on the ring

It is usually unnecessary to hybridize the orbitals of the terminal atoms—just the central atoms.

4. Lewis electron dot structure of the nitrite ion NO_2^-:

    In the valence bond picture, the nitrogen valence orbitals are sp^2 hybridized. The arrangement of electron pairs (including the lone pair) is trigonal planar. In the Lewis diagram, one N–O bond is drawn as a double bond and the other as a single bond. Resonance between the two possibilities leads to a bond order of 1.5 for each.

    In the molecular orbital approach, the p_π orbital on the nitrogen can bond with appropriate p_π orbitals on the oxygen atoms. All three p_π orbitals, the one on the sulfur and the two on the oxygens, combine to produce three molecular π orbitals. One is bonding and holds the bonding pair of electrons, one is nonbonding and is empty, and the third is antibonding and is also empty. The pair of π bonding electrons are delocalized over all three atoms.

5. Count up the number of σ bonds and lone pairs around an atom. Establish the arrangement; then the corresponding hybridization.

You may notice that the hybridization is related to the number of double bonds around the atom. How?

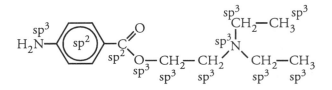

    Note that all the carbon atoms in the benzene ring are sp^2 hybridized. Note also that the lone pairs on oxygen and nitrogen atoms are often omitted!

6.

diatomic molecule	configuration	bond order	magnetism
He_2^+	$(\sigma_{1s})^2(\sigma_{1s}^*)^1$	0.5	paramagnetic
Li_2	$(\sigma_{1s})^2(\sigma_{1s}^*)^2(\sigma_{2s})^2$	1.0	diamagnetic
C_2	$(\sigma_{1s})^2(\sigma_{1s}^*)^2(\sigma_{2s})^2(\sigma_{2s}^*)^2(\sigma_{2p})^2(\pi_{2p})^2$	2.0	paramagnetic
C_2^{2-}	$(\sigma_{1s})^2(\sigma_{1s}^*)^2(\sigma_{2s})^2(\sigma_{2s}^*)^2(\sigma_{2p})^2(\pi_{2p})^4$	3.0	diamagnetic
N_2	$(\sigma_{1s})^2(\sigma_{1s}^*)^2(\sigma_{2s})^2(\sigma_{2s}^*)^2(\sigma_{2p})^2(\pi_{2p})^4$	3.0	diamagnetic
O_2^{2-}	$(\sigma_{1s})^2(\sigma_{1s}^*)^2(\sigma_{2s})^2(\sigma_{2s}^*)^2(\sigma_{2p})^2(\pi_{2p})^4(\pi_{2p}^*)^4$	1.0	diamagnetic
F_2	$(\sigma_{1s})^2(\sigma_{1s}^*)^2(\sigma_{2s})^2(\sigma_{2s}^*)^2(\sigma_{2p})^2(\pi_{2p})^4(\pi_{2p}^*)^4$	1.0	diamagnetic

7.

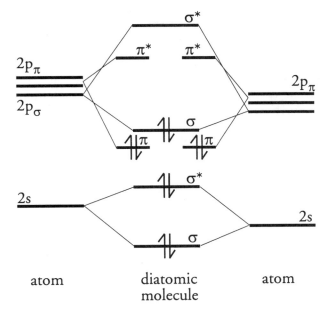

atom   diatomic   atom
              molecule

The energy levels of the two atoms are not quite equal because the ions are heteronuclear. The atom on the right is the more electronegative of the two; the orbitals (and electrons) are slightly lower in energy.

Both ions have the same energy level diagram; they are isoelectronic. Both have a bond order of 3 (4 pairs of bonding electrons and 1 pair of antibonding electrons). The electron configuration of both is:

$$(\sigma_{1s})^2(\sigma_{1s}^*)^2(\sigma_{2s})^2(\sigma_{2s}^*)^2(\sigma_{2p})^2(\pi_{2p})^4$$

8.   In allene $CH_2=C=CH_2$, the hybridization at each end carbon atom is sp^2 (trigonal). This leaves a p_π orbital perpendicular to the trigonal plane of the carbon atom and the two hydrogen atoms. The hybridization at the central carbon atom is sp (linear); the entire 3–carbon chain is linear. There are two p_π orbitals on the central carbon atom—at right angles to one another and the axis of the molecule. One p_π orbital forms a π bond with the carbon at one end of the molecule and the other p_π orbital forms a π bond with the carbon at the other end. Therefore the two π bonds are not in the same plane; they too are at right angles to each other. This means, of course, that the four hydrogen atoms are not in the same plane either.

9.   Superficially, the carbon dioxide molecule has a structure similar to allene in the last question. And, according to valence bond theory the π bonding is the same. The hybridization at the central carbon atom is sp (linear) and the entire molecule is linear. There are two p_π orbitals on the central carbon atom—at right angles to one another and the axis of the molecule. One p_π orbital forms a π bond in one direction and the other p_π orbital forms a π bond in the other direction.

However, unlike in allene, there are two orbitals on each terminal oxygen atom in carbon dioxide that have the correct orientation to π bond, and they do. So the π bonds in carbon dioxide are still at right angles to one another, but they extend the entire length of the molecule. Electrons in them are delocalized over all three atoms. Each π bond contains two electrons so the bond order of both C=O bonds is still 2.

10. Going down the group, from carbon to lead, there is a change in structure. Carbon, silicon, and germanium all have a diamond-like structure; gray tin has a diamond-like structure but it is less stable than white tin which has a six-coordinate structure. And lead, at the bottom, has a typical metallic close-packed structure. If all the elements crystallized in a close-packed structure, they would all be metals—but they don't!

For those elements that have a diamond-like structure, the gap between the valence band and the conduction band decreases down the group:

C	5.47 eV
Si	1.12 eV
Ge	0.66 eV
Sn (gray)	0 eV

The gap is too large in carbon; it is an insulator. The gap is too small in gray tin ; it is too good a conductor. Only for silicon and germanium is the gap about right for semiconducting behavior at room temperature.

11. The band gap for silicon carbide is between that for carbon and silicon, about 3.00 eV, too large for any appreciable current at room temperature. So it is an insulator. Gallium arsenide, on the other hand, has a band gap of about 1.42 eV, about the same as silicon, and in some respects is a better semiconductor than silicon.

Note that an equimolar combination of any Group 3A element and any Group 5A element (for example, GaP (band gap 2.26eV), InAs (band gap 0.36eV)) will have a full valence band and may have semiconducting properties. Or any combination of a Group 2B element and Group 6A element (for example, CdS).

CHAPTER 11

Introduction

Carbon is an amazing element. Along with hydrogen it is involved in more compounds than any other element. It is the basis for life on this planet. In this chapter we take a look the covalent compounds of carbon and provide examples of the chemical bonding discussed in the last two chapters. Carbon is said to be electron perfect; it has four electrons in four valence orbitals and can form four bonds. There are no lone pairs of electrons on carbon and this allows carbon atoms to link together in long chains producing molecules of tremendous biological importance.

In inorganic diatomic or triatomic molecules and ions, (e.g. CO, CN⁻, C_2^{2-} or CNO⁻) carbon has one lone pair.

In larger molecules, carbon always forms four bonds:
  four σ, or
  three σ and one π, or
  two σ and two π

Contents

11.1   The Uniqueness of Carbon

An atom of carbon can—

form four bonds
bond to up to four different other atoms
form single, double or triple bonds
participate in relatively stable and kinetically inert carbon compounds
exist in a linear, trigonal, or tetrahedral geometries
form optical and geometrical stereoisomers

Carbon itself exists as three different allotropes—diamond, graphite and the fullerenes (buckyballs). In diamond the hybridization is sp³; in graphite and fullerene, the hybridization is sp². Fullerenes and the related nanotubes were unknown until fairly recently.

Carbon–carbon bonds are relatively strong and difficult to break under normal conditions. In single, double, and triple bonds, the overlap of the carbon orbitals is very efficient leading to high bond energies. Carbon compounds are also kinetically inert, so that even if a thermodynamically favorable reaction is possible, it is often difficult to get it started. For example, most organic compounds will burn given the chance, but most will coexist with oxygen for ever unless something happens to start the reaction.

Carbon can form an extremely diverse collection of compounds!

These three allotropes are illustrated in Figure 11.1 on page 481.

Nanotubes are illustrated in Figure 11.2 on page 483 of the text.

Examples are the initiation of the polymerization of ethylene, or lightning initiating a forest fire.

Organic chemistry is the chemistry of carbon compounds; and there are millions of different ones. Fortunately, it is possible to classify at least the simpler compounds of carbon by the various characteristic groups of atoms that occur in the molecule. Just a few will be examined in this chapter:

C–H   hydrocarbons: alkanes, alkenes, alkynes, aromatics
C–O   alcohols, ethers
C–N   amines
C=O   aldehydes, ketones, acids, esters, amides

11.2   Hydrocarbons

Alkanes have the general formula C_nH_{2n+2} where n is an integer. The series involves a longer and longer chain of carbon atoms as n increases. When n is 4 or greater, structural isomers are possible because the chain may or may not branch. Each carbon in an alkane is sp^3 hybridized; so the bonds at each carbon atom radiate toward the corners of a tetrahedron.

When the alkane joins one end to the other, eliminating two hydrogen atoms, a cycloalkane results. Cycloalkanes have the general formula C_nH_{2n} and cyclohexane C_6H_{12} is a well known example. Because the angles at the carbon are 109.5°, the cyclohexane ring is not planar; it can adopt two forms—the chair and the boat. Smaller rings become more and more strained as the size decreases.

As the molar mass of the alkane increases, the stable state changes from gas, to liquid, and then to solid. Methane CH_4, for example, is a gas, hexane C_6H_{14} is a liquid, and eicosane $C_{20}H_{42}$ is a solid.

Alkanes burn readily in air; they are widely used as fuels. Other reactions involve substitution of the hydrogen atoms by other elements (for example, chlorine).

11.3   Alkenes and Alkynes

Alkenes (C_nH_{2n}) contain a carbon–carbon double bond. An example is ethylene C_2H_4. The presence of a double bond leads to stereoisomerism—specifically geometrical isomerism. For example, 2–butene can exist as the cis–isomer or the trans– isomer. Carbon atoms involved in the double bond are sp^2 hybridized; the arrangement of bonds around the carbon is trigonal planar.

Hydrocarbons can contain more than one double bond; butadiene contains two double bonds. Alkenes can also be cyclic; an example is cyclohexene.

Alkynes (C_nH_{2n-2}) contain a carbon–carbon triple bond. An example is acetylene C_2H_2. Cis–trans geometrical isomerism is impossible about a triple bond and alkynes possess fewer isomers than alkenes. Carbon atoms involved in the triple bond are sp hybridized; the arrangement of bonds is linear.

The chemical properties of the alkenes and alkynes center on the double or triple bond. Compounds possessing multiple C–C bonds are often referred to as unsaturated. Addition of hydrogen to the double bond produces an alkane—a process called hydrogenation. Other molecules can be added across the bond.

Members of the alkanes:

methane CH_4
ethane C_2H_6
propane C_3H_8
butane C_4H_{10}
pentane C_5H_{12}

illustrated in Figure 11.3 on page 487.

Be able to name the simpler alkanes. The procedure is listed in Appendix E. of the text.

The chair and boat forms of cyclohexane are illustrated on page 491 of the text.

An old name for alkene is olefin.

Structural isomers are molecules with the same molecular formula but with different arrangements of atoms.

Stereoisomers have the same sequence of atoms in the molecule, but their arrangement in space is different. There are two types: geometrical and optical.

Geometrical isomers have different geometrical arrangements of the same atoms.

Cis-and trans- isomers of 2-butene are illustrated on page 493 of the text.

For example, addition of water produces alcohols, chlorine produces a dichloroalkane, a hydrogen halide produces a haloalkane, etc.

Aromatic compounds contain a benzene ring. Benzene C_6H_6 is the simplest aromatic compound. Other examples are toluene (a benzene ring with a methyl group) and naphthalene (two benzene rings fused together).

The bonding in the six–membered benzene ring involves sp^2 hybridization at each carbon atom. p_π orbitals on the carbon atoms (perpendicular to the molecular plane) interact to form π bonds that extend over all six carbon atoms. Electrons in the π bonds are delocalized over the entire benzene ring. This delocalization of the π electrons confers extra stability. Energy equivalent to this stability—about 184 kJ mol^{-1}—is required any time the π bonding is broken up—for example in the hydrogenation of benzene to form cyclohexane. The extra energy is called resonance energy because in valence bond theory, resonance is required to explain the bonding.

As a result of the delocalization of the π electrons, benzene does not behave like an alkene. For example, unlike an alkene, benzene does not readily add reactants across the double bond. Instead, reactants substitute the hydrogen atoms, keeping the aromatic ring intact.

If there are 2 or more substituents on the ring, then structural isomerism is possible. For example, ortho–, meta–, and para– xylene.

Kekule was the first to propose the cyclic structure for benzene.

The xylenes are dimethyl benzenes; this one is para–xylene.

ortho-, meta-, and para- isomers are illustrated on page 500 of the text.

11.4 Alcohols and Amines

Organic chemistry can be organized very logically by considering any organic compound to be a hydrocarbon in which one or more hydrogen atoms have been substituted by other atoms or groups or atoms. These other groups of atoms are called **functional groups**. In general formulas, the hydrocarbon chain is represented by R or R'. Some examples of functional groups are:

R–Cl	chloroalkane
R–OH	alcohol
R–O–R'	ether
R–NH$_2$	amine
R–CHO	aldehyde
R–CO–R'	ketone
R–CO$_2$H	carboxylic acid
R–CO$_2$–R'	ester
R–CO$_2$–NH$_2$	amide

Alcohols contain the functional group –OH. Common examples are methanol CH_3OH, ethanol CH_3CH_2OH, and isopropanol $CH_3CH(OH)CH_3$. Alcohols can contain more than one –OH group. Examples are ethylene glycol (antifreeze) $CH_2(OH)CH_2OH$ and glycerol $CH_2(OH)CH(OH)CH_2OH$.

Just as the parent alkanes can exhibit isomerism, so can the derived alcohols. The hydrocarbon chain can have structural isomers, and then the position of the –OH group can vary as well. For example, for butanol C_4H_9OH, there are four isomers: two alcohols based upon the straight 4-carbon chain and two alcohols based upon the branched 2-methylpropane chain.

Solubility in all proportions is referred to as miscibility.

Like all hydrocarbon derivatives, alcohols possess properties of both parts of the molecule. Properties of the hydrocarbon chain are, for example, combustion and insolubility in water. Properties of the –OH functional group are, for example, strong intermolecular bonding and solubility in water. Which has precedence depends upon the relative size of the two parts. For example, methanol is soluble in water in any proportion but butanol is much less soluble.

Ethers have two alkyl groups bonded to oxygen:

Without the –OH functional group, ethers differ sharply in their properties compared to alcohols.

Amines are organic compounds based upon ammonia just as alcohols are based upon water. The functional group is –NH_2 for a primary amine, –NH– for a secondary amine, and –N– for a tertiary amine. As expected, the properties of the amine functional group parallel the properties of ammonia. For example, amines can be protonated; the amines are bases just like ammonia.

11.5   Compounds with a Carbonyl Group

The carbonyl functional group is C=O. It occurs in aldehydes, ketones, carboxylic acids, amides, and esters:

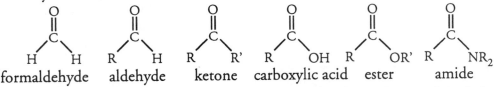

The aldehydes, ketones, and acids are prepared by the oxidation of alcohols. The esters and amides are derivatives of the acids.

Oxidation of a primary alcohol yields an **aldehyde** and then a **carboxylic acid**. For example, ethanol produces acetaldehyde and acetic acid:

Vinegar is a 5% solution of acetic acid.

$$CH_3CH_2OH \quad \rightarrow \quad CH_3CHO \quad \rightarrow \quad CH_3CO_2H$$

A secondary alcohol produces a **ketone** but a tertiary alcohol is not so easily oxidized. Aldehydes and ketones often have pleasant smells.

Carboxylic acids are the end product of the oxidation of primary alcohols. These acids are readily soluble in water provided that the hydrocarbon chain is not too long. The hydrogen of the carboxylic acid functional group is partially ionized to produce a weakly acidic solution. For example, acetic acid is a weak acid.

Some carboxylic acids are listed in Tables 11.10 and 11.11 on page 513 of the text.

Some esters are listed in Table 11.12 on page 515.

Carboxylic acids react with alcohols to produce esters and with amines to produce amides:

$$\begin{array}{lll} CH_3CO_2H & + \quad HOCH_2CH_3 & \rightarrow \quad CH_3CO_2CH_2CH_3 \\ \text{acetic acid} & \quad\text{ethanol} & \quad\text{ethyl acetate} \end{array}$$

$$\begin{array}{lll} CH_3CO_2H & + \quad H_2NCH_2CH_3 & \rightarrow \quad CH_3CONHCH_2CH_3 \\ \text{acetic acid} & \quad\text{ethylamine} & \quad\text{N–ethylacetamide} \end{array}$$

11.6   Fats and Oils

Fats (solids) and oils (liquids) are triesters made from the alcohol glycerol and long chain carboxylic acids. The hydrocarbon chain may be saturated, unsaturated, or polyunsaturated:

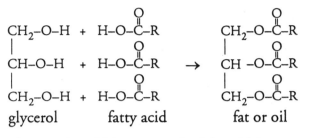

These acids are often referred to as fatty acids.

For example, stearic acid is saturated, oleic acid is mono-unsaturated, and linolenic acid is polyunsaturated. All three have C_{18} chains. In general, the more saturated the fatty acid is, the more solid the fat. **Hydrogenation**, therefore, is used to convert oils into fats.

Linolenic acid has three double bonds.

Hydrolysis of these esters in the presence of base breaks the ester link and produces glycerol and the sodium salt of the fatty acid. This sodium salt of a long chain fatty acid is a soap and therefore the hydrolysis process is often called **saponification**.

Saponification means soap-making.

11.7   Synthetic Organic Polymers

Polymers are giant molecules with molar masses up to several millions. They are prepared by joining together many thousands of small molecules called **monomers**. There are different ways to classify polymers. A chemical classification divides polymers into **condensation** and **addition** polymers. An industrial classification divides polymers into **thermoplastics** and **thermosetting** plastics. A classification based upon the use of the polymers divides them into **plastics**, **fibers**, **elastomers**, **coatings**, and **adhesives**.

Addition polymers are built by adding simple alkenes together. The prototype, formed by adding molecules of ethylene together, is called polyethylene or polythene. The process takes place at a temperature above 100°C and a pressure between 1000 and 3000 atm in the presence of a catalyst. The polymer is a chain of $-CH_2-$ units, thousands of them, so is in fact a giant alkane. Different conditions lead to different polymers. Unbranched long chains form a high density polymer (HDPE). When the chain is branched, the chains cannot pack together so closely, and a low density polymer is formed (LDPE). Cross linking the chains lead to a more rigid and inflexible polymer. Other examples of polyethylenes are polystyrene, polyvinyl alcohol (PVA), polyacrylonitrile (PAN), and teflon.

Some addition polymers are listed in Table 11.15 on page 521 of the text.

Natural rubber is a polymer of isoprene (2-methyl-1,3-butadiene). This diene can exist as a cis– or as a trans– isomer and in natural rubber it is always the cis– isomer. Vulcanization crosslinks the polymer chains leading to a more resilient polymer—one that can stretch and reform repeatedly. Polymers able to do this are called **elastomers**. Elastomers are often copolymers, formed by the polymerization of two or more different monomers.

Condensation polymers involve a chemical reaction between two molecules in which a small molecule (such as water) is eliminated. If the two molecules are both difunctional (contain two functional groups—one at each end), then polymerization can occur. There are two common types: polyesters and polyamides.

Polyesters include polyethylene terephthalate (PET) which is an ester of terephthalic acid and ethylene glycol. This polymer is sold as terylene, dacron, or mylar. Nylon is a **polyamide** made from adipoyl chloride (from adipic acid) and hexamethylenediamine.

Many polymers are used in composite materials. Fibers of glass or graphite are embedded in a polymer matrix just like steel reinforced concrete. These low density materials are used in sporting goods and aircraft construction where low weight and high strength are at a premium.

Review Questions

1. Describe, and give an example of, structural isomerism in an alkane.

2. Describe geometrical isomerism in 2–butene. Are there any structural isomers of 2–butene?

3. Explain the difference in the hybridization of the orbitals on carbon when it is involved in four single bonds, two single bonds and a double bond, two double bonds, and one single bond and one triple bond.

4. How is the π bonding in benzene different from the π bonding in an alkene?

5. Describe what functional groups are. Make a table of, and name, some of the functional groups you have encountered in this chapter.

6. Describe esterification, hydrolysis, hydrogenation, and saponification.

7. Describe the two major classes of polymerization processes.

8. Describe other ways to classify polymers.

9. What is the difference between a thermoplastic and a thermosetting resin?

10. What is an elastomer?

11. Would you describe polyethylene as an alkene or an alkane? If the latter, why is it called polyethylene and not polyalkane?

12. What is a composite material?

13. What is the difference between an amide and a peptide?

Answers to Review Questions

1. Structural isomerism occurs when there are alternative ways of connecting the atoms in a molecule. The simplest alkane that exhibits structural isomerism is butane. The carbon atoms can be arranged in a straight chain or they can be arranged in a branched chain:

$$CH_3-CH_2-CH_2-CH_3 \quad \text{or} \quad CH_3-CH-CH_3$$
$$\underset{\displaystyle CH_3}{|}$$

2. Geometrical isomerism involves the same basic connections between atoms in the molecule but different arrangements in space.

   cis–   trans–

   A structural isomer is:

3.
Four single bonds:	sp^3
Two single bonds, one double bond:	sp^2
Two double bonds:	sp
One single and one triple bond:	sp

   Note that the total number of bonds for one carbon atom always equals 4.

4. The π bonding in an alkene can be considered in valence bond theory as a simple localized sharing of a pair of electrons between two atoms. The valence bond description of the π bonding in benzene requires resonance. All bonds in the benzene ring are identical and the interaction of the six p_π orbitals of the carbon atoms is best described by molecular orbital theory. There are six p_π orbitals and therefore six molecular orbitals. There are three pairs of bonding electrons delocalized in orbitals extending over the entire carbon ring. The π bonding in benzene is stronger than a set of three simple π bonds. The difference in energy is sometimes referred to as the resonance energy.

   When the alkene is a conjugated diene, or polyene, then delocalization can occur and the valence bond picture requires resonance.

5. Functional groups are atoms, or groups or atoms, that add functionality (a particular set of properties) to a molecule.

R–Cl	chloroalkane	R–OH	alcohol
R–O–R'	ether	R–NH₂	amine
R–CHO	aldehyde	R–CO–R'	ketone
R–CO₂H	carboxylic acid	R–CO₂–R'	ester
R–CO₂–NH₂	amide		

6. esterification— the condensation of a carboxylic acid and an alcohol to form an ester.

   hydrolysis— the reverse of the ester formation, a process in which the ester is broken down to the acid and the alcohol.

   hydrogenation— adding hydrogen across a double bond, producing a saturated compound from an unsaturated compound.

   saponification— soap making, breaking down a fat or oil to glycerol and the sodium salt of the fatty acid.

7. Addition (eg. polyethylene) and condensation (e.g. polyamide).

8. An industrial classification divides polymers into thermoplastics and thermosetting plastics. A classification based upon how the polymers are used divides them into plastics, fibers, elastomers, coatings, and adhesives.

9. A thermoplastic resin softens and flows when heated and then rehardens when cooled. A thermosetting resin sets when it is made and cannot be softened when heated.

10. An elastomer is a rubber that has a high coefficient of reconstitution. It can be stretched but is able to reform without degradation. An example is a 1:3 copolymer of styrene and butadiene called SBR (styrene–butadiene rubber) used to make tires.

11. Polyethylene is a giant alkane molecule $-(CH_2)_n-$ where n is in the thousands; it is called polyethylene because it is made from ethylene. Polystyrene is made from styrene, polyacrylonitrile from acrylonitrile, etc.

12. A composite material is a combination of two or more materials differing in form or composition on a macroscopic scale. The composite is heterogeneous. One phase, the reinforcement, provides the desired strength, toughness, or other physical characteristic, while the other phase, the matrix, provides the adhesive that unites and protects the reinforcement. The matrix is the polymer and the reinforcement is the graphite or glass fiber.

    There are other types of composites: metal–matrix and ceramic–matrix are alternatives to polymeric–matrix composites. Some composites employ polymers as both reinforcement and matrix. For example, aramid-epoxy composites are made with kevlar fibers in an epoxy resin.

Epoxy resin is a polymeric ether. Kevlar is a polyamide made from *para*-phenylenediamine and terephthalic acid.

13. There is no difference between amide and peptide. Both refer to the link –CO–NH–. Nylon, for example, is a polyamide. Peptide often refers to the link between amino acids in a polypeptide chain or protein. In other words, peptide is a biological term whereas amide is a chemical term for the same thing.

Study Questions and Problems

The important thing in science is not so much to obtain new facts as to discover new ways of thinking about them.

Sir William Lawrence Bragg
(1890-1971)

1.   Draw the structural formulas for:

   a.   2–methylpentane

   b.   2,2-dimethylpropane

2.   Classify the following as alkanes, alkenes, or alkynes:

   C_2H_4 C_6H_{10} C_3H_8 C_4H_{10} C_5H_{12}

3.   How many isomers of $C_4H_{10}O$ are there? Classify the isomers you draw.

4.   How many isomers of the alkene pentene C_5H_{10} are there?

5.   Draw three examples of the structural isomers of the alcohol $C_5H_{11}OH$ that are primary, secondary, and tertiary.

6.   What condensation products would be formed from:
   a.   ethanol and formic acid
   b.   acetic acid and dimethylamine
   c.   propanol and propanoic acid
   d.   butyric acid and aniline

7.   Organic compounds have names with characteristic endings. To what classes of organic compounds do the following endings belong?

   a.   –oic acid   d.   –one
   b.   –ene   e.   –ol
   c.   –al   f.   –oate

8.   What products result from the oxidation of
   a.   ethanol
   b.   2–propanol (isopropanol)
   c.   acetaldehyde (ethanal)
   d.   methanol

9.   Draw:
   a.   an aldehyde with the molecular formula C_4H_8O
   b.   a ketone with the molecular formula C_4H_8O
   c.   an ester with the molecular formula $C_3H_6O_2$
   d.   an ether with the molecular formula C_2H_6O
   e.   an aromatic amide with the molecular formula C_8H_9NO

10.   Write an equation illustrating the condensation of p–diaminobenzene with the acid chloride of oxalic acid $(COCl)_2$.

Answers to Study Questions and Problems

1. a. 2–methylpentane

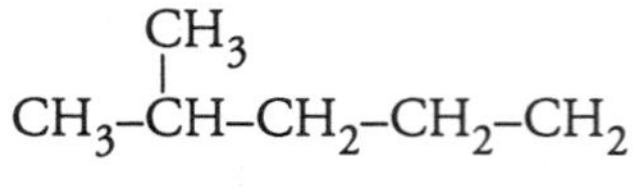

$$CH_3$$
$$|$$
$$CH_3-CH-CH_2-CH_2-CH_2$$

   b. 2,2-dimethylpropane

$$CH_3$$
$$|$$
$$CH_3-C-CH_3$$
$$|$$
$$CH_3$$

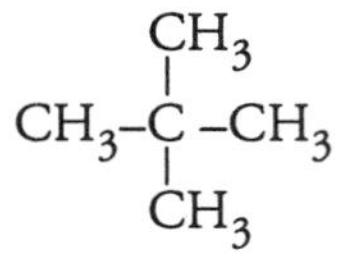

2. Alkanes: C_nH_{2n+2} C_2H_4 alkene

   Alkenes: C_nH_{2n} C_6H_{10} alkyne

   Alkynes: C_nH_{2n-2} C_3H_8 alkane

                            C_4H_{10} alkane

                            C_5H_{12} alkane

3. Isomers of $C_4H_{10}O$ (all are structural isomers):

   alcohols: ethers:

   $CH_3-CH_2-CH_2-CH_2-OH$ $CH_3-CH_2-CH_2-O-CH_3$

   $CH_3-CH_2-CH(OH)-CH_3$ $CH_3-CH_2-O-CH_2-CH_3$

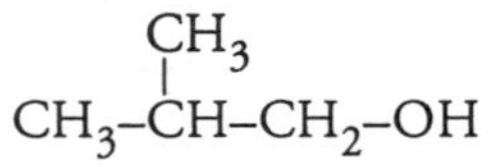

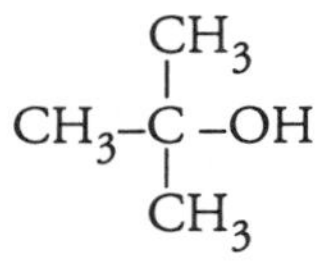

   $$CH_3$$
   $$|$$
   $$CH_3-CH-CH_2-OH$$

   $$CH_3$$
   $$|$$
   $$CH_3-CH-O-CH_3$$

   $$CH_3$$
   $$|$$
   $$CH_3-C-OH$$
   $$|$$
   $$CH_3$$

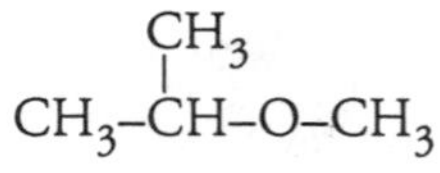

4. Isomers of pentene C_5H_{10}:

   $CH_3-CH_2-CH_2-CH=CH_2$ 1–pentene (no geometrical isomerism)

   $CH_3-CH_2-CH=CH-CH_3$ 2–pentene (both cis and trans)

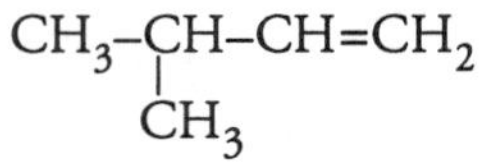

   $$CH_3-CH-CH=CH_2$$
   $$|$$
   $$CH_3$$
   3–methyl–1–butene
   (no geometrical isomerism)

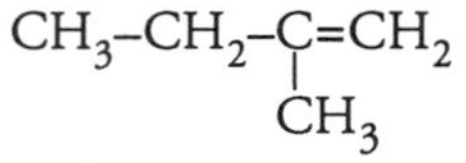

   $$CH_3-CH_2-C=CH_2$$
   $$|$$
   $$CH_3$$
   2–methyl–1–butene
   (no geometrical isomerism)

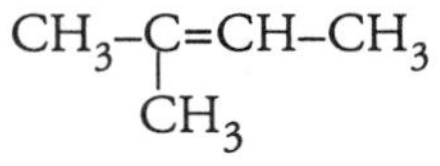

   $$CH_3-C=CH-CH_3$$
   $$|$$
   $$CH_3$$
   2–methyl–2–butene
   (no geometrical isomerism)

5. $CH_3–CH_2–CH_2–CH_2–CH_2–OH$ primary

$$CH_3–CH_2–CH_2–\overset{\overset{\displaystyle CH_3}{|}}{CH}–OH$$ secondary

$$CH_3–CH_2–\overset{\overset{\displaystyle CH_3}{|}}{\underset{\underset{\displaystyle CH_3}{|}}{C}}–OH$$ tertiary

6. a. ethanol and formic acid → ethyl formate (or methanoate)
    b. acetic acid and dimethylamine → N,N–dimethylacetamide
    c. propanol and propanoic acid → propylpropionate (or propanoate)
    d. butyric acid and aniline → N-phenylbutyrate (or butanoate)

7. a. –oic acid   carboxylic acid   d. –one   ketone
    b. –ene   alkene   e. –ol   alcohol
    c. –al   aldehyde   f. –oate ester

8. a. ethanol → acetaldehyde → acetic acid
    b. 2–propanol → acetone (propanone)
    c. acetaldehyde → acetic acid
    d. methanol → formaldehyde (methanal) → formic acid

9. Draw:

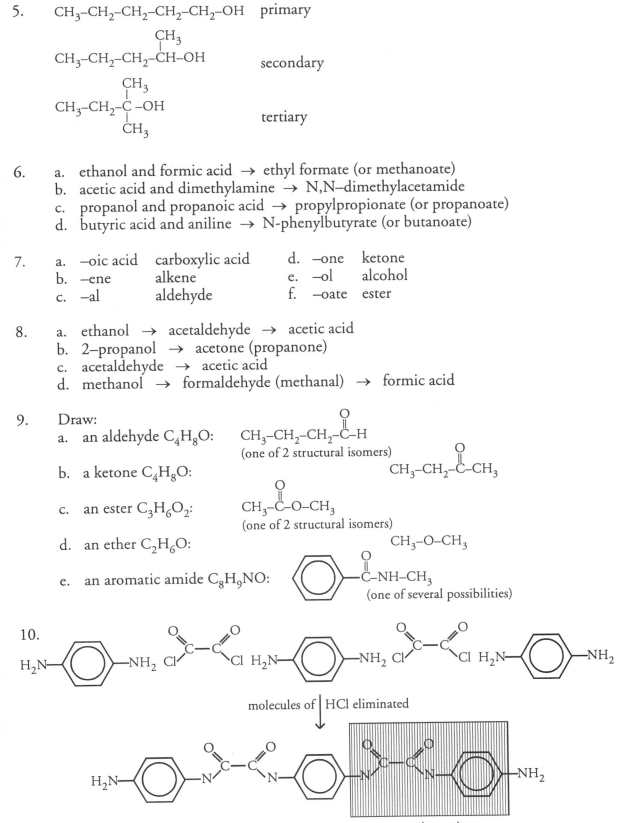

    a. an aldehyde C_4H_8O: $CH_3–CH_2–CH_2–\overset{\overset{\displaystyle O}{\|}}{C}–H$
        (one of 2 structural isomers)

    b. a ketone C_4H_8O: $CH_3–CH_2–\overset{\overset{\displaystyle O}{\|}}{C}–CH_3$

    c. an ester $C_3H_6O_2$: $CH_3–\overset{\overset{\displaystyle O}{\|}}{C}–O–CH_3$
        (one of 2 structural isomers)

    d. an ether C_2H_6O: $CH_3–O–CH_3$

    e. an aromatic amide C_8H_9NO: $\overset{\overset{\displaystyle O}{\|}}{C}–NH–CH_3$
        (one of several possibilities)

10. molecules of | HCl eliminated

repeating unit

Challenge 5

Examine the following structure and list different functional groups present:
(note that not all these amino acids occur naturally!)

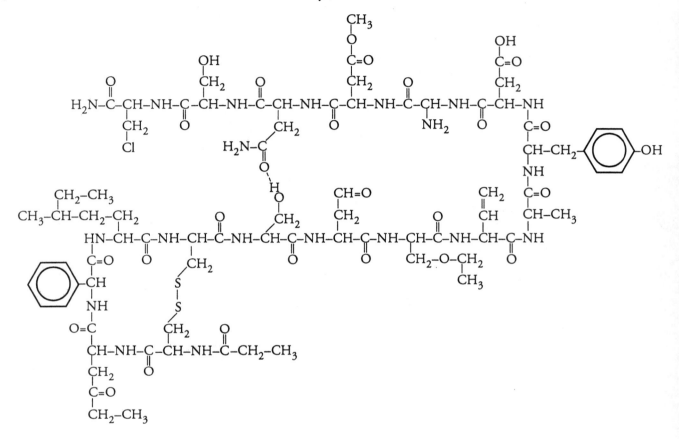

EXAMINATION 2

Introduction

This examination tests your knowledge and understanding of the chemistry in Chapters 7 through 11 of Kotz & Treichel. The questions are again formatted as true–false questions and multiple choice questions. Try the exam before looking at the answers provided at the end of this study guide.

True–false questions

1. Red light has a lower frequency than blue light.

2. The maximum allowed value for m_ℓ is $+\ell$.

3. Diamagnetic materials are drawn into a magnetic field.

4. A fluorine atom is larger than a boron atom.

5. The number of electrons that can be placed in a subshell of orbitals = $2(2\ell+1)$.

6. The most stable ion of sodium is the Na^- ion.

7. A lone pair of electrons is a nonbonding pair of electrons.

8. The bond order can be determined from the number of pairs of electrons shared in the bond.

9. Electronegativity is a measure of the ability of an atom in a compound to attract electrons to itself.

10. An ionic bond is the result when the both participating elements have low electronegativity.

11. The see-saw molecular geometry is derived from the trigonal bipyramidal arrangement of electron pairs.

12. Bonding pairs of electrons repel each other more than nonbonding pairs of electrons.

13. A triple bond indicates that three atoms are involved in the bond.

14. There's no such thing as a 100% ionic bond.

15. If an ion has a charge it must be polar.

16. The geometry of a set of sp hybrid orbitals is linear.

17. The number of hybrid orbitals always equals the number of atomic orbitals used to make the set.

18. Cis-trans isomerism is possible about any multiple bond.

19. An antibonding orbital is always higher in energy than the corresponding bonding combination.

20. p orbitals can form σ bonds and π bonds.

21. Because there are 3 π bonding pairs of electrons in a benzene molecule, the bond order is 3.

22. Resonance is required in valence bond theory when π electrons are delocalized.

23. A single bond is a σ bond.

24. Polyamides result from the condensation of difunctional amines with difunctional acids (or acid chlorides).

25. Copolymers result from the polymerization of two or more different monomers, often in an established ratio.

Multiple choice questions

1. What is the wavelength in cm of an infrared wave whose velocity is 3.0×10^8 m s^{-1} and whose frequency is 9.0×10^{11} s^{-1}?

   a. 27×10^{-1} cm   c. 3.0×10^{-21} cm   e. 3.0×10^{-21} s^{-1}
   b. 3.3×10^{-2} cm   d. 30 cm   f. 3.3×10^{-2} cm s^{-1}

2. What is the energy of a photon with a wavelength of 5.00×10^{-8} m in the ultraviolet region of the electromagnetic spectrum? (h = 6.624×10^{-34} Js; c = 3.0×10^8 m s^{-1})

   a. 3.97×10^{-18} J   c. 3.31×10^{-41} J   e. 2.52×10^{17} J
   b. 7.55×10^{-25} J   d. 3.02×10^{40} J   f. 1.33×10^{-26} J

3. What is the energy of this radiation (question 2) in kJ mol^{-1}?

   a. 1200 kJ   c. 3310 kJ   e. 4430 kJ
   b. 2390 kJ   d. 3650 kJ   f. 6590 kJ

4. The secondary (angular momentum) quantum number ℓ designates

   a. the main energy level in which an orbital is found.
   b. whether an orbital is an s, p, d, or f orbital.
   c. the number of orbitals of a given type (s, p, d, f, etc.).
   d. the orientation of an electron in a magnetic field.
   e. the number of spin orientations allowed for an electron.

5. When $n = 4$, the secondary quantum number ℓ must be less than or equal to what number?

   a. 0   c. 1   e. 3   g. 5
   b. 1/2   d. 2   f. 4   h. 7

6. For any one atom, what is the total number of electrons that can have a principal quantum number of 5 and an secondary (angular momentum) quantum number of zero?

   a. 2   c. 5   e. 7   g. 14
   b. 4   d. 6   f. 10   h. 18

7. What is the ground state electron configuration of aluminum?

   a. $1s^2\,2s^2\,2p^5\,3s^2\,3p^2$ d. $1s^2\,2s^2\,2p^6\,3s^2\,3p^2$
   b. $1s^2\,2s^2\,2p^6\,3s^2\,3p^1$ e. $1s^2\,2s^2\,2p^4\,3s^2\,3p^3$
   c. $1s^2\,2s^2\,2p^6\,3s^2\,4s^1$ f. $1s^2\,2s^2\,2p^6\,3s^2\,3d^1$

8. According to the aufbau principle, which electron energy sublevel is filled after the 4s sublevel?

   a. 4p   c. 4d   e. 3d   g. 5p
   b. 5s   d. 4f   f. 3g   h. 3f

9. According to Hund's Rule, how many singly occupied orbitals are there in the valence shell of a iron atom in its lowest energy state?

   a. none   b. 1   c. 2   d. 3   e. 4

10. The electron configuration of an element is illustrated below. Which, if any, of the three principles that determine the most stable electronic configuration of an element does this configuration disobey?

   a. Pauli's exclusion principle   c. Hund's rule of maximum multiplicity
   b. Bohr's Aufbau principle   d. none; the configuration is correct

11. Printed below is a table of ionic radii. The radii are given in picometers. One of the values listed is incorrect. Which one?

   a. N^{3-} : 171 pm   d. O^{2-} : 140 pm   g. F^- : 136 pm
   b. P^{3-} : 212 pm   e. S^{2-} : 184 pm   h. Cl^- : 181 pm
   c. As^{3-} : 230 pm   f. Se^{2-} : 180 pm   i. Br^- : 195 pm

12. If two elements combine to form a compound, and both elements have relatively high (greater than 2.0) but different electronegativities, what kind of bond is formed between the two elements?

   a. covalent
   b. ionic
   c. polar covalent
   d. metallic

13. The illustration on the right represents the Lewis electron dot symbol for an element in Period 2 or Period 3 of the Periodic Table. Which element?

   a. Be   c. C   e. O
   b. P   d. Si   f. F

14. Which of the following molecules does not obey the octet rule?

   a. HF   b. PF_3 c. OF_2 d. XeF_2 e. CHF_3

15. What is the formal charge on the phosphorus atom in the orthophosphate ion PO_4^{3-} if all the bonds between the phosphorus and the oxygen atoms are drawn as single bonds?

   a. +1   c. +3   e. +5   g. −2
   b. +2   d. +4   f. −1   h. −3

16. In which area of the Periodic Table are the least metallic elements found?

   a. upper left   c. upper center   e. upper right
   b. lower left   d. lower center   f. lower right

17. What is the arrangement of electron pairs, bonding and nonbonding, around the central atom in the ion ICl_4^- ?

   a. linear   c. trigonal   e. tetrahedral
   b. trigonal bipyramidal   d. square planar   f. octahedral

18. The V–shaped and trigonal pyramidal molecular geometries are associated with which one of the five VSEPR electron pair arrangements?

   a. linear   c. tetrahedral   e. octahedral
   b. trigonal   d. trigonal bipyramid

19. Which one of the following molecules or ions is polar?

   a. XeF_4 b. I_3^- c. SF_6 d. PCl_3 e. NO_3^-

20. A set of sp^3 hybrid orbitals is formed by combination of

   a. one s and one p orbital   d. three s and one p orbitals
   b. three s and three p orbitals   e. one s and three p orbitals
   c. one s and two p orbitals

21. How many σ bonds are there in an ordinary triple bond?

   a. 1   c. 3   e. 5
   b. 2   d. 4   f. 6

22. If the internuclear axis between two bonded atoms is the x axis, a π bond could form by the interaction of which orbitals on the two atoms?

   a. p_x and p_x c. p_z and p_z e. s and p_x g. p_y and p_z
   b. s and s   d. p_y and s   f. p_y and p_x h. s and p_z

23. Estimate the heat of formation ΔH_f° of methane in kJ/mol if

   the bond energy of the H–H bond = 436 kJ/mol
   the bond energy of the C–H bond = 413 kJ/mol
   the sublimation energy of graphite = 716 kJ/mol

   a. +1565   c. +1652   e. +652   g. −64
   b. +739   d. −780   f. +808   h. −500

24. How many pairs of electrons on the central atom in the molecule BrF_5 determine the VSEPR arrangement and hence the geometry of the molecule?

   a. 1   c. 3   e. 5   g. 7
   b. 2   d. 4   f. 6   h. 8

25. How many isomers, including geometrical cis–trans isomers, of the alkene pentene C_5H_{10} are there?

   a. 5   b. 6   c. 7   d. 8   e. 9

26. Orbital hybridization is the

   a. creation of bonding and antibonding orbitals from atomic orbitals
   b. combination of atomic orbitals to form a new set of orbitals with directional properties more appropriate for bonding
   c. formulation of many Lewis dot resonance structures for a molecule
   d. filling of molecular orbitals with electrons according to the aufbau principle
   e. creation of a set of molecular orbitals in which the electrons can be delocalized over the entire molecule

27. Which of the following molecular orbital configurations is correct for the ion NO^+?

   a. $(\sigma_{1s})^2(\sigma_{1s}^*)^2(\sigma_{2s})^2(\sigma_{2s}^*)^2(\pi_{2px})^2(\pi_{2py})^2(\sigma_{2pz})^2(\pi_{2px}^*)^1(\pi_{2py}^*)^1$
   b. $(\sigma_{1s})^2(\sigma_{1s}^*)^2(\sigma_{2s})^2(\sigma_{2s}^*)^2(\pi_{2px})^2(\pi_{2py})^2(\sigma_{2pz})^2(\pi_{2px}^*)^2$
   c. $(\sigma_{1s})^2(\sigma_{1s}^*)^2(\sigma_{2s})^2(\sigma_{2s}^*)^2(\pi_{2px})^2(\pi_{2py})^2(\sigma_{2pz})^2$
   d. $(\sigma_{1s})^2(\sigma_{1s}^*)^2(\sigma_{2s})^2(\sigma_{2s}^*)^2(\pi_{2px})^2(\pi_{2py})^2(\sigma_{2pz})^2(\pi_{2px}^*)^2(\pi_{2py}^*)^2$
   e. $(\sigma_{1s})^2(\sigma_{1s}^*)^2(\sigma_{2s})^2(\sigma_{2s}^*)^2(\pi_{2px})^2(\pi_{2py})^2(\pi_{2px}^*)^1(\pi_{2py}^*)^1$

28. The carbonate ion CO_3^{2-} has a delocalized π system. How many orbitals are there in the delocalized π system?

   a. 1   b. 2   c. 3   d. 4   e. 5

29. An organic hydrocarbon was named logically but incorrectly by a student as 2–isopropylpropane. What is its correct name?

   a. 2,3–dimethylbutane   d. 2–ethylbutane
   b. 2–methyl–2–isopropylethane e. 2–methylpentane
   c. isohexane   f. di(isopropane)

30. A cyclic alkane has the general formula

    a. C_nH_{2n-2} b. C_nH_{2n-4} c. C_nH_{2n} d. C_nH_{n+2} e. C_nH_{2n+4}

31. The monomers used to produce the addition polymers, teflon $-(CF_2)_n-$, orlon $-(CH_2CHCN)_n-$, PVC, lucite $-(CH_2C(CH_3)CO_2CH_3)-$, polyethylene, etc. all have one thing in common. What is it?
    a. a halogen that can be easily lost to form a free radical
    b. a $-C=C-$ double bond
    c. only two carbon atoms
    d. a high melting point
    e. low density
    f. the ability to crosslink

32. What compound lies in the missing space in the sequence of oxidations:

    methane $\rightarrow$ methanol $\rightarrow$? $\rightarrow$ formic acid $\rightarrow$ carbon dioxide

    a. HCO_2H c. CH_3CH_2OH e. CH_3OH
    b. CH_3CHO d. CH_3OCH_3 f. $HCHO$

33. Esters are formed as a result of the condensation of

    a. an acid and an amine   d. an acid and an alcohol
    b. an alcohol and an aldehyde   e. an alcohol and a ketone
    c. an acid and a ketone   f. an alcohol and an amine

CHAPTER 12

Introduction

Of the three states of matter, the gas state is perhaps the simplest to understand from a molecular point of view. It is possible to describe the behavior of molecules in the gas state under ordinary conditions in terms of relatively simple mathematical equations. In this chapter we will examine the properties of gases and how they are characterized and described. The kinetic molecular theory is reintroduced.

Contents

12.1   The Properties of Gases

A sample of gas can be described using only four quantities: pressure, volume, temperature, and amount.

Pressure is exerted by a gas when molecules of the gas strike a surface. The force exerted per unit area is the pressure of the gas. A barometer is a device used to measure atmospheric pressure. At sea level, the pressure exerted by the atmosphere—the result of the collisions of the molecules in the atmosphere with the surface of the liquid—can support a column of mercury 76 cm high or a column of water about 1025 cm high.

12.2   Gas Laws

Gases are compressible; if you increase the pressure, the volume occupied by the gas decreases. Robert Boyle discovered that for a fixed quantity of gas at a constant temperature, the pressure and volume are inversely proportional. This is known as **Boyle's law**. Mathematically, the law is written as PV = constant (at constant temperature).

Jacques Charles discovered that the volume of a fixed quantity of gas increased if the temperature was increased at constant pressure. If plots of volume *vs.* temperature are extrapolated to zero volume, then all the lines meet at a

---

Pressure = force per unit area
Pa = Nm^{-2}

Force = mass × acceleration
N = kg ms^{-2}

So Pa = kg m^{-1}s^{-2}

The Pa is a small unit and ordinary pressures are expressed in kPa.

1 atm pressure = 101.325 kPa

1 bar = 100 kPa exactly

1 atm pressure = 14.7 psi

1 atm pressure = 760 torr

A torr is another name for mm Hg.

Be able to convert between the common units of pressure.

See Figure 12.6 on page 546 of the text.

$K = °C + 273.15$

common temperature: −273.15°C. **Kelvin** proposed a new temperature scale—now known as the Kelvin scale—with a value of zero at this common temperature. **Charles' law** can now be expressed as: the volume of a given quantity of gas at constant pressure is proportional to its temperature in K. Mathematically, this is written as V = constant × T (or V = kT where k is a constant).

A general gas law combines both Boyle's law and Charles' law: PV = kT where the temperature is in K and the quantity of gas is constant.

Another relationship was discovered by a Frenchman Guillaume Amontons in the late 1600s that stated that the pressure of a gas is directly proportional to its temperature:

$P = kT$

Amonton's data led to the first suggestion of an absolute zero temperature.

Joseph Gay-Lussac discovered that the ratio of the volumes of gases involved in a chemical reaction was always a small whole number—provided that the gas volumes were measured at the same temperature and pressure. This is now known as **Gay-Lussac's law** of combining volumes. For example, the volume of hydrogen in the reaction of hydrogen and oxygen to produce water is always twice the volume of oxygen. Amedeo Avogadro explained this law with his hypothesis that at the same temperature, volume, and pressure, a gas always contains the same number of molecules. **Avogadro's law** is expressed as V is proportional to n at constant pressure and temperature.

n = the number of moles of gas.

12.3   The Ideal Gas Law

You should be able to do calculations involving the ideal gas law; be able to solve for any one of the four variables, pressure, volume, temperature, and quantity, if you are given the other three.

The laws discussed in the previous section can be combined into a single law called the **ideal gas law**. If

Boyle: PV = constant, at constant T and n
Charles: V = constant × T, at constant P and n   }   then PV = nRT
Avogadro: V = constant × n, at constant T and P

Be sure to use the correct value of R in your problems—it depends upon the units of pressure and volume that you are using.

The combined proportionality constant R is called the gas constant. It has a value equal to 0.08206 L atm K⁻¹ mol⁻¹ or 8.315 J K⁻¹ mol⁻¹.

The density of a gas decreases if the temperature is increased. Jacques Charles, a balloonist of some renown, knew this. The relationship of the density to the temperature, pressure, and molar mass of the gas can be derived from the ideal gas law:

The other common mistake when doing gas law problems is to forget to use the Kelvin temperature.

$$\text{density} = \frac{PM}{RT} \qquad \text{where M is the molar mass of the gas.}$$

There's no such thing as an ideal gas; all gases are real. However, under ordinary conditions, most gases behave almost ideally. Some are more ideal than others.

Standard temperature and pressure (STP) describes the conditions of 1.00 atm pressure and 273.15K temperature (0°C). The volume of an ideal gas at STP is 22.414 liters.

12.4   Gas Laws and Chemical Reactions

Problems involving the ideal gas law, and problems in calculating density or molar mass, are provided in the section at the end of this study guide. Solutions are provided but try the problems before looking at the solutions!

The stoichiometry of a chemical reaction (the coefficients in the equation) provides information about the relative numbers of moles of reactants required and moles of products produced. The ideal gas law relates the number of moles of a gas to the pressure, temperature, and volume of the gas. Therefore stoichiometric calculations for reactions involving gases can be done.

The determination of the limiting reactant, the calculation of theoretical and percent yields, and the calculation of how much excess reactant remains at the end of the reaction, all apply in exactly the same way as they did earlier.

See Figure 12.11 on page 555 of the text.

12.5   Gas Mixtures and Partial Pressures

The pressure exerted by a gas is due to the millions of collisions that the gas molecules have with the surface of the container every second. If the gas is a mixture of different gas molecules, then the different gases in the mixture will all contribute to the total pressure because they will all collide with the surface of the container. The contribution that each gas makes to the total pressure will depend upon how much of that gas is present.

John Dalton observed that the pressure of a mixture of gases is the sum of the partial pressures of the different gases in the mixture. This is known as **Dalton's law of partial pressures.**

An ideal gas is one in which the individual molecules behave quite independently of one another. They collide without interaction; they have zero attraction for one another. The identity of the gas is therefore immaterial; a mixture of ideal gases behaves just like a single ideal gas. Therefore, each gas in an ideal mixture can be considered separately:

For gas a, for example, $P_aV=n_aRT$ and likewise for gas b, or c, etc.

For the total mixture, $P_{total}V=n_{total}RT$.

Dividing the first expression by the second: $\dfrac{P_a}{P_{total}} = \dfrac{n_a}{n_{total}}$

or $P_a = (n_a/n_{total})P_{total}$

or $P_a = X_aP_{total}$ where X_a is called the **mole fraction** of a in the mixture.

An application of the use of Dalton's law occurs when gases are collected or prepared in the presence of water. Water vapor is just like any other gas; it contributes to the total pressure exerted by a gas mixture. Therefore the total pressure is the sum of the partial pressure of water vapor at that particular temperature and the partial pressure of the gas being collected.

> The partial pressure of a gas in a mixture is the contribution that gas makes to the total pressure.

> The sum of the mole fractions of all components in the mixture must equal 1.

> If the mole fraction equals one, then there must be only one gas present, and the partial pressure is equal to the total pressure. In other words, the partial pressure of a gas equals the pressure it would exert if it were the only gas present. This is a particularly useful thing to know in calculations involving partial pressures.

12.6   The Kinetic-Molecular Theory of Gases

At a particulate level, molecules in the gas state are in constant motion, travelling at high velocities and in constant collision with each other and the container. This picture of the gaseous state is described by the kinetic-molecular theory. The average kinetic energy of the gas molecules depends only upon the temperature—the higher the temperature, the higher the average kinetic energy, and the faster the molecules move.

Because kinetic energy is $\frac{1}{2}mu^2$, the average kinetic energy of many molecules is related to the average of the squares of the speeds of all the molecules. This average is referred to as the **mean square speed** and is written $\overline{u^2}$.

The principal features of the kinetic-molecular theory are:

- Gases consist of molecules with intermolecular distances much greater than their size.
- Molecular motion is continual, random, and rapid.
- The average KE is proportional only to the temperature of the gas. All gases have the same KE at the same temperature.
- No energy is "lost" during the collisions with each other and the container.

> At room temperature, the average speed of molecules of oxygen and nitrogen in the air is 1000 mph.

> The square root of the mean square speed is called, logically enough, the root mean square speed. The significance of the rms speed is that it is the speed of a molecule possessing average kinetic energy.

> Maxwell's equation relates the root mean square speed to the temperature and the molar mass:
> $$\sqrt{\overline{u^2}} = \sqrt{\frac{3RT}{M}}$$

All the experimental gas laws can be derived from the kinetic molecular theory. Pressure is caused by molecules colliding with the container walls. If the temperature is increased, the kinetic energy of the molecules is increased, and the molecules hit the container with greater force. Therefore pressure is proportional to temperature. Increasing the number of molecules, without changing the temperature or pressure, means that the molecules need more room, so volume is proportional to n. The other laws can be explained in the same way.

Not all molecules in a gas sample move at the same speed; some have low speeds (and low KE), whereas others are moving very fast (with high KE). The distribution of molecular speeds is called a **Boltzmann distribution**. The distribution shifts to higher speeds as the temperature is increased.

Boltzmann distribution curves are illustrated in Figure 12.18 on page 566 and in Figure 12.19 on page 567 of the text.

12.7   Diffusion and Effusion

Effusion is the movement of a gas through a small opening or a porous barrier.

Diffusion is the mixing of gases.

See review question 14.

Thomas Graham (1805–1869) studied the effusion of gases. He found that the rate at which a gas effused was inversely proportional to the square root of its molar mass. This law, called **Graham's law of effusion**, can be derived directly from the kinetic molecular theory. If the average kinetic energies of all gases are the same at the same temperature, then a gas with a greater molar mass must have a lower speed (so that $\frac{1}{2}mu^2$ is the same).

12.8   Some Applications of Gas Laws

Rubber balloons leak because the molecules inside the balloon can escape through small holes in the rubber latex. They will escape until the pressures on both sides of the balloon are equal. Hydrogen and helium escape more easily because their molar masses are low, they are smaller particles, and their molecular speeds are higher.

When deep sea diving, the pressure of air in your lungs is higher the deeper you go. At 33 feet, the pressure is 2 atm. When using compressed air, the higher partial pressure of nitrogen at depths below 33 feet can cause nitrogen narcosis. High partial pressures of oxygen can cause oxygen toxicity (the partial pressure of oxygen in air at 150 feet is equivalent to breathing pure oxygen).

Oxygen is mixed with helium for deep sea diving.

12.9   Nonideal Behavior

At ordinary temperatures and pressures the ideal gas law works well. However, at low temperatures, where the molecules are moving much more slowly and attract one another, and at high pressures, where the molecules are squeezed together, deviation from ideal behavior is observed.

Two assumptions made in the derivation of the ideal gas law from the kinetic molecular theory do not hold at low temperatures and high pressures. The first, the more important, is that molecules do attract one another; because of this attraction a gas will eventually liquefy as the temperature is lowered. The effect of the attraction is to reduce the observed pressure exerted by the gas; the molecules, because of their attraction for one another, are held back in their collisions with the surface of the container, thus reducing the pressure.

The second, the less important, is that the actual volume of the molecules is not zero. Although the volume is very small, it becomes more and more significant the higher the pressure and the lower the temperature.

Johannes van der Waals developed a modification of the ideal gas law to accommodate the facts that there is attraction between molecules and that molecules do have a finite volume. This modified gas law is called the **van der Waals equation (of state):**

$$[P + a(\tfrac{n}{V})^2][V - nb] = nRT$$

The correction factor for the pressure term contains $\tfrac{n}{V}$, which is a concentration term. Squared, and multiplied by a, it is a measure of what proportion of the molecules stick together. The van der Waals constant a is a measure of the strength of the intermolecular attraction. The correction factor for the volume term is nb, where the van der Waals constant b is a measure of the volume of one mole of molecules. The two constants, a and b, are determined experimentally. Notice that the two correction terms have opposite signs; they do to some extent compensate for one another.

One mole of water vapor occupies 22.4 L at 25°C and 1 atm pressure. One mole of water liquid occupies 18 mL – very small in comparison.

The equation is often referred to as an equation of state because it describes the behavior of the gaseous state.

Table 12.3 on page 572 shows values for a and b for some gases.

Review Questions

1. What is the difference between a gas and a vapor?

2. Write an expression for Boyle's law. If the pressure on a gas is tripled at constant temperature, what happens to the volume?

3. Write an expression for Charles' law. Explain why the temperature must be in K.

4. What is the combined gas law? Write a mathematical expression for the law.

5. Write statements of Gay-Lussac's law and Avogadro's law.

6. To what does the 'ideal' refer in the ideal gas law? Is there such a thing as an ideal gas?

7. Derive the expression for the density of a gas $d = \dfrac{PM}{RT}$ from the ideal gas equation.

8. One property that distinguishes the gas state from the other two states is its compressibility. Why can gases be compressed whereas liquids and solids cannot?

9. Derive the expression $P_a = X_a P_{total}$ for the partial pressure of component a in a mixture of gases. What is X_a in this expression?

10. Is air containing a high proportion of water vapor (high humidity) less dense or more dense than dry air?

11. How is the speed of a molecule in the gaseous state related to the temperature and the molar mass of the gas?

12. Does a single molecule have a temperature?

13. Describe the principal features of the kinetic-molecular theory.

14. Derive Graham's law from the result you obtained in question 11.

15. Why does a helium filled mylar balloon last longer if you put it in the freezer?

16. Describe the conditions under which a gas might start to deviate from ideal behavior. Explain why the deviation occurs.

Answers to Review Questions

1. There is no significant difference between a gas and a vapor. A gas is a substance that is normally a gas at ordinary temperatures and pressures (for example, oxygen, carbon dioxide). A vapor is a substance that is normally a liquid or solid at ordinary temperatures and pressures (for example, water, naphthalene).

2. Boyle's law can be expressed in several ways; one is PV = constant at constant temperature; another is $P_1V_1 = P_2V_2$ for two states 1 and 2 at the same temperature. If the pressure exerted on a gas is tripled, then the volume must decrease to one third of its initial volume to keep the product PV constant (at constant temperature).

3. Charles' law can be expressed as $V = kT$ at constant pressure. Alternatively, it can be written as $V_1/T_1 = V_2/T_2$ for two states 1 and 2 at the same pressure. The temperature must be in K for the volume to be directly proportional (i.e. a slope with an origin = zero). You could write the law using °C but you would have to include the intercept −273.15°C.

4. The combined gas law is a combination of Boyle's law and Charles' law. It can be written as $PV = kT$ where the temperature is in K and the quantity of gas is constant. Or it can be written as $P_1V_1/T_1 = P_2V_2/T_2$ for two states 1 and 2 where the temperature is again in K and the quantity of gas is constant.

k is a constant.

5. Gay-Lussac's law states that the volumes of gases involved in a chemical reaction is always a small whole number ratio—provided that the volumes are measured at the same temperature and pressure. Avogadro's law states that the volume occupied by a gas at constant pressure and temperature is directly proportional to the number of molecules of gas in the sample.

6. The 'ideal' in the ideal gas law means the attraction between the molecules is zero and the actual volume of the molecules is zero. There is no such thing as a molecule with zero volume! However, most gases at ordinary temperatures and pressures behave ideally.

7. The number of moles n equals the mass of the gas divided by its molar mass:

    $$n = \frac{m}{M} \quad \text{therefore } PV = \frac{m}{M}RT \quad \text{or} \quad PM = \frac{m}{V}RT$$

    $\frac{m}{V}$ is density, so density $= \frac{PM}{RT}$

8. Liquids and solids are called the condensed states. Molecules in these states are next to one another, held together by intermolecular forces of attraction. As a result, both liquids and solids have surfaces. A gas, on the other hand, will always fill whatever volume is available. Molecules in the gas state behave independently and are on average far apart from one another. So the volume occupied by a certain quantity of gas can change quite dramatically whereas the same quantity of liquid or solid occupies a comparatively small volume which changes little.

    Gases on earth are held by gravitational attraction, they don't, fortunately, expand to fill the entire space of the universe. Gases released from space craft will however expand for ever until individual molecules are trapped by the gravitational fields of various stars, planets, etc.

9. The ideal gas law can be written for any component in a mixture of gases, or it can be written for the entire mixture.

    For gas a, for example: $\qquad P_aV=n_aRT$
    For the total mixture $\qquad P_{total}V=n_{total}RT \qquad \dfrac{P_a}{P_{total}} = \dfrac{n_a}{n_{total}}$

    Dividing the first expression by the second:

    or $P_a = (n_a/n_{total})P_{total}$

    or $P_a = X_aP_{total}$ where X_a is the mole fraction of a in the mixture.

10. The molar mass of water (18 g mol^{-1}) is lower than the molar masses of nitrogen and oxygen (28 and 32 g mol^{-1}). Air containing water vapor (high humidity) is therefore less dense than dry air—even though it may feel more oppressive.

11. The average kinetic energy of the molecules of any gas at the same temperature is always the same. If the molar mass of one gas is greater than the molar mass of another gas, then it must have a proportionately lower mean square speed. The average kinetic energy is proportional to the temperature:

    $$\tfrac{1}{2}m\overline{u^2} = \text{constant} \times T$$

12. The temperature of a gas is proportional to the average kinetic energy of the molecules of the gas. It is a macroscopic property that refers to the entire sample. However, in a gas, there are molecules travelling fast with high speeds and high kinetic energies and there are molecules travelling at slow speeds with low kinetic energies. The distribution of molecular

speeds is illustrated by a Boltzmann distribution curve; this curve shows that most molecules have speeds somewhere in the center. In the sense that kinetic energy and temperature are related, the molecules in a sample that have high kinetic energies can be described as "hot"; those with low kinetic energies as "cold".

13.   The principal features of the kinetic-molecular theory are:

Gases consist of molecules with intermolecular distances much greater than the size of the molecules.
Molecular motion is continual, random, and rapid.
The average kinetic energy of the molecules in a gas is proportional only to the temperature of the gas. All gases have the same average kinetic energy at the same temperature.
No energy is lost during the collision of one molecule with another or in a collision of a molecule and its container.

14.   The result from question 11 was: $\frac{1}{2}m\overline{u^2} =$ constant $\times$ T for all gases. In other words, if T is the same, then

for two gases A and B: $\frac{1}{2}m_A\overline{u_A^2} = \frac{1}{2}m_B\overline{u_B^2}$

Multiplying through by 2, collecting the m's and u's on either side:

$$\frac{\overline{u_A^2}}{\overline{u_B^2}} = \frac{m_B}{m_A}$$

The rate at which molecules effuse is proportional to their rms speeds and the masses of the individual molecules can be multiplied by Avogadro's number to obtain the ratio of molar masses:

$$\frac{\text{Rate of effusion of A}}{\text{Rate of effusion of B}} = \frac{\sqrt{\overline{u_A^2}}}{\sqrt{\overline{u_B^2}}} = \sqrt{\frac{M_B}{M_A}}$$

15.   If the temperature is lowered, the rate at which a gas effuses or diffuses must decrease, so a helium filled mylar balloon will stay inflated longer in the freezer.

16.   The ideal gas law does not hold well at low temperatures and high pressures. Molecules do attract one another and molecules do have a finite volume. The effect of the attraction is that molecules are held back in their collisions with the surface of the container. The observed pressure is less than the ideal pressure. And although the volume of molecules is very small, it becomes more and more significant the higher the pressure and the lower the temperature. The observed volume is greater than the ideal volume. Johannes van der Waals developed an alternative gas law to accommodate the intermolecular attraction and the finite volume of molecules—a real gas law.

Study Questions and Problems

The formulation of a problem is far more often essential than its solution, which may be merely a matter of mathematical or experimental skill.

Albert Einstein
(1879-1955)

1. Convert the following pressures to atm.

    a. 726 torr
    b. 2.31 bar
    c. 98 kPa
    d. 16.33 psi

2. Consider the following changes imposed upon a sample of gas, assuming the variables not mentioned remain constant:

    a. What happens to the pressure if the temperature in K is doubled?

    b. What happens to the volume if the pressure is tripled?

    c. What happens to the volume if the temperature decreases from 300K to 200K?

    d. What happens to the temperature if one-half of the gas is removed?

    e. What happens to the pressure if volume decreases from 4 liters to 2 liters and the temperature increases from 25°C to 323°C?

3. Methane burns in air to produce carbon dioxide and water:

    $$CH_4(g) \ + \ 2O_2(g) \ \rightarrow \ CO_2(g) \ + \ H_2O(l)$$

    What volume of carbon dioxide, at 1 atm pressure and 112°C, will be produced when 80.0 grams of methane is burned?

4. What is the volume of 6 moles of helium gas at 0.34 atm pressure and 33°C? What is the density of the helium gas under these conditions?

5. Jacques Charles used the reaction of hydrochloric acid on iron to produce the hydrogen for one of his balloons. For one flight in 1783 he used 1000 lbs of iron and excess acid. What volume of hydrogen gas (in cubic meters) did he produce for this flight? Assume the pressure is 1 atm and the temperature is 22°C.

    $$Fe(s) \ + \ 2HCl(aq) \ \rightarrow \ FeCl_2(aq) \ + \ H_2(g)$$

6. If 1.0 liter of oxygen at 2.0 atm pressure, 2.0 liters of nitrogen at 1.0 atm pressure, and 2.0 liters of helium at 2.0 atm pressure, are all mixed in a 3.0 liter vessel with no change in temperature, what is the final pressure of the mixture in the 3.0 liter vessel?

7. What is the partial pressure of oxygen in the atmosphere at the top of Mt Everest? Atmospheric pressure at the summit of Mt Everest is 253 torr. The partial pressure of oxygen in air at 1 atm pressure is 0.20946 atm.

8. An ideal gas occupies a volume of 10 liters at 27°C. If the pressure on the gas is tripled at this temperature, the volume changes. To what value must the temperature be changed to restore the volume to the initial 10 liters at the new pressure?

9. Using the Maxwell equation, calculate the root mean square speed of nitrogen gas at 25°C. What happens to the rms speed if the temperature is doubled to 50°C?

10. Imagine three automobiles travelling down the road at 20 mph, 32 mph, and 68 mph. Calculate the average speed and the rms speed. What is the significance of the rms speed?

11. A gas diffuses 5/3 times faster than carbon dioxide. Which gas might it be?
    a. O_2 b. N_2 c. CO   d. He   e. CH_4

12. For nitrogen, the van der Waals constants a and b have values 1.39 and 0.0391 respectively. Calculate the pressure of 5 moles of nitrogen gas confined to a 1.0 liter vessel at a temperature of 300K using the ideal gas equation and the van der Waals equation of state. Comment on the difference.

Answers to Study Questions and Problems

1. Conversion to atmospheres:

    a. 726 torr × (1 atm / 760 torr) = 0.955 atm
    b. 2.31 bar × (1 atm / 1.01325 bar) = 2.28 atm
    c. 98 kPa × (1 atm / 101.325 kPa) = 0.97 atm
    d. 16.33 psi × (1 atm / 14.696 psi) = 1.111 atm

2. Use the ideal gas law to establish the relationships between the four variables pressure, volume, temperature, and number of moles. Variables not mentioned are constant.

$\underline{P}V = nR\underline{T}$ a. Pressure is proportional to temperature in K, so if the temperature in K is doubled, the pressure must also double.

$\underline{PV} = nRT$ b. Volume and pressure are inversely proportional (Boyle's law), so if the pressure is tripled, the volume must decrease to one-third of the initial volume.

$P\underline{V} = nR\underline{T}$ c. A decrease from 300K to 200K is a decrease to two-thirds of the initial temperature. Volume is proportional to temperature, so the volume too must decrease to two-thirds of its initial value.

$PV = \underline{n}R\underline{T}$ d. Temperature and the number of moles are inversely proportional. If one-half of the gas is removed, the temperature must double—if the remaining gas is to occupy the same volume at the same pressure.

e. Remember to convert the temperature to K: 25°C is 298K and 323°C is equal to 596K (double the initial temperature). Pressure is inversely proportional to the volume but proportional to the temperature, so if the volume is doubled and the temperature is also doubled, the pressure does not change.

$\underline{PV} = nRT$
$\underline{PV} = nR\underline{T}$

3. $$CH_4(g) \; + \; 2O_2(g) \; \rightarrow \; CO_2(g) \; + \; H_2O(l)$$

80.0 grams of methane is 80.0 g × (1 mol / 16 g mol^{-1}) = 5.0 mol. According to the 1:1 stoichiometry of the equation, 5.0 mol of CO_2 will be produced. The ideal gas law can be used to calculate the volume:

$PV = nRT$ so
1atm × volume in L
   = 5.0 mol × 0.08206 L atm K^{-1}mol^{-1} × (112+273.15)K.
Volume = 158 liters.

Remember to use the correct value for R.

4. $PV = nRT$ so
 0.34 atm × volume in L
   = 6.0 mol × 0.08206 L atm K^{-1}mol^{-1} × (33+273.15)K.
Volume = 443 liters.

Using the expression: density = PM/RT, the density = 0.0542 g/L, or, having already calculated the volume, density = mass/volume, so the density = 24.018g /443 L = 0.0542 g/L again.

Molar mass = 4.003 g mol^{-1}
Mass of 6.0 moles = 24.018 g

5. 1000 lbs of iron = 453.6 kg = 8122 moles of iron.

$$Fe(s) \; + \; 2HCl(aq) \; \rightarrow \; FeCl_2(aq) \; + \; H_2(g)$$

According to the equation, 8122 moles of hydrogen were produced.
$PV = nRT$ so
1 atm × volume in L
   = 8122 mol × 0.08206 L atm K^{-1}mol^{-1}× (22+273.15)K.
Volume = 200,000 liters
       = 200 m^3.

The partial pressure of a gas in a mixture is the pressure it would exert if it were the only gas present.

6. The 1.0 liter of oxygen at 2.0 atm pressure is expanded to occupy 3.0 liters, so its pressure decreases by a factor of 1/3 to 2/3 atm.
Likewise, the pressure of nitrogen is 1.0 atm × 2.0/3.0 = 2/3 atm.
And the pressure of helium is 2.0 atm × 2.0/3.0 = 4/3 atm.
The total pressure is the sum of these partial pressures:
Total pressure = $P_{O_2} + P_{N_2} + P_{He}$

   = 2/3 + 2/3 + 4/3 atm = 8/3 atm = 2.67 atm.

Notice that the partial pressures of both oxygen and nitrogen are the same. This is because P×V is the same for both, and P×V is proportional to n, the number of moles. The number of moles of oxygen = the number of moles of nitrogen and their partial pressures must therefore be the same.

7. The partial pressure of oxygen in the atmosphere at the top of Mt Everest is equal to 253 torr × 0.20946 atm / 1atm = 53 torr.
In atmospheres, this is 0.070 atm, not very much!

$PV = nR\underline{T}$

$\sqrt{\overline{u^2}} = \sqrt{\dfrac{3RT}{M}}$

You must have the correct units —use SI units!
So use the SI value for the gas constant R and use the correct base units, for example kg for mass. Then you know that the units of the answer must be SI also, i.e. ms^{-1}. Check to see if they cancel correctly.
One J = kg m^{2}s^{-2}

8. Use the ideal gas law to establish the relationship between the variables. In this problem, the volume is restored to its original value by adjusting the temperature. So, in effect, the volume doesn't change. The initial temperature is 27°C or 300K. If the pressure is to be tripled, then so must the temperature, to 900K, or 627°C.

9. The Maxwell equation relates rms speed to the molar mass and the temperature of the gas.
   The root mean square speed
   $= (3 \times 8.314 \, \text{JK}^{-1}\text{mol}^{-1} \times 298.15 \, \text{K} / 28.02 \, \text{g mol}^{-1} \times 1\text{kg}/1000\text{g})^{\frac{1}{2}}$
   $= 515 \, \text{ms}^{-1}$
   If the temperature is doubled to 50°C, then the rms speed will increase proportionately—by $\sqrt{(323/298)}$—to 536 ms^{-1}. Use degrees K!

10. The average speed is $(20 + 32 + 68)/3 = 120/3 = 40$ mph.

    To calculate the rms speed, square the speeds, take the average, then take the square root of the average:
    $\text{Rms speed} = [(20^2 + 32^2 + 68^2)/3]^{\frac{1}{2}}$
    $\qquad\qquad\quad = [(400 + 1024 + 4624)/3]^{\frac{1}{2}}$
    $\qquad\qquad\quad = [2016]^{\frac{1}{2}}$
    $\qquad\qquad\quad = 45 \, \text{mph}$

    The rms speed is higher than the average speed. The rms speed is the speed at which an automobile would have to travel if it were to have a kinetic energy equal to the average kinetic energy of the other three cars.

11. According to Graham's law, the rate of diffusion is inversely proportional to the square root of the molar mass. The unknown gas diffuses faster and therefore must have a lower molar mass than carbon dioxide. The ratio of the molar masses must be $(3/5)^2$.

    $44 \, \text{g mol}^{-1} \times (3/5)^2 = 15.8 \, \text{g mol}^{-1}$, the gas is methane, CH_4.

12. Using the ideal gas equation PV = nRT:
    $P = 5.0 \, \text{mol} \times 0.08206 \, \text{L atm K}^{-1} \text{mol}^{-1} \times 300 \, \text{K} / 1.0 \, \text{L} = 123 \, \text{atm}$.

    Using the van der Waals equation of state $[P + a(\frac{n}{V})^2][V - nb] = nRT$:

    $[P + 1.39 \times (5.0/1.0)^2] \times [1 - (5.0 \times 0.0391)] = 5.0 \times 0.08206 \times 300$

    $[P + 34.75] \times [0.8045] = 123.1 \, \text{atm}$

    $P = 118 \, \text{atm}$

    The two pressures agree tolerably well, but the temperature is still quite high, and the pressure is not very high. An increase in pressure by a factor of 10 would lead to a more significant deviation. Notice how the pressure correction and the volume correction tend to compensate one another.

Challenge 6

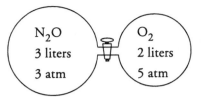

The apparatus shown consists of two glass bulbs connected by a tap which is closed. In the first bulb, volume 3 liters, there is a sample of pure nitrous oxide at 3 atm pressure. In the second bulb, volume 2 liters, there is a sample of pure oxygen at a pressure of 5 atm.

Maintaining a constant temperature throughout, the tap is opened, the gases are allowed to mix, and they react to produce nitrogen dioxide according to the equation written below. Assume that the reaction goes as far as possible to completion:

$$2N_2O(g) \ + \ 3O_2(g) \ \rightarrow \ 4NO_2(g)$$

a. What is the partial pressure of the nitrogen dioxide in the apparatus at the end of the reaction?

b. What is the partial pressure of the excess (nonlimiting) reactant at the end of the reaction?

c. What is the total pressure in the apparatus at the end of the reaction?

People of this chapter and their States

—a crossword puzzle

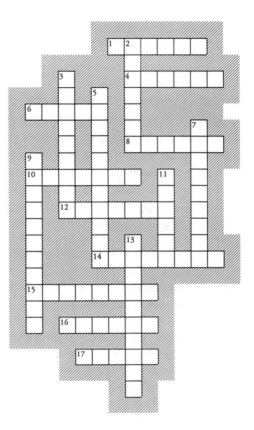

Across:

1. William Thomson's new temperature scale.
4. He formulated the law of effusion.
6. A sceptical chemist, his was the first gas law.
8. His law states that the total pressure is the sum of the partial pressures.
10. Country of 7 down.
12. A balloonist of some renown.
14. Country of 4 across.
15. He is remembered by his number.
16. Country of 12 across, 3 down, and 11 down.
17. Country of 15 across and 13 down.

Down:

2. Country of 8 across.
3. He formulated the law of combining volumes.
5. Country of 9 down.
7. He was concerned with probabilities and distributions.
9. He developed a gas law for real gases.
11. Mathematician for whom the SI unit of pressure is named.
13. He invented the mercury barometer in 1643.

CHAPTER 13

Introduction

The liquid and solid states of matter are condensed states—the molecules, atoms, or ions that make up the matter are close together, held together by strong interparticle forces of attraction. In this chapter we will examine these forces of attraction and investigate the properties of different kinds of liquids and solids. The energies involved in changing from one state to another will be examined.

Contents

13.1   States of Matter and the Kinetic Molecular Theory

Molecules behave independently in the gas state. Their behavior in the liquid and solid states is, however, more complicated. Molecules do attract one another—without such attraction all substances would be gases. In both liquids and solids, molecules are held together to a finite volume by their attraction for one another. This volume is approximately the same for the liquid and solid states, but much smaller than the gas state. For example, 18 mL of liquid water, or 18 g of ice, produce 22.4 liters of water vapor—a dramatic change in volume.

13.2   Intermolecular Forces

Intermolecular forces are the electrostatic forces of attraction between molecules or between ions and molecules. There are a variety of different types. For example, there is the attraction between ions and polar molecules, between polar molecules themselves, and between nonpolar molecules. Those forces not involving ions are generally referred to as **van der Waals forces**.

Intermolecular forces are considerably weaker than intramolecular bonds. A covalent bond has an energy between 100 and 500 kJ mol^{-1}. Multiple bonds are stronger. Typical ionic forces in an ionic compound vary between 700 and 1100 kJ mol^{-1}. The magnitude of intermolecular forces are in general much weaker.

These various forces are listed in Table 13.1 on page 585.

A N$_2$ triple bond has a bond energy of 946 kJ/mol.

The strongest intermolecular force is the force between an ion and a polar molecule. This is the type of interaction that occurs in solution between an ion and the solvent molecules. In this case the process is called **solvation** and the energy released is called the **solvation energy**. The magnitude of the force depends upon the charge on the ion, the dipole of the polar molecule, and the distance between the ion and the polar molecule.

When water is the solvent, the solvation of ions is called **hydration**. Hydrated salts are often encountered in chemistry. The hydration energy varies in a predictable way, depending to a great extent on the charge and size of the ion. The number of water molecules associated with a single ion varies in the same way. A Li^+ ion has a radius 90 pm, an enthalpy of hydration of -515 kJ mol^{-1}, and a hydration number of about 25 water molecules. A Cs^+ ion is much larger, 181 pm, has an enthalpy of hydration of -263 kJ mol^{-1}, and a hydration number of only 10.

When two molecules have dipole moments, they attract one another. The positive end of one molecule attracts the negative end of another molecule and *vice versa*.

Many molecules with the groups O–H, N–H, and F–H have properties indicating an exceptionally strong intermolecular force of attraction. This force is called **hydrogen bonding** and involves a partial sharing of a hydrogen atom attached to O, N, or F, with another N, O, or F. It is the high electronegativity and small size of oxygen, nitrogen, and fluoride that is responsible for the strength of the hydrogen bond. Consider the hydrogen compounds of all the elements in Groups 3, 4, 5, 6, and 7; compounds like SiH_4, PH_3, HBr, B_2H_6, CH_4, H_2S— all are gases except for water which is a liquid. Water is a liquid because of the strong intermolecular hydrogen bonds.

Hydrogen bonding has important implications for any property of a compound that is related to the intermolecular forces of attraction. Examples of such properties are melting and boiling points, heats of fusion, vaporization and sublimation, viscosity, and structure in the liquid and solid states. Examples of some substances in which hydrogen bonding plays an important role in the structure and properties of the substance are proteins, synthetic polymers, and DNA.

Some molecules are nonpolar, and yet at sufficiently low temperatures they will liquefy and solidify. What holds nonpolar molecules together? All molecules are held together by **dispersion forces**. Even the attraction between polar molecules like HCl involves more dispersion force than dipole–dipole attraction. A dispersion force is the attraction between **induced dipoles**—these induced dipoles are temporary displacements in the electron cloud surrounding the atoms in a molecule. The inducement of such temporary dipoles explains how a nonpolar molecule like oxygen can dissolve in polar solvent like water: As the water molecule approaches an oxygen molecule, the dipole of the water molecule distorts the electron distribution in the oxygen molecule. The result is an attraction between the dipole of the water molecule and the induced dipole of the oxygen molecule.

The larger the electron cloud, the more easily **polarized** it is—the easier it is to induce a dipole. Thus, nonpolar iodine is a solid, bromine is a liquid, and

---

The mobility of Li$^+$ ions in solution is less than Cs$^+$ ions because the Li$^+$ ions are dragging all these water molecules around.

Hydrogen bonding involving
N–H...N–
N–H...O–
O–H...O–
O–H...N–
is extraordinarily important in biological and biochemical structures and reactions.

You may ask why ammonia NH$_3$ and hydrogen fluoride HF are not liquids like water. The structure of water, with two H atoms and two lone pairs on the oxygen, is perfect for the formation of four hydrogen bonds from each molecule. Ammonia has two many H, and HF has too many lone pairs, and so they form half as many hydrogen bonds as water and are gases.

Remember:
The hydrogen bond is a bond *between* two water molecules not a bond *in* a water molecule. Generally hydrogen bonding is an intermolecular attraction although there are many examples of hydrogen bonding between two functional groups of the same molecule.

chlorine is a gas. But in these halogens there are no polar molecule to induce the dipole, so how does the induced dipole attraction work? As the temperature is decreased, molecules slow down; the molecules approach one another long enough to allow the electron clouds and nuclei of different molecules to interact; this leads to a mutual distortion of the electron clouds on adjacent molecules; a temporary induced dipole attraction is the result.

13.3   Properties of Liquids

Molecules in the liquid state are in constant motion, just like they are in the gas state, but, unlike in the gas state, they are not independent of one another. The molecules are held together by intermolecular forces of attraction. The speeds and energies of the molecules range from low to high; the distribution is similar to that described for the gas state. Some molecules have sufficient energy to break free of the intermolecular attraction and enter the gas phase. The ability of a molecule to break free depends upon the strength of the intermolecular force and its kinetic energy. The latter depends upon the temperature. The boiling point and the heat of vaporization are indications of the strength of the intermolecular bonding.

> The intermolecular forces were described in the last section.

> The stronger the bonding, the higher the boiling point and the greater the heat of vaporization.

A liquid left in an open container will eventually evaporate; molecules escape from the liquid and diffuse away through the atmosphere. If the container is closed, then the escaping molecules cannot diffuse away and may eventually collide with the surface of the liquid and reenter the liquid phase. After some time, the rate at which molecules leave the liquid will equal the rate at which molecules reenter the liquid. At this point the concentration of the molecules in the vapor phase will be constant. The system is said to be in **dynamic equilibrium**. The partial pressure of the molecules in the gas above the liquid at equilibrium is called the **equilibrium vapor pressure**. This vapor pressure is a measure of how easy it is for the molecules in the liquid to escape—often referred to as the **volatility**.

> The movement of molecules between the liquid and vapor phases does not stop at equilibrium—it's just that the two rates are equal so there is no macroscopic change in the system. The two processes continue unabated.

The vapor pressure increases with temperature—the molecules have greater energy at higher temperatures and therefore find it easier to escape. A mathematical relationship between vapor pressure (P) and temperature (T) was determined by Clausius and Clapeyron:

$$\ln P = -\left(\frac{\Delta H_{vap}}{RT}\right) + C \qquad \text{(where C is a constant)}$$

> A plot of the vapor pressure *vs.* the temperature is called the vapor pressure curve.

If the temperature is increased to the point at which the vapor pressure equals the atmospheric pressure, then the liquid will boil. The pressure exerted by the molecules in the atmosphere colliding with the surface of the liquid is balanced by the force of the molecules escaping from the surface, and bubbles form in the liquid.

The vapor pressure curve does not continue upward without limit. The curve stops at the **critical point**. At this point it is no longer possible to distinguish between liquid and vapor. The combination of high pressure and high temperature means that the molecules are still close to one another but have sufficient kinetic energy to overcome the intermolecular forces. It's like a liquid in which the intermolecular forces are ineffective—just like a gas.

> Supercritical fluids have interesting properties and commercial uses.

Molecules at the surface of a liquid experience an unbalanced attraction. There are molecules within the liquid pulling the surface molecules, but no molecules beyond the surface to balance the pull. As a result there is a net inward force referred to as **surface tension**. This surface tension tends to minimize the surface area—so that drops of water are spherical for example. **Capillary action** is related to surface tension. Water will rise in a small diameter glass tube because of the adhesive forces between the water molecules and the glass surface and the cohesive forces between the water molecules. The water will be pulled up the tube until the cohesive and adhesive forces are balanced by the downward gravitational forces.

Viscosity is the resistance of a liquid to flow. Stronger intermolecular forces will increase the viscosity of a liquid. Physical entangling of long chain molecules will also inhibit flow and increase viscosity.

13.4 Metallic and Ionic Solids

In a solid, the structural units—molecules, atoms, or ions—do not move as they do in liquids and gases. They vibrate and sometimes rotate, but they cannot move. A regular repetitive pattern of structural units is characteristic of the solid state. Identical points (lattice points) in the crystal define what is called a **crystal lattice**.

A **unit cell** is the smallest repeating unit of a crystalline material. The unit cell has all the symmetry characteristics of macroscopic crystal and is drawn so that, as far as possible, the structural units are at the corners of the cell. The shape of the unit cell is defined by the **crystal system**. There are seven crystal systems—seven different shapes for the unit cells. They vary in the angles at the corners and the lengths of the sides—the simplest is the cubic crystal system. For this system the unit cell is a cube.

How the structural units are placed in the unit cell is defined by the **crystal lattice**. There are 14 crystal lattices (sometimes referred to as Bravais lattices). The cubic system has three lattices associated with it—the simple cubic lattice (sc), the body-centered cubic lattice (bcc), and the face-centered cubic lattice (fcc). Many metals adopt a fcc lattice (for example: copper, aluminum, gold), some adopt a bcc lattice (for example: sodium, potassium, iron), and one (polonium) adopts a sc lattice.

Ionic salts are composed of at least two different particles, one positive and the other negative. The lattices of many ionic compounds are built up by creating a close-packed lattice (often cubic) of the larger of the two ions (usually the anion) and then filling the holes in this lattice with the smaller ion (usually the cation). The type of hole (tetrahedral, octahedral, or cubic) is determined by the relative sizes of the ions. The fraction of holes filled is determined by the stoichiometry of the compound. If the unit cell of the ionic compound is determined, then the stoichiometry of the compound can be calculated from the composition of the unit cell. It is important to note that ions on the edge of one unit cell are shared with the adjacent unit cell(s).

Cohesive forces are the forces within the substance, holding the substance together. **Adhesive forces** are the forces between one substance and another.

Crystal system— the shape of the unit cell.

Crystal lattice— the arrangement of structural units within the unit cell.

The seven crystal systems are illustrated in Figure 13.27 on page 612.

The three lattices of the cubic system are illustrated in Figure 13.28 on page 613.

Cubic system:

simple cubic lattice (1 structural unit per cell)

body-centered cubic lattice (2 structural units per cell)

face-centered cubic lattice (4 structural units per cell)

Contribution to a unit cell:

inside:	contribution 1
on a face:	contribution 1/2
on an edge:	contribution 1/4
on a corner:	contribution 1/8

13.5   Other kinds of Solid Materials

The structural units in a crystal can be molecules; for example, the water molecules that make up ice. How the molecules are arranged in the lattice depends upon the shape of the molecules and their intermolecular forces. The molecules tend to pack as densely as possible and are aligned to maximize their attraction for one another.

Another type of solid is a covalent network solid in which the structural units are atoms covalently bonded together. These solids are typically high melting, hard, rigid materials. Examples are diamond, graphite, silicon, carborundum, silica, boron nitride, boron carbide, etc.

True solids are crystalline. However, there are some materials that appear to be solid that are not crystalline. The order within the material is random—just as in a liquid. Such materials are called **amorphous**. Window glass is an example. In glass the molecular order is random but the intermolecular forces are sufficiently strong to prohibit much movement of the molecules. The glass, although apparently solid, is just a very viscous liquid. Many plastics (polymers) are similar in structure.

If you ever come across a very old (>200 years) glass window, examine it to see if the thickness of the glass is greater at the bottom than at the top.

13.6   The Physical Properties of Solids.

Solids are characterized by their regular crystalline lattice structure. The melting point and heat of fusion depend upon the strength of the intermolecular forces in the lattice. The energy required to break up the lattice is often referred to as the lattice energy. Metals vary considerably in their melting points and lattice energies. Mercury is a liquid at room temperature; tungsten has a melting point of 3422°C. Most representative group metals have relatively low melting points. Most transition metals have relatively high melting points.

The melting points of ionic compounds depend upon the lattice energies —which in turn depend upon the charges and sizes of the ions. Covalent network materials (e.g. diamond) generally have very high melting points.

Molecular solids have a range of melting points and heats of fusion. Again, both depend upon the strength of the intermolecular force. Dispersion forces increase as the size of the molecule increases, so within any series, the melting point will increase as the molar mass increases.

Be able to do calculations involving the heats of fusion and vaporization. These have been encountered before in Chapter 6.

Solids made up of small molecules have low melting points. An example is ice. Indeed, most small molecules are liquids or gases.

13.7   Phase Diagrams.

A **phase diagram** is a summary of the conditions of temperature and pressure under which each of the three states of matter exists. The diagram also describes the conditions under which two, or even three, phases can coexist in equilibrium. And it often includes information about the particular phases that can exist within the solid state. On the phase diagram:

- the areas represent conditions under which only one phase will exist,
- the lines represent conditions under which two phases will coexist, and
- the point where the lines meet represents conditions under which three phases will coexist in equilibrium.

The phase diagram for water is illustrated in Figure 13.37 on page 630 of the text.

This is why it is called a phase diagram rather than a state diagram.

This point is called the triple point.

Review Questions

1.   Summarize the differences in the behavior of molecules in the three states of matter.

2.   Describe the different types of interparticle forces that can occur between atoms, molecules, and ions. Distinguish the forces called intermolecular forces. What forces are referred to as van der Waals forces? Draw a flow chart or diagram to summarize these intermolecular forces.

3.   What is a hydrogen bond? Describe the requirements for hydrogen bonding.

4.   Why do larger molecules (higher molar mass) melt at higher temperatures?

5.   Describe the process of dynamic equilibrium that exists between a liquid and its vapor in a closed container. Why does a system in an open container never reach equilibrium?

6.   What is the critical point?

7.   Describe what you understand by surface tension.

8.   What is the difference between a crystal system and a crystal lattice?

9.   Describe how you would calculate the number of ions in the unit cell of an ionic crystal lattice.

10. Is there such a thing as an amorphous solid? Explain.

For the mathematicians amongst you: Can you derive a mathematical relationship between the number of phases that exist or coexist in equilibrium and how the conditions of temperature and pressure can vary?

11. Draw a generic phase diagram and indicate the areas in which the various states are stable. What is the significance of the lines on the diagram and the point at which the lines meet?

12. On the diagram you drew for question 11, draw a horizontal line at a pressure of 1 atm. The line crosses the solid-liquid line and the liquid-vapor line (the vapor pressure curve). What is the significance of these points where the lines cross?

13. What is significant about the direction in which the solid–liquid line on the phase diagram slopes?

Answers to Review Questions

1. Some differences are:

    Solid: very regular array or lattice of molecules
    very effective attraction between molecules
    molecules close together
    molecular motion zero

    Liquid: random order of molecules, some close-range order
    intermolecular forces of attraction intermediate in effectiveness
    molecules close together
    molecules move with respect to one another but are held together

    Gas: totally random order of molecules
    intermolecular forces of attraction totally ineffective
    molecules far apart
    molecules move independently of one another

2. There are three major classes of bonds. These are characterized by the behavior of the valence electrons in the bond which, in turn, depends upon the electronegativity difference between the elements involved:

    covalent: sharing of electrons (need not be equal sharing)
    ionic: transfer of electrons
    metallic: delocalization of electrons

    Covalent bonding leads to the formation of molecules or polyatomic ions. The bonding between these molecules, or between these molecules and ions can be classified as follows:

    Intermolecular forces: ion–molecule forces (e.g. solvation)
    van der Waals forces of attraction

    The van der Waals forces can be subdivided further into:
    hydrogen–bonding
    dipole–dipole attraction
    dipole–induced dipole attraction
    dispersion forces (induced dipole)

    For a diagram summarizing these intermolecular forces that will help you to decide what types of forces are appropriate for a given set of molecules, see Figure 13.14 on Page 597.

3. A hydrogen bond involves a partial sharing of a hydrogen atom attached to O, N, or F, with another N, O, or F. The N, O, or F must have a lone pair of electrons. The high electronegativity and small size of oxygen, nitrogen, and fluoride is responsible for the strength of the hydrogen bond because this polarizes the O–H, N–H, or F–H bond. Strong intermolecular hydrogen bonding has important implications for any property that is related to the intermolecular forces of attraction.

4. All molecules are held together by dispersion forces. A dispersion force is the attraction between induced dipoles—these induced dipoles are temporary polarizations in the electron cloud surrounding the atoms in a molecule. The larger the electron cloud, the more easily polarized it is—

the easier it is to induce a dipole. Larger molecules also move more slowly than smaller molecules, and therefore the interaction between the molecules is more effective.

5.   In a closed container, the molecules escaping from the surface of the liquid cannot diffuse away and eventually collide with the surface of the liquid and reenter the liquid phase. After some time, the rate at which molecules leave the liquid will equal the rate at which molecules reenter the liquid. The system is then said to be in dynamic equilibrium; at this point the concentration of the molecules in the vapor phase is constant. A liquid left in an open container will eventually evaporate; molecules escape from the liquid and diffuse away through the atmosphere—the system can never reach equilibrium.

6.   At the critical point it is no longer possible to distinguish between the liquid and vapor states. The combination of high pressure and high temperature means that the molecules are still close to one another but have sufficient kinetic energy to overcome the intermolecular forces.

7.   Molecules at the surface of a liquid experience an attraction by the molecules within the liquid. This pulls the surface molecules inward; this inward force is referred to as surface tension. The surface tension tends to minimize the surface area because the smaller the area the lower the energy. Drops of water are spherical for example. It's almost as if the surface of the water has a skin; an insect walking on the surface causes a slight indentation in the surface that is resisted by the cohesive forces within the liquid water.

8.   The crystal system defines the shape of the unit cell. There are seven crystal systems—seven different shapes for the unit cells. The unit cells of the seven systems vary in the angles at the corners and the lengths of the sides. The crystal lattice describes how the structural units are placed in the unit cell. There are 14 crystal lattices (sometimes referred to as Bravais lattices). For example, the cubic system has three lattices associated with it—the simple cubic lattice (sc), the body-centered cubic lattice (bcc), and the face-centered cubic lattice (fcc).

Try problems #8, 9, and 10.

9.   The key to determining the number of ions in the unit cell of an ionic crystal lattice is to remember that ions (or any structural unit) on the edge of one unit cell are shared with all adjacent unit cells. Thus, the contributions to a particular unit cell are:
        a structural unit (ion) on the inside contributes 1
        a structural unit (ion) on a face contributes 1/2
        a structural unit (ion) on an edge contributes 1/4
        a structural unit (ion) on a corner contributes 1/8

10.   A true solid is crystalline. It contains a regular lattice of atoms, ions, or molecules. In an amorphous material the array of structural units is random or disordered. By this definition, amorphous materials are not solids. Many modern materials, including polymers, plastics, and composites are amorphous. Some would prefer to define a solid as something hard, as something with sufficiently strong intermolecular bonding to prevent movement of the molecules, or as something that hurts if you kick it.

11.   The lines on a phase diagram represent the conditions under which two states are in equilibrium together. The triple point, where the lines meet, represents the condition of temperature and pressure at which all three states can coexist in equilibrium together.

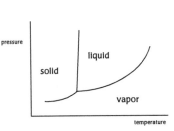

There is a rule, called Gibbs' Phase Rule, that relates the number of phases P, the number of components in the mixture C, and how the pressure and temperature can vary F:

$$F = C - P + 2$$

For a pure substance, $C = 1$:

$$F = 3 - P$$

For one phase (an area), $P = 1$ and therefore $F = 2$ and P and T can vary independently.

12.   The temperatures at which the line at 1 atm crosses the solid-liquid line and the liquid-vapor line (the vapor pressure curve) correspond to the normal melting point and the normal boiling point of the substance. The word normal signifies 1 atm pressure.

For two phases (on a line), $P = 2$ and therefore $F = 1$ and P can change only if T does and *vice versa*.

13.   If the line slopes to the right (as it does for almost all substances), then it means that at a particular temperature, the solid state is more dense than the liquid state. This means that an increase in pressure will cause the liquid to freeze. If the line slopes to the left (as it does for water, bismuth, and antimony), then at a particular temperature, the solid is less dense than the liquid. The solid will float. An increase in pressure at a melting point will cause the solid to melt.

For three phases (at the triple point), $P = 3$, and $F = 0$ and neither P nor T can change.

Study Questions and Problems

1.   Describe the interparticle forces at work in the following:

    a. within a water molecule H_2O
    b. in a crystal of the salt NaCl
    c. in a solution of potassium nitrate KNO_3
    d. in diamond
    e. in a fiber of nylon
    f. in liquid butane
    g. between water molecules in ice
    h. between the two strands in the double helix of DNA
    i. in paraffin wax
    j. between the molecules of carbon dioxide CO_2 in dry ice.
    k. between the molecules of HCl in liquid HCl.
    l. in tungsten metal
    m. in a solution of perchloric acid

Our ideas must be as broad as Nature if they are to interpret Nature.

Sir Arthur Conan Doyle
(1859-1930)

2. Which one of the following pairs of molecules would you expect to have the higher melting point?
   a. Cl_2 or Br_2
   b. C_4H_{10} or C_5H_{12}
   c. NH_3 or PH_3
   d. Na or Mg
   e. BeO or KCl
   f. ICl or Br_2

3. Which states or types of matter would be characterized by each of the following statements?

   a. High individual molecular speeds.
   b. A melting point spread over a wide temperature range.
   c. A regular repeating array of structural units.
   d. Molecules move with respect to one another but are held together in a condensed state.
   e. Molecules close together but having sufficiently high kinetic energies to overcome the intermolecular forces.
   f. Valence electrons delocalized over huge arrays of atoms.
   g. Totally random molecular order with comparatively great distances between individual molecules.
   h. A three-dimensional network of covalent bonds.

4. Acetone and chloroform form an unusually strong intermolecular bond. Why is this? Draw a picture of how the molecules attract each other.

5. a. How much heat is required to melt 15 grams of ice at 0°C?
   b. How much heat is released when 100 grams of steam condenses at 100°C?
   c. If a system of ice and water has a mass of 12 grams, and it is converted completely to water at 0.0°C by supplying 1.33 kJ of heat, how much water was initially present?

   Heat of fusion of ice = 333 J/g
   Heat of vaporization of water = 2250 J/g

6. Silver crystallizes in a face-centered cubic lattice. If the edge of the cube is 407 pm in length, what is the radius of a silver metal atom?

7. From the data provided in question 6, calculate the density of silver metal.

8. What unit cell would you choose for the two–dimensional space lattice shown on the right?

   How would you write the stoichiometry this material?

9.   Draw a unit cell of a body-centered cubic lattice. Show that the number of structural units within the unit cell equals 2.

10.   Examine the unit cell of the ionic crystalline solid $A_aB_bC_c$ illustrated below. The AB_6 unit is a polyatomic ion.

   a.   How many A ions are there in the unit cell?
   b.   How many B ions are in the unit cell?
   c.   How many C ions are in the unit cell?
   d.   How is the lattice structure of the AB_6 ions described?
   e.   How is the lattice structure of the C ions described?
   f.   What is the stoichiometry of the salt?

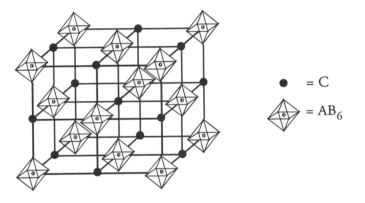

• = C

= AB_6

11.   Examine the data in Table 13.4 on page 601 of the text. Convert the boiling points from °C to K, and then divide the enthalpy of vaporization by the boiling point in K. Order the results and see if you can detect any correlation between the magnitude of the results and the intermolecular forces of attraction.

   Compare the boiling points in K with the magnitude of the van der Waals constants a listed in Table 12.3 on page 572 of the text. The easiest way to detect any correlation between the two is to plot the value of a against the boiling point. Does the existence of some correlation between the two quantities surprise you? The values are listed below with those for some additional substances:

van der Waals a		b. pt. °C		van der Waals a		b. pt. °C
He	0.034	–269		Ar	1.34	–186
H_2	0.244	–253		N_2	1.39	–196
O_2	1.36	–183		Cl_2	6.49	–34
CO	1.49	–192		CH_4	2.253	–162
NH_3	4.17	–33		Ne	0.21	–246
Xe	4.19	–108		C_2H_6	5.49	–89
SO_2	6.71	–10		HCl	3.67	–85
H_2S	4.43	–61		C_2H_2	4.39	–84
HBr	4.45	–66		Kr	2.32	–107

Notice that water is not included!

Answers to Study Questions and Problems

1. a. in a water molecule   covalent bonding
    b. in a crystal of NaCl   ionic bonding
    c. in a solution of KNO_3 ion–molecule interparticle bonding
    d. in diamond   covalent bonding (covalent network)
    e. in a fiber of nylon   covalent bonding within the polymer; inter molecular hydrogen bonding and dispersion forces between the polymer chains
    f. in liquid butane   dispersion forces
    g. in ice   hydrogen bonding
    h. between DNA strands   hydrogen bonding
    i. in paraffin wax   dispersion forces
    j. in dry ice   dispersion forces
    k. in liquid HCl   dipole–dipole attraction and dispersion forces
    l. in tungsten metal   metallic bonding
    m. in perchloric acid   strong ion–molecule interaction between the hydrogen and perchlorate ions and the solvent water molecules

There are K^+ and NO_3^- ions in solution.

2. a. Br_2 greater molar mass; greater dispersion forces
    b. C_5H_{12} greater molar mass; greater dispersion forces
    c. NH_3 hydrogen bonding
    d. Mg   stronger metallic bonding (2 valence electrons *vs* 1)
    e. BeO   charges on the ions greater, higher lattice energy
    f. ICl   polar; molar mass approximately the same as Br_2

It is interesting that ICl was discovered before Br_2. As a result, the first samples of the red-brown Br_2 were mistaken for ICl. The molar masses are almost the same:
Br2   159.82 g/mol mpt −7°C
ICl   162.35 g/mol mpt 27°C

3. a. a gas
    b. an amorphous material (a glass or polymer)
    c. a crystalline solid
    d. a liquid
    e. a supercritical fluid
    f. a metallic solid
    g. a gas
    h. a crystalline covalent network solid like diamond or graphite

4. The chlorine atoms on the carbon of the chloroform $CHCl_3$ are electronegative— they pull the electrons away from the carbon atom, and from the hydrogen attached to the carbon. Effectively, the electronegativity of the carbon is increased. As a result, hydrogen bonding is possible between the H of the $CHCl_3$ and the O of the $(CH_3)_2C=O$.

As a rule, the hydrogen atoms attached to carbon do not participate in hydrogen bonding.

5. a. Heat required to melt 15 grams of ice
        = 333 J/g × 15 g = 5 kJ
    b. Heat released when 100 grams of steam condenses
        = 2250 J/g × 100 g = 225 kJ

        This is why steam can cause severe burns.
    c. Since 4 g × 333 J/g = 1.33 kJ, this heat is sufficient to melt 4 grams
        of ice. So the quantity of water originally present = 8 grams.

6. If the lattice is face-centered cubic, then the
    distance across the diagonal of one face is
    equal to four times the radius of one silver
    atom. If the edge of the cube is 407 pm in
    length, the diagonal = $\sqrt{(407^2+407^2)}$ = 575.6 pm.
    The radius of one atom, therefore is 144 pm.

    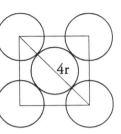

7. The volume of one unit cell = 407^3 = 6.742×10^7 pm^3.
    This equals 6.742×10^{-23} cm^3.

    One pm = 10^{-10} cm

    One pm^3 = 10^{-30} cm^3

    There are four atoms in a fcc unit cell, each having a mass equal to
    107.8682 g mol^{-1}/ 6.02214×10^{23} mol^{-1} = 1.791×10^{-22} grams.
    The density is therefore $4 \times 1.791 \times 10^{-22}$ / 6.742×10^{-23} g cm^{-3}.
        = 10.6 g cm^{-3}.

    The actual density of silver is 10.49 g cm^{-3} at 20°C.

8. The unit cell is outlined in black at the top.
    There are other alternatives shown lower on
    the pattern.

    The stoichiometry of this material is ⊞·

    enlarged

    The unit cell is the smallest repeating unit that has the same symmetry as the crystal. Lattice points should be at the corners as far as possible. The diagonal square does not match the overall sysmmetry as well.

9. In the unit cell of a body-centered cubic lattice, there are structural units
    on each corner and there are 8 corners. Each contributes 1/8 of itself to
    that particular unit cell, the remainder belongs to adjacent unit cells in
    the lattice. There is also a structural unit at the center of the cell that
    belongs entirely to that unit cell. The total number of structural units is
    therefore 8 × 1/8 + 1 = 2.

    See Figure 13.28 on page 613 for pictures of the unit cells of the three cubic lattices.

10. a. The lattice array of the AB$_6$ octahedral ions is face-centered cubic.
        There are ions on each corner (contribution 8 × 1/8) and one in the
        center of each face (contribution 6 × 1/2). The total number of AB$_6$
        octahedra is therefore 4. So there are 4 A in one unit cell.

        You can count the B individually and get the same answer:

    b. There must be 6 times as many B as there are A, so there are 24 B per
        unit cell.

        At each of 8 corners, there is
        1 on a face = 8 × 1/2 = 4
        1 on an edge = 8 × 1/4 = 2
        total = 4 + 2 = 6.
    c. There are C ions halfway along each edge, and there are twelve edges.
        Each C ion on an edge contributes 1/4 to that unit cell, so the total
        contribution form the C ions on the edges is 3. There is also an C ion
        in the center of the unit cell—contribution 1. So the total number of
        C ions in the unit cell = 4.

        On each of 6 faces, there is
        1 inside the cell = 6 × 1 = 6
        4 on the face = 4 × 6 × 1/2 = 12
        total = 18.

        Total number of B = 24.

d. The lattice structure of the AB_6 ions is face-centered cubic.

e. The lattice structure of the C ions is also face-centered cubic.

f. The stoichiometry is CAB_6.

11. Results of the division of ΔH_{vap} by the boiling point in K are listed in the margin. Notice that the value is most often in the range of 70 to 90. The substances with values below 70 are those substances with extremely weak intermolecular forces (helium and hydrogen). The substances with values at the top of the scale are those substances with strong intermolecular hydrogen bonds (ammonia and water). The consistency of the result was first discovered by Trouton and is known as Trouton's Rule. The magnitude of the result is directly related to the disorder present in the liquid state—which in turn is related to how well the molecules are held together.

As the value of van der Waals constant a increases, so does the boiling point. A plot of one against the other indicates a linear relationship. Both quantities are a measure of the strength of the intermolecular attraction and it is not surprising that they correlate reasonably well.

Helium	19
Hydrogen	44
Neon	63
Nitrogen	72
Argon	73
Methane	73
Oxygen	75
Xenon	76
Fluorine	78
Ethane	80
Propane	82
Butane	82
Hydrogen Iodide	83
Chlorine	85
Hydrogen fluoride	86
Bromine	90
Hydrogen bromide	93
Sulfur dioxide	95
Ammonia	97
Water	109

Challenge 7

A question in the Survey of Natural Sciences section on a recent sample Admission Test read as follows:

> Potassium metal crystallizes in a body-centered cubic lattice. It has a density of 0.86 g/cc. What would the density be if potassium metal crystallized in a simple cubic lattice instead?
>
> a. 0.43   b. 0.57   c. 0.69   d. 0.76
>
> The correct answer was listed as a. (0.43 g/cc).

a. How was this "correct answer" obtained?

b. What incorrect assumption was made in determining this answer.

c. What is the correct answer?

CHAPTER 14

Introduction

Solutions are homogeneous mixtures of two or more substances; they are everywhere. The air we breathe, the water we drink, and all sorts of materials encountered every day, are solutions. Chemists perform many reactions in solution because, in the solution state, molecules are free to move about and react. So in this chapter we will look further into the solution state, investigate the solution process itself, and examine the properties of solutions called colligative properties. At the end of the chapter we will look at colloids and surfactants.

Contents

14.1   Units of Concentration

There are many ways to express the concentration of the various components in a solution. Each has its own merits. Molarity is a concentration expressed as the number of moles of solute per liter of solution and is particularly useful in stoichiometric calculations for reactions occurring in solution.

The word "component" refers to a solute or the solvent in the solution. Very often there is more than one solute. Note that components are (pure) substances.

Molarity does not explicitly state the quantity of solvent in the solution. There are other concentrations units that do: molality, mole fraction, mass percent, and parts per million or billion:

Recall that molarity is defined as the number of moles per liter of solution.

$$\text{Molality} = \frac{\text{moles of solute}}{\text{kg of solvent}}$$

$$\text{Mole fraction} = \frac{\text{moles of one component}}{\text{total moles of the solution}}$$

$$\text{Mass percent} = \frac{\text{mass of one component}}{\text{total mass of all components in the solution}} \times 100$$

$$\text{Parts per million} = \frac{\text{mass of one component in mg}}{\text{total mass of the solution in kg}}$$

Do not confuse molality with molarity.

Mass percent is often called weight percent.

The concentration ppm is often expressed as mg/L (approximately the same as mg/kg). The concentration ppb is expressed as μg/kg or μg/L. Although solutions are often thought of as being liquids, remember that solutions can be homogeneous mixtures in any one of the three states (solid, liquid, or gas).

Be able to calculate concentrations using these units and be able to convert between them.

14.2   The Solution Process

When one substance dissolves in another, any intermolecular bonds in the two substances have to be at least partially broken, and new intermolecular bonds between the two substances are made. For example, when sodium chloride salt dissolves in water, the salt crystal is broken up, the intermolecular bonds in the water are partially broken, and the ions are hydrated by the water (ion–polar molecule attraction).

There is a limit to the amount of sodium chloride that will dissolve. Before the limit is reached, the solution is said to be **unsaturated**. At the limit, it is **saturated**, and beyond the limit(!) it is **supersaturated**.

When the two components of a solution are both liquids, and they can be mixed in any proportion, they are said to be **miscible**. If they do not dissolve in each other, they are called **immiscible**. Often two liquids are partially miscible. A common rule that indicates the likely solubility of one substance in another is "like dissolves like". A polar molecular is likely to be soluble in a polar solvent, a nonpolar molecule in a nonpolar solvent. The reason for this is that the character of the bonds in each of the components and those between the components should be similar if the solution process is going to be product–favored.

Whether energy is liberated or absorbed during the solution process depends upon the relative strengths of the bonds broken and formed in the process. The bonds broken are those in the solute (for example, the lattice energy of a NaCl crystal) and in the solvent (for example, some hydrogen bonds in water). The bonds made are those between the solute and the solvent (for example, in the hydration of the sodium and chloride ions). If the disparity in the bond strengths is sufficiently large, the salt may not dissolve even though the process would lead to an increase in disorder. For example, barium sulfate and silver chloride are insoluble. Heats of solution can be measured in a calorimeter.

Temperature affects the solubility of most substances. If the solute is a gas, its solubility is also affected by pressure.

The effect of pressure on the solubility of a gas is described by Henry's law. This law states that the solubility of a gas is directly proportional to the partial pressure of the gas above the solution:

$$S_g = k_H P_g$$

The constant k_H is a characteristic of both gas and solvent. The reason why helium–oxygen mixtures are used in preference to nitrogen–oxygen mixtures for deep water diving is the lower solubility of helium in blood.

The solubility of a gas in water decreases as the temperature is raised. The solution process is a dynamic process—gas molecules are leaving and reentering the solution all the time. If allowed, the process will reach equilibrium when the two rates are equal and the concentration reaches a steady value. Because no intermolecular bonds between the gas molecules need to be broken in a solution process, and solute–solvent bonds are made, processes involving the solution of a gas in water are exothermic:

Gas   +   liquid solvent $\rightleftharpoons$ saturated solution of the gas   +   HEAT

Unsaturated:
concentration < solubility

Saturated:
concentration = solubility

Supersaturated:
concentration > solubility

The supersaturated state is thermodynamically unstable, and given half a chance, a supersaturated solution will precipitate the excess solute with the evolution of considerable energy.

The product in this case is the solution.

There two driving forces for any process:
The first is a decrease in the enthalpy of the system.
The second is an increase in the disorder of the system.

The thermodynamic term for disorder is entropy. It is the increase in disorder, or entropy, that drives many solution processes.

In Chapter 20 you will learn that nothing happens unless the entropy of the universe increases.

Henry's law is valid providing that the solvent and solute do not react.

See also the comments on scuba diving on page 570 in Chapter 12 on gases.

There are no intermolecular bonds in a gas.

According to LeChatelier's Principle, addition of heat (on the product side), drives the reaction to the left so that the heat is absorbed by the system. In other words, if a solution of a gas (e.g. a carbonated drink) is heated, the gas will bubble out of the solution. The effect of temperature on the solubility of solids is less easy to predict.

14.3   Colligative Properties

Solutions have properties different from those of the pure solvent. The freezing point, boiling point, vapor pressure, osmotic pressure are all different for the solution. These properties of the solution, called **colligative properties**, depend upon the relative numbers of solute and solvent particles in the solution.

The equilibrium vapor pressure is the partial pressure of the vapor above a liquid when the system is in equilibrium. Addition of a solute to the liquid results in fewer solvent molecules being able to escape. The disorder in the solution state is increased and the rate at which solvent molecules escape into the vapor state is reduced. As a result the equilibrium vapor pressure is less for the solution.

Raoult's law states that the equilibrium vapor pressure of a solution depends upon two factors: one is the volatility of the solvent (usually expressed as the vapor pressure of the pure solvent $P^\circ_{solvent}$), and the second is the concentration of solvent molecules in the solution (usually expressed as the mole fraction $X_{solvent}$):

$$P_{solvent} = X_{solvent} P^\circ_{solvent}$$

Raoult's law assumes that the solution is an **ideal solution**. An ideal solution is one in which the forces of intermolecular attraction in the two components (solute and solvent) are very similar to one another and to the forces of attraction in the solution. Ideal solutions are therefore ones for which the enthalpy of solution is zero. They occur when solute and solvent are very similar—for example, hexane and heptane.

If the forces of attraction differ in the two components and the solution, the solution is said to deviate from ideal behavior, or to deviate from Raoult's law.

A lowering of the vapor pressure causes an increase in the boiling point. A higher temperature is required to compensate for the lower vapor pressure caused by the presence of the solute particles. The **elevation of the boiling point** ΔT_b is given by the relationship:

$$\Delta T_b = k_b \times m$$

where m is the molality of the solute and k_b is a constant characteristic of the solvent.

The freezing point of a solvent is depressed when a solute is added. This is why salt is spread on an icy road, or ethylene glycol is added to radiator fluid. The solid–liquid interface is altered just as it was for the vaporization process. The rate at which solvent molecules move between the solution and the solid phases is reduced. The **depression of the freezing point** ΔT_f is given by a similar relationship:

$$\Delta T_f = k_f \times m$$

where m is the molality of the solute and k_f is a constant characteristic of the solvent.

LeChatelier's Principle will be discussed again in the chapter on chemical equilibria (Chapter 16).

The principle states that: A system in equilibrium will adjust so as to reduce or accommodate any stress placed on the system.

For example, in this case, the system will adjust to absorb the heat added to the system; the equilibrium will shift to the left.

Recall that at equilibrium, the rate at which molecules leave the liquid phase equals the rate at which molecules reenter the liquid phase; it is a dynamic equilibrium. At equilibrium the partial pressure of the vapor is constant.

An ideal solution is often said to be a solution that obeys Raoult's law.

Dilute solutions generally obey Raoult's law reasonably well.

Positive deviation, when the vapor pressure is higher than Raoult's law would predict, occurs when $\Delta H_{solution}$ is positive. Negative deviation occurs when $\Delta H_{solution}$ is negative.

All colligative properties are the result of the increased entropy (disorder) of the liquid phase. There is no solute in the solid and vapor phases (assuming the solute is nonvolatile). This increase in disorder means that (compared to the pure solvent) the change in entropy during vaporization is less and the change in entropy during freezing is greater. This will be examined further in Chapter 20.

It is possible to determine the molar mass of an unknown solute by measuring the depression in the freezing point of its solution. The constant k_f is established using a known solute and then the experiment is repeated using the unknown solute.

A colligative property depends upon the relative number of solute particles in the solution. It doesn't matter whether the particles are molecules or ions. If a solute breaks up to form ions, it has a greater effect on the freezing point depression or boiling point elevation than a solute that doesn't break up. This effect was discovered by Raoult, studied by van't Hoff and explained by Arrhenius. Arrhenius proposed that salts like sodium chloride dissociate in solution to form two solutes, one sodium ions and the other chloride ions. Thus the effect is twice as great. It is not the identity, but the relative number, of solute particles that matters.

The **van't Hoff factor**, i, represents the extent to which solutes break up to form ions. In dilute solution, NaCl, for example, breaks up to form two ions per formula unit, and therefore $i = 2$. The value of i, which can be determined by experiment, indicates the extent to which ionization occurs. In all but very dilute solution, there is some association between ions, and i is less than the ideal integer. The earlier equations can be modified to include the van't Hoff factor:

$$\Delta T_b = k_b \times m \times i$$
$$\Delta T_f = k_f \times m \times i$$

Osmosis is the movement of solvent molecules through what is called a semipermeable membrane. A semipermeable membrane allows solvent molecules to pass through but not the solute particles. The water molecules tend to move from the side of low solute concentration to the side of high solute concentration. In other words, the system attempts to equalize the concentrations. The interface at the membrane is just the same as the solid–liquid and liquid–vapor interfaces already described. The rate of movement of the solvent molecules from the side of high concentration to the side of low concentration is reduced.

The osmosis continues until the concentrations are equal, or until pressure is applied to the high concentration side to equalize the rates of movement of solvent molecules through the membrane. The pressure required to equalize the rates from one side to the other is called the osmotic pressure Π. The osmotic pressure is related to the concentration of solute by a equation similar to the ideal gas law:

$$\Pi V = nRT \qquad \text{or} \qquad \Pi = cRT \quad \text{where } c = \frac{n}{V}$$

Large osmotic pressures result from relatively small molar concentrations and measurement of the osmotic pressure of a solution is an effective method of determining the molar mass of an unknown solute—particularly for solutes having high molar masses and low solubilities (for example, proteins and polymers).

The values of k_f and k_b are now known for most common solvents. See Table 14.3 on page 661 of the text.

Recall that the depression of the freezing point when a solute is added is a result of the increased disorder in the solution. A solute that breaks up to form two particles in solution creates twice as much disorder.

k_f is usually given as a negative number because the freezing point decreases.
The units of k_f and k_b are Km^{-1}.

Solutions having the same osmotic pressure are said to be isotonic.

Intravenous drips are made isotonic with the patient's blood to prevent rupture of cells.

14.4　Colloids

A solution is a homogeneous mixture; the mixing of solute and solvent particles occurs at the molecular or particulate level. If the solute particle is large in size (about 1 μm) or large in mass (molar mass in the thousands), but not so large that the particles settle out or precipitate, a **colloidal dispersion** forms.

A colloidal dispersion is not a solution, nor is it a suspension; it is somewhere between. The interesting properties of colloids are due to the relatively large surface areas of the particles. A **hydrophobic** colloid is one in which there are weak forces of attraction between the colloid particle and water. Hydrophobic colloidal particles carry charges due to ions absorbed on the surface. The repulsion between the like charges prevents the coagulation of the particles and the formation of a suspension and precipitation. A **hydrophilic** colloid exhibits a strong attraction for water molecules. This is often due to hydrogen bonding. Examples are colloids of proteins and starch.

Emulsions are colloidal dispersions of one liquid in another—for example, milk and mayonnaise. Mayonnaise is a dispersion of oil in water. Lecithin (a protein in egg yolk) is added to stabilize the emulsion, or the oil and water would separate. Lecithin is a **surfactant** or emulsifying agent.

A surfactant is a surface active agent. **Soaps** and **detergents** are examples. A soap is the sodium or potassium salt of a long chain fatty acid (*cf.* Section 11.6). This salt has a polar hydrophilic end and a long chain nonpolar hydrophobic end. It is able to bridge the interface between oil and water. It therefore facilitates the emulsion of the oil and water, enabling the removal of the oil stain. Magnesium and calcium salts of the fatty acids are insoluble and create soap scum. To avoid the problem, synthetic detergents use sulfonate instead of carboxylate salts (the calcium and magnesium sulfonate salts are more soluble). An example of a detergent is sodium lauryl benzenesulfonate (which is biodegradable):

$$CH_3-CH_2-CH_2-CH_2-CH_2-CH_2-CH_2-CH_2-CH_2-CH_2-CH_2-CH_2-\langle \bigcirc \rangle-SO_3^-$$
$$Na^+$$

As the size of the solute particle increases:

solution
↓
colloid
↓
suspension

Colloids have the ability to scatter light (like dust in a sunbeam). This is called the Tyndall effect—see Figure 14.18 on page 673 of the text.

A colloid, like a solution, has two components: the dispersing medium (the "solvent") and the dispersed phase (the "solute").

sol: a colloidal dispersion of a solid in a fluid.
gel: a colloidal dispersion of a liquid in a solid.
aerosol: a colloidal dispersion of a liquid or solid in a gas.
emulsion: a colloidal dispersion of one liquid in another liquid.
foam: a colloidal dispersion of a gas in a liquid or solid.

Review Questions

1. Define solute, solvent, and solution.

2. Define unsaturated, saturated, and supersaturated. How would you make a supersaturated solution?

3. Define the concentration units molarity, molality, mole fraction, mass percent, and ppm.

4. Define the terms miscible and immiscible.

5. What two factors are largely responsible for determining the magnitude and sign of the enthalpy of solution?

6. Given that a decrease in energy favors a reaction, how is it that some solution processes are endothermic? Why does the solution process happen in these cases?

7. Describe how temperature and pressure affect the solubility of a gas.

8. What is Henry's law?

9. Why is ammonia much more soluble than oxygen in water ?

10. Describe Henri LeChatelier's principle.

11. What is Raoult's law?

12. What is an ideal solution?

13. What is a colligative property?

14. Describe how the solid–liquid and liquid–vapor phase interfaces are affected in the same way by the addition of a solute. How do these interfaces compare with the interface across a semipermeable membrane in osmosis?

15. Why does the freezing point *decrease*, and the boiling point *increase*, upon addition of a solute?

16. Explain what the van't Hoff factor *i* is. What does it mean if *i* is less than the ideal integer value for a salt?

17. Describe the terms isotonic, hypotonic, and hypertonic.

18. What is a colloidal dispersion?

19. Soaps and detergents are both surfactants; what is the difference between them?

Answers to Review Questions

1. Solute: the substance dissolved in the solvent.
   Solvent: the component that determines the state of the solution.
   Solution: the homogeneous mixture of solute and solvent.

2. Unsaturated: concentration less than the solubility, more solute can be dissolved.
   Saturated: concentration = solubility at that temperature.
   Supersaturated: concentration exceeds the solubility; this solution is thermodynamically unstable.

A saturated solution is made at a high temperature where the solubility is higher. The solution is then allowed to cool slowly to a temperature where the solubility is lower. If the excess solute does not precipitate out, the solution becomes supersaturated.

The fact that the excess solute does not precipitate out is a kinetic, not a thermodynamic, effect.

3. Molarity (M) = moles of solute / liter of solution
    Molality (m) = moles of solute / kg of solvent
    Mole fraction (X) = moles of solute / total moles in solution
    Mass percent (%) = (mass of solute / mass of solution) × 100
    ppm = mass of solute in mg / mass of solution in kg

Because volume changes with temperature, molarity is temperature dependent also.

4. Miscible means that the two liquids making up the solution can mix (usually in all proportions).
    Immiscible means that two liquids do not mix (for example, oil and water are immiscible).

5. As in any reaction, the energy released or absorbed is a measure of the strength of the bonds that have to be broken, and the strength of the bonds that are made. In the solution process, there are bonds in the solute and bonds in the solvent that are broken. Bonds are formed between the two components.
    For example, when sodium chloride dissolves in water, the lattice energy of the NaCl crystal must be supplied. In addition, some hydrogen bonds in the water are broken. The ions are solvated by the water; strong ion–molecule bonds are formed. The energy required or released is the difference.

6. The predominant driving force for many solution processes is the increase in the disorder (entropy) of the system. This often overcomes any unfavorable enthalpy change.

There two driving forces for any process:
The first is a decrease in the enthalpy of the system.
The second is an increase in the disorder of the system.

7. The solubility of a gas in water decreases as the temperature is raised. Because no intermolecular bonds between the gas molecules need to be broken in a solution process, and solute–solvent bonds are made, processes involving the solution of a gas in water are exothermic. According to LeChatelier's Principle, addition of heat drives the reaction from the product (solution) side to the reactant side to absorb the heat. In other words, if a solution of a gas is heated, gas will bubble out of the solution. The effect of pressure on the solubility of a gas is described by Henry's law. This law states that the solubility of a gas is directly proportional to the partial pressure of the gas above the solution: $S_g = k_H P_g$. The constant k_H is a characteristic of both gas and solvent.

8. Henry's law describes how the solubility of a gas depends upon the partial pressure of the gas above the liquid: $S_g = k_H P_g$.

Carbon dioxide and sulfur dioxide are two other gases that are very soluble in water because of the interaction between these molecules and the water molecules. Carbon dioxide forms carbonic acid H_2CO_3 to some extent and sulfur dioxide forms sulfurous acid H_2SO_3 to some extent.

9. Ammonia interacts with the water to a much greater extent than oxygen does. Hydrogen bonding between the ammonia molecules and the water molecules leads to an exceptionally high solubility. A solution of ammonia in water is sometimes referred to as ammonium hydroxide although the ammonia is present predominantly as ammonia molecules. The interaction between oxygen and water is only a dipole–induced dipole attraction and its solubility is much less.

10. Henri LeChatelier's principle states that any system in equilibrium will adjust so as to reduce or accommodate any stress placed on the system.

11. Raoult's law states that the equilibrium vapor pressure of a solution depends upon two factors: one is the volatility of the solvent (expressed as the vapor pressure of the pure solvent $P^\circ_{solvent}$), and the second is the concentration of solvent molecules in the solution (expressed as the mole fraction $X_{solvent}$):

$$P_{solvent} = X_{solvent}P^\circ_{solvent}$$

Raoult's law assumes that the solution is an ideal solution.

12. An ideal solution is a solution that obeys Raoult's law! It is a solution in which the forces of intermolecular attraction in the two components (solute and solvent) are very similar to one another and to the forces of attraction in the solution. Ideal solutions are therefore ones for which the enthalpy of solution is zero. They occur when solute and solvent are very similar—for example, hexane and heptane, benzene and toluene, or pentanol and hexanol. If the forces of attraction differ in the two components and the solution, the solution is said to deviate from ideal behavior, or to deviate from Raoult's law.

13. A colligative property is a property that depends only upon the relative numbers of solute and solvent particles in the solution and not their identity.

There are of course many instances where the solute is volatile.

14. When a solute is dissolved in a solvent to make a solution, in most cases the solute exists only in solution. The solute is nonvolatile. The vapor state and the solid state contain only solvent molecules. It is the increase in the disorder of the liquid state that affects the colligative properties of the solution. In both the vaporization process and the freezing process, the movement of solvent molecules from the liquid phase to the solid or vapor phases is inhibited by the disorder created by the solute particles in solution. In terms of probability, there is a lower probability of the solvent particles being at the surface. The situation is the same across the semipermeable membrane in osmosis.

15. Freezing is a process in the direction of increased order; vaporization (boiling) is a process in the direction of increased disorder. The directions are opposite. A lower temperature is required to create order from a more disordered solution (compared to the pure solvent). A higher temperature is required to create the disordered vapor from a more disordered solution.

From a molecular point of view, the solute particles in the solution interfere with the freezing or boiling process. At equilibrium, at the phase interface, the numbers of solvent molecules travelling in both directions must be equal. Because of the presence of the solute particles in the liquid phase, the temperature must be lower at the freezing point, and must be higher at the boiling point, to achieve the equal rates.

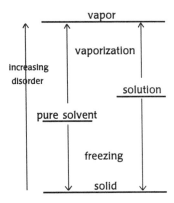

cf. Chapter 20:

If $T\Delta S$ is constant, and ΔS is greater, then T (freezing point) must be lower.
If $T\Delta S$ is constant and ΔS is larger, then T (boiling point) must be higher.

16. The van't Hoff factor i is an indication of how many particles are produced when a solute is dissolved. If the solute breaks up to form individual ions, then the van't Hoff factor i is greater than 1. For example, sodium chloride NaCl produces two ions (Na^+ and Cl^-) in solution and therefore i might be expected to be 2. In a very dilute solutions it is. In more concentrated solutions, the oppositely charged Na^+ and Cl^- ions associate to form ion-pairs so that i is less than 2. In very concentrated solutions, the degree of association between ions is high and i is correspondingly lower.

Values of i are, for example:

Concentration	NaCl	$MgSO_4$
0.10 m	1.87	1.21
0.01 m	1.94	1.53
0.001 m	1.97	1.82
infinitely dilute	2.00	2.00

Notice the greater attraction between the doubly charged ions.

17. Isotonic: equal osmotic pressures (equal solute concentrations).
Hypotonic: the osmotic pressure is less (than a standard).
Hypertonic: the osmotic pressure is greater (than a standard).

18. A colloidal dispersion is not a solution, nor is it a suspension; it is somewhere between. If the solute particle is large in size (about 1 μm or larger) or large in mass (molar mass in the thousands), but not so large that the particles settle out, a colloidal dispersion is produced.

19. A soap is a sodium or potassium salt of a long chain carboxylic acid (a fatty acid). A detergent is a sodium or potassium salt of a long chain sulfonic acid. The magnesium and calcium salts of the sulfonic acids are more soluble than those of the carboxylic acids and so the detergents do not form insoluble soap scums like the soaps do.

Study Questions and Problems

Nothing is interesting if you're not interested.

Helen Clark MacInnes
(1907-1984)

1. A solution of a salt (molar mass 90 g mol⁻¹) in water has a density of 1.29 g/mL. The concentration of the salt is 35% by mass.

   a. Calculate the molality of the solution.
   b. Calculate the molarity of the solution.
   c. Calculate the total number of moles in the solution.
   d. Calculate the mole fraction of the salt in the solution.

2.　Ethylene glycol ($C_2H_4(OH)_2$; 150 grams) is added to ethanol (C_2H_5OH; 250 grams).

    a.　Calculate the mass % of ethylene glycol in the solution.
    b.　Calculate the molality of ethylene glycol in the solution.
    c.　Calculate the mole fraction of ethylene glycol in the solution.

3.　Concentrated sulfuric acid contains very little water, only 5.0% by mass. It has a density of 1.84 g/mL. What is the molarity of this acid?

4.　The lattice energy of a salt is 350 kJ/mol and the solvation energies of its ions add up to 320 kJ/mol for the preparation of a 0.50 M solution. In the preparation of this solution would the solution get colder or warmer? What is the driving force for this solution process?

5.　Addition of excess sodium nitrate to water to form a saturated solution results in the following equilibrium. The solution process is endothermic.

$$NaNO_3(s) \quad \rightleftharpoons \quad Na^+(aq) \ + \ NO_3^-(aq)$$

How could the concentration of sodium nitrate in the solution be increased?

    a.　add more $NaNO_3(s)$
    b.　increase the pressure on the solution
    c.　increase the temperature
    d.　stir the solution more vigorously

6.　The value of Henry's law constant k_H for oxygen in water at 25°C is 1.66×10^{-6} M/torr.

    a.　Calculate the solubility of oxygen in water at 25°C when the total external pressure is 1 atm and the mole fraction of oxygen in the air is 0.20.
    b.　Calculate the solubility at the same temperature with the same atmospheric composition but at an increased pressure of 2 atm.
    c.　What would happen to the solubility of the oxygen gas if the temperature was increased?

7.　Ethanol and methanol form an almost ideal solution. If 64 g of methanol is mixed with 69 g of ethanol, what is the total vapor pressure above the solution?
The vapor pressure of pure methanol at this temperature = 90 torr.
The vapor pressure of pure ethanol at the same temperature = 45 torr.

8.　A 3.0 molal solution of naphthalene in cyclohexane boils at 89.4°C. What is the boiling point of pure cyclohexane? Although solid naphthalene is slightly volatile, assume its volatility is zero in this calculation. The constant k_b for cyclohexane is +2.80 Km^{-1}.

9. Which of the following solutions would you expect to have the lowest freezing point? Assume that the values of i are ideal.

    a. 0.010m NaCl
    b. 0.100m sugar
    c. 0.070m KNO_3
    d. 0.050m glycerol
    e. 0.060m $Ca(NO_3)_2$
    f. 0.075m KCl

10. In order to depress the freezing point of water to $-12°C$, how much magnesium nitrate would you have to add to 500 grams of water? Assume that the van't Hoff factor i is the ideal value.
    k_f for water $= -1.86$ Km^{-1}.

11. An unknown protein (350 mg) was dissolved in water to produce 10 ml of solution. The osmotic pressure of the solution at 25°C was determined to be 10.8 torr. What is the molar mass of the unknown protein?

12. Classify the following materials as foam, aerosol, emulsion, gel, or sol.

    a. homogenized milk
    b. red jelly
    c. cheese
    d. whipped cream
    e. atmospheric dust
    f. mayonnaise
    g. pumice stone
    h. fog

Answers to Study Questions and Problems

1. The concentration of the salt is 35% by mass; this means that in 100 g of solution there is 35 g of salt. So there must be 65 g of water.

    Number of moles of salt = 35 g / 90 g mol^{-1} = 0.389 mol
    Molality = moles of solute / kg of solvent = 0.389 mol / 0.065 kg water

    a. The solution is 6.0 m.

    100 g of solution = 100 g /1.29 g mL^{-1} = 77.52 mL solution
    Molarity = moles of solute / liter of solution = 0.389 mol / 0.07752 L

    b. The solution is 5.0 M.

    Number of moles of solute in 100 g of solution = 0.389 mol
    Number of moles of water = 65 g / 18 g mol^{-1} = 3.61 mol

    c. Total number of moles = 4.0 mol.

    d. Mole fraction of the salt = 0.389 / 4.0 = 0.10.

2. Mass of ethylene glycol = 150 grams
    Mass of ethanol = 250 grams
    a. Mass % of ethylene glycol = 150/400 × 100% = 37.5%

    b. Number of moles of ethylene glycol = 150 g / 62 g mol^{-1} = 2.42 mol.
        Molality = 2.42 mol / 0.250 kg solvent = 9.68 m.

    c. Mole fraction = 2.42 / (2.42 + 5.43) = 0.31.

    Moles of ethanol
    = 250 g / 46 g mol^{-1}
    = 5.43 mol

3. In 100g there are 95g of sulfuric acid.
The volume is 100g / 1.84 g mL^{-1} = 54.35 mL.
The number of moles is 95g / 98 g mol^{-1}= 0.969 mol.
Molarity = 0.969 mol / 0.05435 L = 17.8 M.

4. More energy is required (lattice energy 350 kJ/mol) than is released (sol-vation energies 320 kJ/mol). So the process is endothermic; energy is absorbed and the solution will get cold. The process is not driven by a decrease in enthalpy! As in many cases, the solution process is driven by the increase in disorder.

5. $$NaNO_3(s) \quad \rightleftharpoons \quad Na^+(aq) \ + \ NO_3^-(aq)$$

 a. adding more solid $NaNO_3$ has no effect, the solution is already satu-rated.

 b. changing the pressure on a solution of a solid in a liquid has virtually no effect on the position of equilibrium.

This is LeChatelier's principle; the system will adjust to accommodate any stress placed upon the system. Heat is a reactant in this solution process; it is an endothermic process.

 c. increasing the temperature will shift the equilibrium to the right; the concentration of the sodium and nitrate ions will increase.

 d. stirring the solution more vigorously may allow the equilibrium state to be reached more quickly (a kinetic effect) but the position reached will be the same.

6. The value of Henry's law constant k_H for oxygen in water at 25°C is 1.66×10^{-6} M/torr.

1 atm = 760 torr.

 a. $S_g = k_H P_g = 1.66 \times 10^{-6}$ M/torr $\times (0.20 \times 760)$ torr $= 2.52 \times 10^{-4}$ M.

 b. The solubility will be twice as high $= 5.05 \times 10^{-4}$ M.

 c. At a higher temperature the solubility of the oxygen gas will be less.

7. 64 g of methanol = 2.0 moles
69 g of ethanol = 1.5 moles

This is Raoult's law.

The total vapor pressure above the solution is the sum of the partial vapor pressures, and the partial vapor pressures depend upon the mole fractions of the components in the solution.

The vapor pressure due to methanol = (2.0/3.5) × 90 torr = 51.4 torr
The vapor pressure due to ethanol = (1.5/3.5) × 45 torr = 19.3 torr
Total vapor pressure = 51.4 + 19.3 torr = 70.7 torr.

8. The elevation of the boiling point = k_f × m × i. (i = 1 in this case)
$\Delta T = 2.80 \times 3.0 \times 1 = 8.40$°C.
The boiling point of pure cyclohexane = 89.4 – 8.40°C = 81.0°C.

9.　The solution with the lowest freezing point is the solution with the greatest concentration of solute particles. Calculate $m \times i$ for each solution:

　　a.　0.010m NaCl　　　　$m \times i = 0.010 \times 2 = 0.020$ molar
　　b.　0.100m sugar　　　　$m \times i = 0.100 \times 1 = 0.100$ molar
　　c.　0.070m KNO_3　　　$m \times i = 0.070 \times 2 = 0.140$ molar
　　d.　0.050m glycerol　　　$m \times i = 0.050 \times 1 = 0.050$ molar
　　e.　0.060m $Ca(NO_3)_2$　$m \times i = 0.060 \times 3 = 0.180$ molar — lowest f.pt.
　　f.　0.075m KCl　　　　 $m \times i = 0.075 \times 2 = 0.150$ molar

10.　$\Delta T = k_f \times m \times i$ where $\Delta T = -12°C$, $k_f = -1.86$, and $i = 3$ (ideally).

　　Molality $m = 12 / 1.86 \times 3 = 2.15$ moles / 1000 g water.

　　In 500 grams of water, you would dissolve 1.075 moles $Mg(NO_3)_2$.
　　= 159 grams $Mg(NO_3)_2$.

This solution is very concentrated and the value of *i* will be considerably less than 3 in reality. In other words, this concentration of magnesium nitrate would lead to less of a depression in the freezing point.

11.　$\Pi = cRT$ where c is the concentration of the solute in moles per liter.

　　The osmotic pressure = 10.8 / 760 atm = 0.0142 atm.
　　The temperature = 25 + 273.15 K = 298.15 K.
　　R = 0.082057 L atm K^{-1} mol^{-1}.
　　Therefore c = 0.0142 / (0.082057 × 298.15) = 5.808×10^{-4} M.

　　The number of moles in 10 mL = 5.808×10^{-6} moles.
　　The mass was 350 mg or 0.350 grams, so the molar mass is
　　0.350g / 5.808×10^{-6} moles = 60,300 g mol^{-1}.

A large molar mass can be determined by measurement of the osmotic pressure.

12.　a.　homogenized milk　liquid in liquid　emulsion
　　b.　red jelly　　　　　liquid in solid　　gel
　　c.　cheese　　　　　　liquid in solid　　gel
　　d.　whipped cream　　gas in liquid　　　foam
　　e.　atmospheric dust　solid in gas　　　aerosol
　　f.　mayonnaise　　　　liquid in liquid　emulsion
　　g.　pumice stone　　　gas in solid　　　foam
　　h.　fog　　　　　　　liquid in gas　　aerosol

Challenge 8

Suppose that you have 10 grams of ice and 100 grams of water at 0°C in a well-insulated container. You add to the container 40 grams of ammonium nitrate at 0°C and stir gently to dissolve. What happens? Does ice melt due to the depression of the freezing point? Or does water freeze due to the endothermic solution process?

$\Delta H_{solution}$ (NH$_4$NO$_3$) = +25.7 kJ mol^{-1}

heat of fusion of ice = +333 J g^{-1}

k_f for water = −1.86 Km^{-1}

molar mass of ammonium nitrate = 80.04 g mol^{-1}

specific heat capacity of water = 4.184 JK^{-1}g^{-1}; assume that the specific heat of the solution is the same.

specific heat of ice = 2.06 JK^{-1}g^{-1}

Assume that i, the van't Hoff factor, for the concentrated solution of ammonium nitrate is 1.6.

EXAMINATION 3

Introduction

This examination tests your knowledge and understanding of the chemistry in Chapters 12 through 14 of Kotz & Treichel—the states of matter. The questions are again formatted as true–false questions and multiple choice questions. It is essential to try the exam before looking at the answers provided at the end of this study guide.

True–false questions

1. There's no such thing as an ideal gas.

2. The only way to increase the pressure of a fixed volume of an ideal gas, without changing the amount of gas present, is to increase the temperature.

3. The partial pressure of a gas in a mixture of gases is the pressure the gas would exert under the same conditions if it were the only gas present.

4. The larger a molecule in the gas phase, the faster its speed at the same temperature.

5. There are exactly 100 kPa in one atmosphere.

6. The stronger the intermolecular forces, the more a gas deviates from ideal behavior.

7. If nitrogen gas and hydrogen gas react to form ammonia gas in a constant volume vessel, and the initial and final temperatures are the same, the pressure inside the vessel would increase.

8. The rms speed of a gas is the speed of a molecule possessing average kinetic energy.

9. It's possible to see the color of a gas only if it condenses to form a liquid.

10. The liquid and solid states are condensed states of matter.

11. The strongest intermolecular force is the force of attraction between ions and polar molecules.

12. Dispersion forces are not important if the molecules are polar.

13. Doubly–charged metal ions are more strongly solvated than singly–charged metal ions.

14. Hydrogen bonding cannot occur between ethanol molecules.

15. HF has the lowest boiling point of HF, HCl, HBr, and HI.

16. Most substances are less dense in the solid state than the liquid state.

17.   If the heat of vaporization for a compound is higher than for another compound, its boiling point will also be higher.

18.   The vapor pressure curve for a liquid starts at the triple point and ends at the critical point.

19.   Surface tension is due to the accumulation of impurities at the surface of the liquid.

20.   There are only seven crystal systems.

21.   There are 14 crystal lattices because there are 2 lattices for each system.

22.   There are 9 structural units within a body–centered unit cell; and 14 within a face–centered unit cell.

23.   Molality is a temperature-independent concentration unit.

24.   It is impossible for a solute concentration to exceed its solubility at that temperature.

25.   When gases dissolve in a solvent, heat is usually released.

26.   If a solution process has a positive ΔH, then the process does not occur, the solute does not dissolve.

27.   According to Raoult's law, the vapor pressure of the solvent over a solution must be lower than over the pure solvent.

28.   The freezing point of a solution is depressed, compared to that of the pure solvent, because the vapor pressure of the solution is reduced.

29.   The van't Hoff factor i is an integer only in extremely-dilute solutions.

30.   Immiscible liquids will mix in any proportion with one another to form a homogeneous solution.

Multiple choice questions

1.   The SI derived unit for pressure is the

    a. bar   c. torr   e. Pa
    b. psi   d. atm   f. $kg\ m^{-2}$

2.   A sample of ammonia gas at 10°C and 380 torr is contained in a 2.50 L vessel. How many moles of the ammonia gas are there in the vessel?

    a. 5.6   c. 0.24   e. 0.12
    b. 0.054   d. 38   f. 1.52

3.   If an ideal gas in a balloon has a volume of 3.00 liters at 1.00 atm pressure, and the pressure is increased to 1013 torr at constant temperature, what does the volume become?

    a. 1.33 L   c. 2.33 L   e. 2.67 L
    b. 2.25 L   d. 2.50 L   f. 4.0 L

4. Propane burns in air to produce carbon dioxide and water. How many liters of oxygen at standard temperature and pressure are required to oxidize 2.2 grams of propane?

$$C_3H_8(g) + 5O_2(g) \rightarrow 3CO_2(g) + 4H_2O(l)$$

   a. 1.12 L   c. 5.60 L   e. 22.4 L
   b. 2.24 L   d. 11.20 L   f. 44.8 L

5. A 2.0 L vessel of hydrogen gas at 1 atm pressure and a 4.0 L vessel of nitrogen gas at 1.5 atm pressure are connected and the gases allowed to mix at constant temperature. What is the final pressure?

   a. 1.0 atm   c. 1.67 atm   e. 2.33 atm
   b. 1.33 atm   d. 2.0 atm   f. 2.5 atm

6. If argon diffuses through a porous barrier at a rate of 2.0 mol/min, at what rate would nitrogen gas diffuse?

   a. 0.70 mol/min   c. 1.4 mol/min   e. 2.4 mol/min
   b. 1.2 mol/min   d. 1.67 mol/min   f. 3.4 mol/min

7. Which of the following phase changes are endothermic (mark all that apply)?

   a. melting   c. sublimation   e. freezing
   b. vaporization   d. condensation   f. deposition

8. On a phase diagram for a substance, the lines represent the conditions under which

   a. only one phase is stable
   b. only the solid state can exist; liquids being the stable state between the lines.
   c. any or all three phases can coexist in equilibrium
   d. two phases coexist in equilibrium
   e. the substance is at its normal boiling point

9. What property of a liquid explains that liquids have an associated vapor pressure?

   a. liquids exhibit surface tension
   b. liquids have a low compressibility
   c. gases are soluble only to a slight extent in liquids and escape with little difficulty
   d. some molecules in a liquid have sufficient energy to escape intermolecular attraction
   e. molecules attract one another
   f. a liquid adopts the shape of its container

10. In a cubic lattice, what fraction of a structural unit at the corner of a unit cell belongs to that unit cell?

    a. 1/2   c. 1/4   e. 1/8   g. all
    b. 1/3   d. 1/6   f. 1/12

11. Copper crystallizes in a cubic lattice. If the side dimension of the unit cell is 362 pm and the density of copper is 8.93 g cm^{-3}, how many copper atoms are there in each unit cell?

    a. 1   c. 4   e. 8
    b. 2   d. 6   f. 12

12. Within a group of metal ions, the one most strongly solvated is the one with

    a. the smallest charge and the smallest size   c. the smallest charge and the largest size
    b. the largest charge and the smallest size   d. the largest charge and the largest size

13. What happens to the concentration of oxygen in water if the partial pressure of oxygen above the water is tripled at the same temperature, without changing the total pressure?

    a. it triples   c. it decreases to one-third
    b. it stays the same   d. it increases by 1.5

14. At room temperature, the vapor pressure of acetone is 176 torr, and the vapor pressure of ethanol is 50 torr. What is the mole fraction of ethanol in the solvent vapor above an equimolar solution?

    a. 0.11   c. 0.28   e. 0.72
    b. 0.22   d. 0.50   f. 0.78

15. How many grams of sodium hydroxide must be dissolved in 200 grams of water to make a 0.15 m solution?

    a. 0.03 g   c. 0.75 g   e. 6.0 g
    b. 0.15 g   d. 1.20 g   f. 30 g

16. What is the molality of a 20% solution of methanol in ethanol?

    a. 0.78 m   c. 6.25 m   e. 25 m
    b. 2.5 m   d. 7.8 m   f. 32 m

17. A 2.0 m solution of sugar in pure acetic acid has a boiling point of 124°C. What is the boiling point of pure acetic acid? k_b for acetic acid = –3.0 Km^{-1}.

    a. 118°C   c. 121°C   e. 126°C
    b. 119°C   d. 122°C   f. 130°C

18. What is the molar mass of a substance if a 50 mg sample dissolved in 25 mL of water has an osmotic pressure of 9.8 torr at 20°C?

    a. 920   c. 3730   e. 5730
    b. 1850   d. 4290   f. 14900

19. An aerosol is a colloidal dispersion of

    a. a liquid in a gas
    b. a liquid in another liquid
    c. a gas in a liquid
    d. a gas in a solid
    e. a solid in a liquid

20. Which one of the following solutions has the lowest vapor pressure?

    a. 0.010m NaCl   c. 0.050m sugar   e. 0.020 $Mg(NO_3)_2$
    b. 0.100 glycerol   d. 0.075 KCl   f. 0.040 Na_2SO_4

CHAPTER 15

Introduction

Chemical thermodynamics provides information about whether or not a reaction will go, and if so, how far it will go. What thermodynamics doesn't say anything about is how fast the reaction will happen. Indeed, many spontaneous reactions occur extremely slowly.

It is the chemical kinetics of a reaction that provides information about how fast the reaction goes (the rate of the reaction) and how the rearrangement of atoms actually takes place (the mechanism of the reaction). We will look at the rates and mechanisms of chemical reactions in this chapter.

The spontaneity of a reaction is a thermodynamic property of the system—a spontaneous reaction is one that happens.

Do not confuse the word spontaneous with the word instantaneous. Spontaneous does *not* mean fast.

The spontaneity of a reaction and the 2nd law of thermodynamics will be examined in Chapter 20.

Contents

15.1   Rates of Chemical Reactions

The rate of a chemical reaction is the rate at which the concentration of a substance changes per unit time. It may be the rate at which reactants are used up, or the rate at which a product appears. Consider the decomposition of N_2O_5 in solution:

$$N_2O_5 \rightarrow 2NO_2 + 1/2\,O_2$$

The progress of the reaction can be followed by monitoring the change in concentration of any one of the three components. The relative rates of appearance of products and disappearance of reactants depend upon the stoichiometry of the reaction. In this case NO_2 appears at twice the rate that N_2O_5 disappears —the stoichiometric ratio is 1:2. However, oxygen appears at a rate one-quarter of the rate of production of NO_2. The reaction rate can be expressed by an equation:

$$\text{Rate of reaction} = \frac{\text{change in } [N_2O_5]}{\text{change in time}} = \frac{-\Delta[N_2O_5]}{\Delta t}$$

The square brackets represent the concentration of the species inside the brackets.
The negative sign indicates that the concentration of N_2O_5 decreases as the reaction proceeds.

The rate of a reaction decreases as the reaction proceeds because the reactants are used up and their concentrations decrease. A plot of rate *vs.* time is a curve. The instantaneous rate at a particular time is the slope of the curve at that point.

The slope of the curve at any point can be obtained by drawing the tangent at that point. See Figure 15.3 on page 689 of the text.

15.2   Reaction Conditions and Rate

A chemical reaction is a rearrangement of atoms. For the rearrangement to take place, the reactant molecules must get together. For this reason, reactions are often carried out in the gas phase or in solution where the molecules are free to move about. Several factors influence the rate of the reaction under these conditions:

- the concentrations of reactants
- the temperature
- the presence of a catalyst
- the nature of the reactant (state, surface area, particle size)

15.3   Effect of Concentration on Reaction Rate

Changing the concentration of a reactant often changes the rate of the reaction. In the earlier example of the decomposition of dinitrogen pentoxide, doubling the concentration doubles the rate. Data are, for example:

Concentration of N_2O_5, $[N_2O_5]$	Rate: $-\Delta[N_2O_5]/\Delta t$
0.34 mol L^{-1}	0.0014 mol L^{-1} min^{-1}
0.68 mol L^{-1}	0.0028 mol L^{-1} min^{-1}

Most often, increasing the concentration increases the rate, but this doesn't always happen.

In this reaction, the rate is directly proportional to the concentration of dinitrogen pentoxide. The rate equation is:

Rate of reaction = $k[N_2O_5]$ where k is a proportionality constant called the rate constant

For a general reaction:

$$aA + bB \xrightarrow{C} xX \text{ where C is a catalyst, the rate equation is:}$$

Rate = $k[A]^m[B]^n[C]^p$

The rate of the reaction may depend upon the concentration of any of the reactants and upon the concentration of any catalyst. It is important to note that the exponents in the rate equation are not necessarily the stoichiometric coefficients in the chemical equation. The **rate constant** k relates the rate and the concentrations at a specific temperature.

The exponents in the rate equation are called the **orders** of the reaction with respect to the corresponding reactants. For example, in the above rate equation the order with respect to the reactant A is m. The order m describes how the reaction rate depends upon the concentration of A. The order may be zero, it may be an integer, it may be a non-integer, and it may be positive or negative.

The sum of the orders of all the components in the rate equation is called the **overall order** of the reaction. The overall order in the example above = m + n + p.

If the order with respect to a reactant is 2, for example, it means that doubling the reactant concentration leads to a quadrupling of the rate, because 2^2 = 4; the order is an exponent, the power to which the concentration term is raised. An order of zero means that the rate is independent of that reactant.

The relation between rate and concentration must be determined experimentally. One method is the **method of initial rates** in which the rate is determined before very much of the reactants have been consumed. Different experiments, with different initial concentrations of reactants, are run and the rates determined. Comparison of the rates allows the determination of the orders of reaction for the various reactants.

See Problem number 5.

Once the orders are known, the rate equation can be written, and the value of k, the rate constant, can be determined.

It is impossible to write the rate equation just by looking at the chemical equation.

15.4   Relationships between Concentration and Time

The rate equation described in the last section is a convenient way of expressing the relationship between rate and concentration. However, in some cases, it is more convenient to use an equation that expresses the relationship between concentration and time. These equations are called the **integrated rate equations**.

See Problem numbers 6 & 7.

The rate equation for a first-order reaction is: where [R] is the reactant concentration.

$$\text{Rate} = \frac{-\Delta[R]}{\Delta t} = k[R]$$

The corresponding integrated rate equation is: where $[R]_0$ is the concentration of R at time zero and $[R]_t$ is the concentration of R at time t

$$\ln \frac{[R]_t}{[R]_0} = -kt$$

The integrated equation expresses the relation between concentration and time—this is useful when you wish to know how much reactant or product will exist after a certain length of time. It is also an easy way to determine k, the rate constant.

The integrated rate equations are somewhat different in appearance for zero order and second order reactions:

Zero order: $\text{Rate} = \dfrac{-\Delta[R]^0}{\Delta t}$ integrated: $[R]_0 - [R]_t = kt$

First order: $\text{Rate} = \dfrac{-\Delta[R]^1}{\Delta t}$ integrated: $\ln \dfrac{[R]_t}{[R]_0} = -kt$

Second order: $\text{Rate} = \dfrac{-\Delta[R]^2}{\Delta t}$ integrated: $\dfrac{1}{[R]_t} - \dfrac{1}{[R]_0} = kt$

Graphical methods are often employed to establish the order of the reaction and then to determine the value of k, the rate constant. All three integrated rate equations can be rearranged to yield graphs with slopes that depend upon k.

How to use your calculator to calculate the logarithms of numbers is explained in Appendix A.

For a first-order reaction, the half-life $t_{1/2}$, is the time required for the concentration of a reactant to decrease to one-half its initial value. If this condition is substituted into the integrated rate equation for a first order reaction, then $\ln(1/2) = -kt_{1/2}$, or $t_{1/2} = 0.693/k$. The half-life is independent of the concentration for a first order reaction.

See Problem Solving Tips and Ideas 15.1 on page 706 of the text on how to establish reaction order from the integrated rate equations.

15.5   A Microscopic View of Reactions

A rate equation describes how the rate depends on the concentrations of the reactants. It does not explain why the rate depends upon, for example, the square of one concentration but is directly proportional to another, or why it does not depend at all upon the concentration of another reactant; it doesn't explain why temperature affects the rate, and how a catalyst participates in a reaction but is not included in the chemical equation. In order to understand these things, it is necessary to examine the reaction at the particulate level—to examine what actually happens in the reaction.

The **collision theory** of reaction rates states three conditions that must be met for a reaction to occur:

- The molecules must collide.
- They must collide with sufficient energy.
- They must collide with the correct orientation with respect to each other.

The frequency of collisions depends in part upon the concentrations of the reactants; the higher the concentrations the higher the rate of reaction. If the temperature is increased, the kinetic energies of the molecules are higher and the more likely the collision is to result in a successful reaction.

All reactions require some minimum initial energy. In the collision between molecules, if a successful reaction is to take place, bonds need to be broken, the atoms must rearrange. The minimum energy required in a successful collision is called the **activation energy**. This energy can be represented as a barrier between reactants and products. At low temperatures there are few collisions that result in enough energy to get over the top of the barrier; at higher temperatures more collisions have sufficient energy.

In a collision, the molecules must be oriented correctly, so that the correct bonds are broken and the desired rearrangement takes place. This steric requirement is sometimes quite critical. And the lower the probability of attaining the correct orientation, the slower the reaction.

A summary of the effects of temperature, activation energy, and orientation is provided by the **Arrhenius equation** for the rate constant k:

$$k \text{ (reaction rate constant)} = A \, e^{-\frac{E_a}{RT}}$$

The pre-exponential factor A is a parameter that is related to the collision frequency and the orientation requirements. The exponent is interpreted as that fraction of molecules having the minimum energy required for reaction. The Arrhenius equation is useful in determining the activation energy E_a.

Catalysts speed up reactions; sometimes by several orders of magnitude. Reactions that are painfully slow can occur with explosive speed in the presence of a catalyst. A catalyst is involved in the reaction but does not appear in the chemical equation; it is not used up. The function of a catalyst is to provide an alternative path for the reaction, one with a much lower activation energy. The use of catalysts allow reactions to be run at lower temperatures, with greater yields, with fewer side reactions, and much more efficiently and cheaply.

The top of the barrier is referred to as the activated complex. This complex represents the molecular situation between reactants and products at the moment of collision when bonds are being broken and new bonds are being formed. See for example Figure 15.14 on page 712 of the text.

The increase in temperature also leads to an increase in the collision frequency but by far the more important result is the increased energy involved in the collision.

The reason why reaction rates vary so much is that the reactions have widely different activation energies.

One of the main reasons why enzymes are so efficient in increasing the rate of biochemical processes is that they orient the reacting molecules precisely.

R in the Arrhenius equation is 8.314 $JK^{-1}mol^{-1}$.

A homogeneous catalyst is a catalyst in the same phase as the reactants.
A heterogeneous catalyst is a catalyst in a different phase than the reactants.

15.6 Reaction Mechanisms

The reaction mechanism is the sequence of bond breaking and making, the rearrangement of the atoms, that occurs in a chemical reaction.

Most reactions occur as a sequence of steps, particularly those involving a catalyst. For example, consider the reaction of bromine with nitric oxide:

$$Br_2(g) + 2NO(g) \rightarrow 2BrNO(g)$$

It is possible, but unlikely, that all three molecules would collide at the same time in the correct orientation. A more probable mechanism would be one that involves two sequential steps:

Step 1: $Br_2(g) + 2NO(g) \rightarrow Br_2NO(g)$

Step 2: $Br_2NO(g) + NO(g) \rightarrow 2BrNO(g)$

The sum of these two steps produces the overall equation. Each step in the multistep process is called an **elementary step**—a single molecular event. Each elementary step has its own activation energy, rate constant k, and rate equation.

There are three types of elementary steps. They are classified by the number of molecules, or reacting species, that come together in the reaction step. This number is called the **molecularity** of the elementary step.

unimolecular   one molecule involved (typically a decomposition)
bimolecular   two molecules involved
termolecular   three molecules involved

The rate equation for an elementary step is defined by the stoichiometry of the step; the exponents in the rate equation equal the coefficients in the chemical equation.

Fitting the observed experimental rate equation to a possible sequence of elementary steps is not easy. It requires considerable knowledge, insight, and intuition, and even then it may be incorrect! Two features at least must be catered for, one is that the sum of the elementary steps must result in the overall chemical equation, the second is that the proposed mechanism must fit the experimentally observed rate. This latter requirement is helped by the observation that in a sequence of steps, the slowest step determines the rate. This step is called the **rate-determining step**.

The rates of the various steps making up a reaction do vary considerably. Some are very fast, others very slow. Often a fast step reaches a state of equilibrium quickly and is followed by a relatively slow rate-determining step. In this case, the first step leads to the formation of a steady concentration of **a reaction intermediate**. This reaction intermediate is then used up in the second (rate-determining) step, but at a rate which allows the first step to maintain its concentration at a steady value. The overall rate equation can be derived by writing an expression for the equilibrium reached in the first step and then substituting this in the rate equation for the second step.

The molecularity of an elementary step is the same as the order of that step—but has nothing to do the the order of the overall reaction.

The exponents for the concentration terms in the rate equation *for an elementary step* are equal to the coefficients in the chemical equation for that step.

Recall that it is impossible to write the overall rate equation from the stoichiometry of the overall chemical equation.

A reaction cannot go faster than the slowest step.

This is sometimes referred to as the steady state approximation.

Chemical equilibria is the subject of the next chapter.

Review Questions

1.   How would you define the rate of a chemical reaction?

2.   How is the rate at which a reactant is used up related to the rate at which a product is formed in a reaction?

3.   Summarize the ways in which the rate of a chemical reaction can be changed.

4.   How does the reaction order describe the dependence of the reaction rate upon the concentrations of the reactants?

5.   Explain the method of initial rates for determining the order of reaction with respect to the reactants.

6.   What is the instantaneous rate? How is it determined?

7.   What is the difference between a rate equation and an integrated rate equation?

8.   Explain how the value of the rate constant k can be obtained graphically using the integrated rate equation for a first order reaction.

9.   What is a half-life? Explain why the half-life for a first order reaction is concentration independent.

10.   List the three requirements for a successful collision between molecules.

11.   Describe what the activation energy is.

12.   What is the Arrhenius equation and what does it describe?

13.   Describe three ways to speed up a reaction and why the reaction does speed up when these three things are done.

14.   What is the difference between a homogeneous catalyst and a heterogeneous catalyst? What does a catalyst actually do?

15.   Describe what an elementary step of a reaction is; is it possible to write the rate equation for an elementary step from its stoichiometry?

16.   What is the molecularity of a reaction step?

17.   What is a reaction intermediate?

18.   Define the rate-determining step of a reaction mechanism.

Answers to Review Questions

1. The rate of a chemical reaction is the rate at which the concentration of a substance changes per unit time. It may be the rate at which reactants are used up, or the rate at which the product appears.

2. The relative rates of appearance of products and disappearance of reactants depend upon the stoichiometry of the reaction. If two reactant molecules yield three product molecules—a stoichiometric ratio of 2:3—the rate of appearance of the product is 1.5 times faster than the rate of disappearance of the reactant.

3. Several factors influence the rate of the reaction: the concentrations of reactants, the temperature, the presence of a catalyst, and the nature of the reactants. A reaction is a rearrangement of atoms, molecules must be in contact for this rearrangement to take place, and anything that facilitates this will increase the rate of the reaction.

4. The rate equation includes concentration terms for the reactants, each raised to some power. An exponent of a concentration term is called the order of reaction with respect to that particular reactant. If the order is one, then the rate is directly proportional to that concentration. If the order is 2, then the rate is proportional to the square of the concentration, and so on. Orders can be non-integer and may even be negative.

    > The order may be 0, in which case the rate does not depend upon the concentration of that reactant.

5. Because the concentrations of the reactants decrease as soon as a reaction starts—the reactants are used up as the reaction proceeds—the best way to determine the rate dependence on reactant concentration is to plot the rate *vs.* the reactant concentration over a period of time. However, this can be quite laborious and sometimes difficult to do. A simpler method is the method of initial rates. In this method the rate is determined as soon as the reaction has started and has not progressed very far. The advantage of the method is that the initial concentrations are known, and they can be easily varied to determine the rate equation.

    > It is only necessary to monitor the change in one reactant or product; the other concentrations are related by the stoichiometry of the reaction (see problems # 1 and 2 in the next section).

6. The instantaneous rate is the rate at any particular time during the reaction. It is the tangent to the curve of reaction rate *vs.* time.

7. The rate equation relates the rate and the concentrations of the reactants. The integrated rate equation relates the time of reaction and the concentrations of the reactants.

8. The integrated rate equation for a first order reaction is:

$$\ln \frac{[R]_t}{[R]_0} = -kt \quad \text{which rearranges to} \quad \ln[R]_t = \ln[R]_0 - kt$$

    Plotting $\ln[R]_t$ *vs.* t yields a straight line with a slope equal to $-k$.

9.   A half-life is a period, the time taken, for the reactant in a reaction to reduce in concentration to one-half of its initial concentration. Substituting $[R]_t = \frac{1}{2}[R]_0$ into the integrated rate equation for a first order reaction yields the relationship: $\ln 2 = kt_{1/2}$ or $t_{1/2} = 0.693/k$ which is concentration independent. This means that for any first order reaction, the concentration of the reactant is always one-half of what it was $t_{1/2}$ ago, no matter what starting time you choose.

$\ln 2 = kt_{1/2}$
is the same as
$\ln(\frac{1}{2}) = -kt_{1/2}$

10.   The molecules must first collide so the rearrangement of atoms is possible; they must collide with sufficient energy to overcome the activation energy; and they must collide with the correct orientation with respect to one another.

11.   All reactions require some minimum initial energy. If a successful reaction is to take place between molecules, bonds need to be broken so that the atoms can rearrange. This takes energy. The minimum energy required in a successful collision is called the activation energy. This energy can be represented as a barrier between reactants and products. At low temperatures there are few collisions that result in enough energy to get over the top; at higher temperatures more collisions have sufficient energy.

12.   The Arrhenius equation is an expression for k (the rate constant) that includes all the factors that affect a reaction other than concentration.

$$k \text{ (reaction rate constant)} = A\, e^{-\frac{E_a}{RT}}$$

A is a parameter that is related to the collision frequency and the orientation requirements. The exponent is interpreted as that fraction of molecules having the minimum energy required for reaction.

13.   a. Increase the concentrations of the reactants. Usually this increases the rate because collisions between the molecules increase in frequency. The exact influence the concentration has is described by the rate equation.
      b. Increase the temperature. Although this also increases the collision frequency, the predominant effect is to increase the effectiveness of those collisions that do occur. More energy is available at the higher temperature and the fraction of molecules possessing sufficient energy to overcome the activation energy barrier is higher.
      c. Add a catalyst. This usually has the most profound effect. The catalyst provides a different mechanism by which the reaction can happen.

14.   A homogeneous catalyst is a catalyst in the same phase as the reactants. A heterogeneous catalyst is a catalyst in a different phase than the reactants. The catalyst provides a different route for the reaction.

15. An elementary step of a reaction is a single step in a sequence of steps making up a chemical reaction; it is a single molecular event. The rate equation for an elementary step is defined by the stoichiometry of the step; the exponents in the rate equation equal the coefficients in the chemical equation. Recall that this is not the case for the overall rate equation; it is impossible to write the rate equation from the stoichiometry of the overall chemical equation.

16. The molecularity of an elementary step is the number of molecules or reacting species that come together in the reaction step. There are three types of elementary steps: unimolecular (one molecule involved), bimolecular (two molecules involved), and termolecular (three molecules involved).

17. A reaction intermediate is a molecule or some species formed in an elementary step during a reaction that is subsequently used up in another elementary step. It is not a product of the reaction nor does it appear in the overall chemical equation.

18. In a sequence of elementary steps making up a chemical reaction, it is the slowest step that determines the rate. This step is called the rate-determining step. The reaction cannot go more quickly than the slowest step.

Study Questions and Problems

1. In the following reaction, what is the relationship between the rate at which the nitrous oxide is used up, the rate at which the oxygen is used, and the rate at which the nitrogen dioxide is produced?

$$2N_2O(g) + 3O_2(g) \rightarrow 4NO_2(g)$$

2. Ammonia can be oxidized by oxygen to produce nitrogen dioxide according to the equation:

$$4NH_3(g) + 7O_2(g) \rightarrow 4NO_2(g) + 6H_2O(g)$$

If, in this reaction, water is formed at a rate of 36 mol L^{-1} min^{-1},
a. at what rate is the ammonia used?
b. at what rate is the oxygen used?
c. at what rate is the nitrogen dioxide formed?

3. If a reactant is used up according to a first order rate equation, and the initial concentration of the reactant is 3.2 mol L^{-1}, what is the concentration of the reactant after two half-lives have passed, and after six half-lives have passed?

Science is a way of thinking much more than it is a body of knowledge.

Carl Sagan
(1934-1996)

4. For each of the following rate equations, describe what would happen to the rate if the concentration of reactant A was tripled and the concentration of reactant B is halved.

   a. Rate = k [A][B]

   b. Rate = k $[A]^2[B]$

   c. Rate = k $[A]^2[B]^2$

   d. Rate = k $[A][B]^3$

5. The reaction of ^{t}butyl-bromide $(CH_3)_3CBr$ with water is represented by the equation:

   $$(CH_3)_3CBr + H_2O \rightarrow (CH_3)_3COH + HBr$$

   The following data were obtained from three experiments using the method of initial rates:

	Initial $[(CH_3)_3CBr]$ mol L^{-1}	Initial $[H_2O]$ mol L^{-1}	Initial rate mol L^{-1}min^{-1}
Experiment 1	5.0×10^{-2}	2.0×10^{-2}	2.0×10^{-6}
Experiment 2	5.0×10^{-2}	4.0×10^{-2}	2.0×10^{-6}
Experiment 3	1.0×10^{-1}	4.0×10^{-2}	4.0×10^{-6}

   a. What is the order with respect to $(CH_3)_3CBr$?
   b. What is the order with respect to H_2O?
   c. What is the overall order of the reaction?
   d. Write the rate equation.
   e. Calculate the rate constant k for the reaction.

6. At 150°C the decomposition of acetaldehyde CH_3CHO to methane is a first order reaction. If the rate constant for the reaction at 150°C is 0.029 min^{-1}, how long does it take a concentration of 0.050 mol L^{-1} of acetaldehyde to reduce to a concentration of 0.040 mol L^{-1}?

7. The decomposition of hydrogen iodide into hydrogen and iodine is a second order reaction. The rate constant k = 0.080 L mol^{-1} s^{-1}. How long does it take an initial concentration of 0.050 M to decrease to half this concentration?

8. Describe some industrial uses of catalysts.

9. The gold-198 isotope has a half-life of 2.7 days. If you start with 10 mg at the beginning of the week, how much remains at the end of the week, seven days later?

10. If the rate of a reaction increases by a factor of 10 when the temperature is increased by 35°C from 300K to 335K, what is the activation energy E_a for the reaction?

11. The decomposition of ozone in the upper atmosphere to dioxygen occurs by a two-step mechanism. The first step is a fast reversible step and the second is a slow reaction between an oxygen atom and an ozone molecule:

    Step 1: $O_3(g) \rightleftharpoons O_2(g) + O(g)$ Fast, reversible, reaction

    Step 2: $O_3(g) + O(g) \rightarrow 2O_2(g)$ Slow

    a. Which is the rate-determining step?

    b. Write the rate equation for the rate-determining step.

    c. Write the rate equation for the overall reaction

12. The rate equation for the reaction of nitrogen dioxide and carbon monoxide in the gas state to form carbon dioxide and nitric oxide is represented by the equation:

    $$NO_2(g) + CO(g) \rightarrow NO(g) + CO_2(g)$$

    The following data were collected at 125°C:

	Initial $[NO_2]$ mol L^{-1}	Initial $[CO]$ mol L^{-1}	Initial rate mol $L^{-1}min^{-1}$
Experiment 1	5.0×10^{-4}	1.6×10^{-2}	1.7×10^{-7}
Experiment 2	5.0×10^{-4}	3.2×10^{-2}	1.7×10^{-7}
Experiment 3	1.5×10^{-3}	3.2×10^{-2}	1.5×10^{-6}

    a. What is the order with respect to NO_2?
    b. What is the order with respect to CO?
    c. What is the overall order of the reaction?
    d. Write the rate equation.
    e. How do you know that this is not a single step reaction?
    f. Suggest a mechanism for the reaction.

Answers to Study Questions and Problems

1. The relative rates at which reactants are used up and products are formed is provided by the stoichiometry of the balanced equation. In this reaction, the relative rates are 2:3:4; the oxygen is used at a rate 1.5 times faster than the rate at which the nitrous oxide is used, and the nitrogen dioxide is formed at a rate 2 times faster than the rate at which the nitrous oxide is used.

    $$2N_2O(g) + 3O_2(g) \rightarrow 4NO_2(g)$$

2. Water is formed at a rate of 36 mol L^{-1} min^{-1}:

    $$4NH_3(g) + 7O_2(g) \rightarrow 4NO_2(g) + 6H_2O(g)$$

    a. the rate that ammonia is used = 36 × 4/6 = 24 mol L^{-1} min^{-1}
    b. the rate that oxygen is used = 36 × 7/6 = 42 mol L^{-1} min^{-1}
    c. the rate that NO_2 is formed = 36 × 4/6 = 24 mol L^{-1} min^{-1}

    The factors 4/6 and 7/6 are the mole ratios in the equation—the stoichiometric factors.

3. The initial concentration of the reactant is 3.2 mol L^{-1}

An expression for the final concentration is $3.2 \times (\frac{1}{2^n})$ or $3.2/64 = 0.05$.

After two half-lives have passed, the concentration has halved twice, to 1.6 and then to 0.8 mol L^{-1}. After six half-lives, the concentration has halved four more times, to 0.4, to 0.2, to 0.1, and then to 0.05 mol L^{-1}.

4. The concentration of reactant A is tripled and the concentration of reactant B is halved:

    a. Rate = k [A][B], rate is multiplied by 3, divided by 2;   × 1.5

    b. Rate = k [A]^2[B], rate is multiplied by 3^2, divided by 2;   × 4.5

    c. Rate = k [A]^2[B]^2, rate is multiplied by 3^2, divided by 2^2;   × 2.25

    d. Rate = k [A][B]^3, rate is multiplied by 3, divided by 2^3;   × 0.375

5.

	Initial [(CH$_3$)$_3$CBr] mol L^{-1}	Initial [H$_2$O] mol L^{-1}	Initial rate mol L^{-1}min^{-1}
Experiment 1	5.0 × 10^{-2}	2.0 × 10^{-2}	2.0 × 10^{-6}
Experiment 2	5.0 × 10^{-2}	4.0 × 10^{-2}	2.0 × 10^{-6}
Experiment 3	1.0 × 10^{-1}	4.0 × 10^{-2}	4.0 × 10^{-6}

If experiments 1 and 2 are examined, it is apparent that the concentration of (CH$_3$)$_3$CBr remains unchanged and the concentration of H$_2$O doubles. The rate stays the same; in other words, changing the concentration of water has no effect; the order with respect to water is zero. Comparison of experiments 2 and 3 indicates a doubling of the concentration of (CH$_3$)$_3$CBr. The water concentration has already been determined to be immaterial. The rate doubles. Therefore the order with respect to (CH$_3$)$_3$CBr is 1. The overall order is 1.

    a. The order with respect to (CH$_3$)$_3$CBr = 1.

    b. The order with respect to H$_2$O = 0.

    c. The overall order of the reaction = 1 + 0 = 1.

    d. The rate equation is: Rate = k[(CH$_3$)$_3$CBr].

    e. Use any experiment to calculate the value of the rate constant k: For example, experiment 1: Rate = 2.0 × 10^{-6} = k × 5.0 × 10^{-2}. k = 4.0 × 10^{-5}.

6. The relationship between time and concentration is provided by the integrated rate equation. For a first order reaction:

    ln([R]$_t$/[R]$_0$) = –kt.

    ln(0.040/0.050) = –0.029 × t

    t = 7.7 minutes

7. Unlike for a first order reaction, the half-life of a second order reaction is concentration dependent. Again, use the integrated rate equation:

$1/[R]_t - 1/[R]_0 = kt$

$1/0.025 - 1/0.050 = 0.080 \times t$

$20 - 40 = 0.080 \times t$

$t = 250$ seconds $= 4$ min 10 sec

8. Catalysts are used in most industrial chemical manufacture. Thirteen of the top twenty synthetic chemicals produced in this country are made by catalytic processes. Some examples are:

    Ostwald Pt-Rh in oxidation of ammonia.
    Ziegler-Natta catalyst in the manufacture of polyethylene.
    Wilkinson's $RhCl(PPh_3)_3$ catalyst for the hydrogenation of alkenes.
    Monsanto $RhI_2(CO)_2^-$ catalyst for oxidation of methanol to acetic acid.

    Notice how much rhodium is used! Other metals often used are palladium and platinum.

9. If the half-life is 2.7 days, then k, the rate constant $= \ln2/2.7 = 0.257$. Then, since $\ln([R]_t/[R]_0) = -kt$

    Note that $-\ln(1/2) = \ln2$ and $\ln(1/2) = -kt_{1/2}$ so $k = \ln2/t_{1/2}$

$\ln([R]_t/10) = -0.257 \times 7$

$\ln[R]_t - \ln(10) = -0.257 \times 7$

    $\ln(a/b) = \ln a - \ln b$

$\ln[R]_t = 0.51$

$[R]_t = 1.66$ mg; 1.66 mg remains after seven days.

10. A useful alternative form of the Arrhenius equation is:

$$\ln\left(\frac{k_2}{k_1}\right) = -\frac{E_a}{R}\left(\frac{1}{T_2} - \frac{1}{T_1}\right)$$

$\ln(10) = -E_a/8.314 \times (1/325 - 1/300)$

$2.303 = -E_a/8.314 \times (-2.564 \times 10^{-4})$

$E_a = 74.7$ kJ/mol

11. Step 1: $O_3(g) \rightleftharpoons O_2(g) + O(g)$ Fast, reversible, reaction

    Step 2: $O_3(g) + O(g) \rightarrow 2O_2(g)$ Slow

    a. Step 2 is the slower step; so this step is the rate-determining step.

    b. The rate equation for the rate-determining step is:

        Rate $= k[O_3][O]$

    c. The rate for the overall reaction is, by definition, equal to the rate of the rate-determining step. So start by writing the rate equation for that step:

        Rate $= k[O_3][O]$

    Remember that you can write the rate equation for an elementary step based upon the stoichiometry of that step.

It is necessary to substitute the concentration terms for any reaction intermediates in this equation. The only concentration terms should be those of the starting reactants. What is [O]? We must express it in terms of O_3 and/or O_2. Write an equilibrium expression for the first step:

Chemical equilibria are the
subject of the next chapter.

$$K = \frac{[O_2][O]}{[O_3]} \qquad \text{and rearrange to obtain an expression for [O]:}$$

$[O] = K[O_3]/[O_2]$ and substitute in the rate equation above:

$$\text{Rate} = k[O_3]K[O_3]/[O_2] = k'[O_3]^2[O_2]^{-1}$$

12. The reaction of nitrogen dioxide and carbon monoxide in the gas state to form carbon dioxide and nitric oxide:

$$NO_2(g) \quad + \quad CO(g) \quad \rightarrow \quad NO(g) \quad + \quad CO_2(g)$$

	Initial $[NO_2]$ mol L^{-1}	Initial [CO] mol L^{-1}	Initial rate mol L^{-1}min^{-1}
Experiment 1	5.0×10^{-4}	1.6×10^{-2}	1.7×10^{-7}
Experiment 2	5.0×10^{-4}	3.2×10^{-2}	1.7×10^{-7}
Experiment 3	1.5×10^{-3}	3.2×10^{-2}	1.5×10^{-6}

a. Tripling the NO_2 concentration increases the rate 9 times; so the order with respect to NO_2 is 2.
b. A change in the CO concentration has no effect; the order is zero.
c. The overall order of the reaction is 2.
d. The rate equation is Rate = $k[NO_2]^2$.
e. If it was a single step elementary reaction the rate equation would be:

$$\text{Rate} = k[NO_2][CO]$$

f. The rate equation suggests a bimolecular collision between two nitrogen dioxide molecules as a slow rate-determining step. The question remains: what is the product of this collision? Experimental evidence for the existence of various reaction intermediates is usually required at this point. Two possibilities are N_2O_4 or NO_3 and NO. The latter is an attractive choice because NO is one of the required products and the NO_3 could react in a fast second step with the CO to produce the other product CO_2:

This is the generally accepted
mechanism.

$$NO_2(g) \quad + \quad NO_2(g) \quad \rightarrow \quad NO_3(g) \quad + \quad NO(g) \qquad \text{slow}$$
$$NO_3(g) \quad + \quad CO(g) \quad \rightarrow \quad NO_2(g) \quad + \quad CO_2(g) \qquad \text{fast}$$

alternatively perhaps, if the collision is N to N:

$$NO_2(g) \quad + \quad NO_2(g) \quad \rightarrow \quad N_2O_4(g) \qquad \text{slow}$$
$$N_2O_4(g) \quad + \quad CO(g) \quad \rightarrow \quad N_2O_3(g) \quad + \quad CO_2(g) \qquad \text{fast}$$
$$N_2O_3(g) \quad \rightarrow \quad NO_2(g) \quad + \quad NO(g) \qquad \text{fast}$$

Challenge 9

Figuring out a possible mechanism for a reaction is difficult—even when the rate equation has been established. Fitting the observed rate equation to a possible sequence of elementary steps requires considerable knowledge, insight, and chemical intuition, and even then suggestions often lead to intense arguments among chemists. So try your hand at the following problem—it's not easy.

Remember that two features must be catered for. The sum of the elementary steps must result in the overall chemical equation and the proposed mechanism must fit the experimentally observed rate. Recall that in a sequence of steps, the slowest step determines the rate. And if there is a preceding step that quickly reaches a steady state or equilibrium, this often simplifies the analysis.

The rate equation for the reaction:

$$I^- + ClO^- \rightarrow Cl^- + IO^-$$

in basic aqueous solution is

$$Rate = k\, \frac{[I^-][ClO^-]}{[OH^-]}$$

Suggest a mechanism.

Kinetics

—a crossword puzzle

Across:

2. What the square brackets [ ] represent.
4. The inverse of rate.
5. One of the platinum group metals used as a catalyst.
6. Required by a system for the reaction to go.
8. Must be correct for an effective collision.
9. Slope.
10. The order that is sum of the exponents in the rate equation.
11. A transient species produced and then consumed in a reaction sequence.
12. The sequence of bond-making and bond-breaking that is the way the reaction happens.
14. *and 19 across:* The largest growth in catalyst use is in this area.
19. *See 14 across.*
20. One molecular event in a reaction sequence.
21. His equation relates *1 down*, *3 down*, and *6 across*.
22. The time taken for a reactant concentration to be reduced by 50%.
25. The slowest step in a *12 across*.
28. The common one is the exponent of 10.
31. The instantaneous rate is the *9 across* of this.
33. Only for an *20 across* is the order equal to this.
36. These must do it with sufficient *35 down* and the correct *8 across* if they are to react.
37. A *17 down* in a biological system.
38. *See 29 down.*

Down:

1. The k in the rate equation.
3. Increase this and the kinetic energy of the molecules increases.
5. The relationship between *2 across* and *4 across*.
7. The exponent of a concentration term in the rate equation.
12. The number of reacting species that come together in an elementary reaction step.
13. The order when the *16 down* is 1.
15. Its magnitude is responsible for dust explosions.
16. The power to which a term is raised.
17. It changes the route of a reaction.
18. The order when the *16 down* is 2.
23. The factor A in the equation of *21 across*.
24. This effect rather than a thermodynamic effect determines the rate of a reaction.
26. Only these collisions result in reaction.
27. Reactions are often done in this because the molecules are free to move about.
29. *and 38 across:* A method used to determine reaction order.
30. A *17 down* in the same phase as the reactants.
32. An explanation.
34. Kinetically unreactive.
35. Used when work is done.

CHAPTER 16

Introduction

All chemical reactions are reversible, at least in principle. Hydrogen and oxygen combine to form water. Water can be decomposed to form hydrogen and oxygen. Most reactions continue until they reach an equilibrium—a balance between the forward and reverse reactions. In this chapter, and the remaining chapters in this part of the text, we will examine chemical equilibria.

Contents

16.1   The Nature of the Equilibrium State

A system in equilibrium is a **dynamic system**, reactants are combining together to form products and the products are combining to reform the reactants. The two reactions, the forward one and the reverse one, occur at the same rate so that there is no net change in the concentrations.

It takes a finite amount of time to reach the equilibrium state, and the state can be reached from either direction, from the reactant side or from the product side. Consider the ionization of acetic acid in water:

$$CH_3CO_2H(aq) + H_2O(l) \rightleftharpoons H_3O^+(aq) + CH_3CO_2^-(aq)$$

The system reaches equilibrium from the reactant side when pure acetic acid is added to water as indicated in the equation above. However, exactly the same equilibrium is established when a strong acid is added to an acetate salt (for example hydrochloric acid and sodium acetate) but in this case the equilibrium is approached from the product side:

$$H_3O^+(aq) + CH_3CO_2^-(aq) \rightleftharpoons CH_3CO_2H(aq) + H_2O(l)$$

If the number of moles of acetic acid used in the first experiment is equal to the number of moles of sodium acetate and hydrochloric acid used in the second experiment, and the volume of solution is the same for both experiments, then the two systems are identical—the position of equilibrium is exactly the same.

16.2   The Equilibrium Constant

A system in equilibrium is characterized by the relative concentrations of products and reactants at equilibrium. If the amount of product is large compared to the reactant, then the equilibrium lies toward the product side. On the other hand, if there is a lot of reactant at equilibrium, and not much product, then the system is said to lie toward the reactant side. The relative concentrations of reactant and product present at equilibrium is represented by the **equilibrium constant**, K.

$$K = \frac{[Products]}{[Reactants]}$$

Always $\dfrac{products}{reactants}$ over

For example, consider the equilibrium system:

$$H_2(g) \ + \ I_2(g) \ \rightleftharpoons \ 2HI(g)$$

Experiments have shown that, at the same temperature, the ratio of the square of the HI concentration to the product of the H_2 and I_2 concentrations is always the same—at 425°C it is always 56. This ratio is always the same regardless of the quantities of HI, H_2, and I_2 present at the start.

$$K = \frac{[HI]^2}{[H_2][I_2]} = 56$$

Each concentration term is raised to a power equal to its stoichiometric coefficient in the balanced equation.

Equilibrium expressions do not contain concentration terms for species in a phase different from that in which the equilibrium exists. For example, equilibria in solution do not include solids or gases in the expression; equilibria in the gas state do not include liquids or solids. Also, a solution equilibrium expression does not include the solvent.

K is temperature dependent—otherwise for any particular reaction it is constant.

The equilibrium constant can be written in terms of the component concentrations as K_c, or it can be expressed in terms of the partial pressures of the components as K_p.

The two are related by the expression:

$K_p = K_c(RT)^{\Delta n}$

where Δn = moles of gaseous products − moles of gaseous reactants

If the equation is changed, the expression for K also changes, because the exponents of the concentration terms equal the stoichiometric coefficients. The expression of the new K equals the expression for the original K raised to a power equal to the factor by which the coefficients were changed. If the reaction is reversed, the new K is the reciprocal of the old K.

If two equations are added to yield a net equation, the K for the net equation is the product of the values of K for the two added equations. Read the Problem Solving Tips and Ideas 16.1 on page 756 of the text.

A large value for K means a proportionately large concentration of product, i.e. a product-favored process. On the other hand, a smaller value for K means less product at equilibrium.

16.3   The Reaction Quotient

At a particular temperature, the value of K for a particular reaction is constant. In other words, the ratio of products to reactants is fixed and cannot change. Suppose, for a general equilibrium system, K = 3:

$$\text{Reactants} \ \rightleftharpoons \ \text{Products} \qquad\qquad K = \frac{[Products]}{[Reactants]} = 3$$

The constant K indicates that the ratio of the concentration of product to the concentration of reactant at equilibrium must be 3; the concentrations can vary infinitely as long as the ratio is 3:1; the system will still be in equilibrium. This condition can be represented by a line, slope K, on a graph of [Product] *vs.* [Reactant].

If the ratio is not 3, the system lies off the line, and it is not in equilibrium! The ratio of the concentrations is called the **reaction quotient Q**; if the value for the reaction quotient Q is equal to K, then the system is in equilibrium. If the reaction quotient Q is larger than K, the system contains too much product and in order to get into equilibrium, some product must reform reactant. If the reaction quotient Q is less that K, then there is not enough product and the reaction must proceed further to the product side.

16.4   Calculating an Equilibrium Constant

If the equilibrium concentrations are known, the equilibrium constant can be calculated by simply substituting these concentrations in the equilibrium constant expression. More often, however, the initial concentrations are known, one of the equilibrium concentrations is measured, and the other equilibrium concentrations must be calculated. The general procedure is to set up what is called an ICE box, where I stands for initial, C for change, and E for equilibrium. Consider, for example, the system:

$$2SO_2(g) \ + \ O_2(g) \ \rightleftharpoons \ 2SO_3(g)$$

Suppose initially 2 moles of sulfur dioxide and 2 moles of oxygen gas are placed in a 1 liter vessel. After reaching equilibrium, the concentration of sulfur trioxide was found to be 1.84 moles. What is the equilibrium constant for this system at this temperature? Set up the box and figure out the unknown equilibrium concentrations:

		$2SO_2(g)$ +	$O_2(g)$ $\rightleftharpoons$	$2SO_3(g)$
I	Initial	2.00	2.00	0.00
C	Change	−1.84	−1.84 × 1/2	+1.84
E	Equilibrium	0.16	1.08	1.84

Now the equilibrium constant can be calculated:

$$K = \frac{[SO_3]^2}{[SO_2]^2[O_2]} = \frac{(1.84)^2}{(0.16)^2(1.08)} = 122$$

16.5   Using Equilibrium Constants in Calculations

Quite often the equilibrium constant is already known. In this case, the initial concentrations might be known, and you might be required to calculate the equilibrium concentrations. There are two things to keep in mind:

The quantities of reactants and products in these calculations must be expressed as concentrations or pressures (not as amounts).

It is sometimes possible to assume that a change in a concentration of a reactant is so small that it can be ignored. If it can be ignored, it makes the

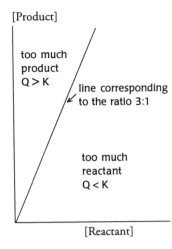

[Product]

too much product Q > K

line corresponding to the ratio 3:1

too much reactant Q < K

[Reactant]

If Q = K, equilibrium!
if Q > K, there is too much product, and the system must move left to get into equilibrium.
if Q < K, there is too much reactant, and the system must move right to get into equilibrium.

Don't forget always to use concentrations, not just the number of moles. (In this case the volume is 1 liter.)
See also some of the problems at the end of this study guide.

The change in the concentrations is always in the same ratio as the stoichiometry: if the change in SO_3 is +1.84, then the change in SO_2 has to be −1.84, and the change in O_2 must be −1.84 × 1/2.

Usually K is given without units.

There are a variety of problems involving chemical equilibria—they are all basically the same! Do not try to memorize a set procedure—think about the question and then set about solving it in a logical way. Often a little algebra is required. Using a table to organize your data is highly recommended!

calculation much easier. A sensible procedure is to assume the simplification is valid, and then once the calculation is done, check to see if it was. The rule of thumb is that the change in the concentration of a reactant can be ignored if it is less than 5% of the smallest quantity initially present. Consider this example:

The ionization of acetic acid is represented by the equation:

$$CH_3CO_2H(aq) \quad + \quad H_2O(l) \quad \rightleftharpoons \quad H_3O^+(aq) \quad + \quad CH_3CO_2^-(aq)$$

Suppose that a 1.0 M solution of acetic acid is prepared, what is the $[H_3O^+]$, the hydronium ion concentration, in the solution? $K = 1.8 \times 10^{-5}$.

Always remember what the unknown x is. Just because you calculate x, don't assume this is the answer required. It is however in this case. Read the Problem Solving Tips and Ideas 16.2 on page 767 of the text.

Notice the absence of $[H_2O]$ from the equilibrium constant expression.

Does this assumption make much difference?—it certainly makes the calculation easier. If the quadratic is solved in this case, x = 4.234 (as opposed to 4.243 with the simplification), not much difference.

See Table 16.2 on page 770 of the text.

		$CH_3CO_2H(aq)$ $\rightleftharpoons$ $H_3O^+(aq)$ $+$ $CH_3CO_2^-(aq)$		
I	Initial	1.00	0.0	0.0
C	Change	−x	+x	+x
E	Equilibrium	1.00 − x	x	x

Assume that 1−x is near enough to 1 that the x can be ignored. Substituting in the expression for the equilibrium constant:

$$K = \frac{[H_3O^+][CH_3CO_2^-]}{[CH_3CO_2H]} = \frac{(x)^2}{(1-x)} = x^2 = 1.8 \times 10^{-5}$$

$x = 4.24 \times 10^{-3}$ M — this is the hydronium ion concentration (it is also the acetate ion concentration).

x is less than 1% of 1, so the assumption that (1−x) = 1 is valid.

16.6 Disturbing an Equilibrium; LeChatelier's Principle

There are three common ways to disturb a chemical equilibrium:

- Change the temperature.
- Change the concentration of a reactant or product.
- Change the volume (or pressure).

LeChatelier's Principle, met briefly in Chapters 13 and 14, states that if a system in equilibrium is subjected to a stress by altering the conditions that determine the state of equilibrium, the system will tend to adjust itself to accommodate the stress placed upon it.

If a reaction is exothermic, like the combination of two nitrogen dioxide molecules to form dinitrogen tetroxide:

$$2NO_2(g) \quad \rightleftharpoons \quad N_2O_4(g) \qquad \Delta H = -57.2 \text{ kJ}$$

Then heat can be regarded as a product of the reaction. Addition of heat will drive the system to the left, breaking the N_2O_4 molecule to form two NO_2 molecules. The movement of the system to the left is a way to accommodate the applied stress; as the system moves to the left the added heat is absorbed. Addition of heat—an increase in temperature—will always move an exothermic reaction back toward the reactants and an endothermic reaction forward toward the products.

Addition or removal of a reactant or product can be predicted in the same way. Think of an equilibrium system as a system in balance; disturb the balance and the system has to regain balance by moving one way or the other. Addition of a product moves the system toward the left; addition of a reactant moves the system toward the product side. Removal of product drives the system to form more product; removal of reactant drives the system back toward the reactant side. The movement will occur until the value of the reaction quotient again becomes equal to the equilibrium constant.

Volume and pressure changes affect equilibria in which the number of moles of gas on the two sides is different. The system will shift to the side occupying the smaller volume to accommodate an increase in pressure or a reduction in volume.

16.7   Is There Life after Equilibrium?

Thermodynamics allows us to establish which reactions are spontaneous—which reactions will happen. It also allows us to determine how far the reactions will go. The answer to the latter question is that a reaction that is spontaneous will go until the system reaches equilibrium. A battery that has run down has reached equilibrium. Many important reactions never reach equilibrium—for example a product may be continually removed from a system. The reactions are said to run under nonequilibrium conditions.

Review Questions

1. What is a chemical equilibrium?

2. What sort of process never reaches equilibrium?

3. How is the expression for an equilibrium constant derived?

4. What sorts of reactants or products are not included in a equilibrium constant expression?

5. What is the difference between K_c and K_p?

6. What happens to the equilibrium constant when the stoichiometric co-efficients in the chemical equation are all doubled? What is the general relationship between what happens to K and what is done to the coefficients?

7. What happens to the value of K when the reaction is reversed?

8. What does the numerical value of K tell you about the state of the system at equilibrium?

9. What is the reaction quotient? What does a comparison of Q and K tell you about the state of the system?

10. When is it reasonable to assume that change in a concentration x is negligible compared to the original concentration. In other words, when can you assume that $(1-x) = 1$.

11. What three disturbances are possible for a system in equilibrium?

12. Is K always constant regardless of the disturbance of the system?

13. Describe in your own words the meaning of LeChatelier's principle.

14. Does addition of a catalyst change the position of a system in equilibrium?

Answers to Review Questions

1. A chemical equilibrium is a dynamic system; reactants are combining to form products and the products are combining to form reactants. The two reactions, the forward one and the reverse one, occur at the same rate so that there is no net change in the concentrations.

2. There are many nonequilibrium systems, systems which never reach equilibrium. A typical system is water evaporating from an open beaker, it will all eventually evaporate as the water vapor diffuses away from the open system. In order to reach equilibrium, the system must be closed—a lid must be put on the beaker.

3. A system in equilibrium is characterized by the relative concentrations of products and reactants at equilibrium. These relative concentrations of reactant and product present at equilibrium are represented by the equilibrium constant, K.

$$K = \frac{[\text{Products}]}{[\text{Reactants}]}$$

4. Expressions for the equilibrium constant K do not contain concentration terms for species in a phase different from that in which the equilibrium exists. For example, equilibria in solution do not include solids or gases in the expression; equilibria in the gas state do not include liquids or solids. Also, a solution equilibrium expression does not include the solvent—the concentration is considered constant.

5. K_c is the equilibrium constant written in terms of the component concentrations (in mol liter^{-1}). K_p is the equilibrium constant written in terms of the partial pressures of the components. There are related by the expression shown on page 236 of this study guide.

6. The equilibrium constant is squared when the stoichiometric coefficients in the chemical equation are all doubled. In general the value of the new K equals the value for the original K raised to a power equal to the factor by which the coefficients were changed.

7. The value of K when the reaction is reversed is the reciprocal of the original value.

8. A large value for K means a proportionately large concentration of product, i.e. a product-favored process. A value greater than 1, in fact, means that the concentration of product is higher than the concentration of reactant at equilibrium. On the other hand, a smaller value for K means less product at equilibrium.

9. The reaction quotient $Q = \dfrac{[Products]}{[Reactants]}$

   If the concentrations [Products] and [Reactants] are equilibrium concentrations then Q = K and the system is at equilibrium. If the reaction quotient Q is larger than K, the system contains too much product and in order to get into equilibrium, some product must form reactant. If the reaction quotient Q is less that K, then there is not enough product and the reaction must proceed further to the product side. The comparison of Q and K tells you which way the system will move in order to get into equilibrium.

10. It's reasonable to assume that the change in a concentration x is negligible compared to the original concentration (i.e. (1−x) = 1), when x is less than 5% of the smallest initial concentration.

11. There are three common ways to disturb a chemical system in equilibrium:
    Change the temperature.
    Change the concentration or partial pressure of reactant or product.
    Change the volume (or pressure) for a system involving gases.

12. No, K will change if the temperature changes. Note that K will remain unchanged if the temperature stays the same (regardless of concentration, volume, or pressure changes).

13. LeChatelier's Principle states that if a system in equilibrium is subjected to a stress by altering the conditions that characterize that state of equilibrium, the system will tend to adjust itself to accommodate the stress placed upon it.

14. No, the position of equilibrium will not change if a catalyst is added. The equilibrium will probably be reached more quickly, but the final position will be the same.

Ce que je rêve, c'est un art d'équilibre, de pureté, de tranquillité, sans sujet inquiétant ou préoccupant, que soit...

What I dream of is an art of balance, of purity and tranquility, without troubling or depressing subject matter...

Henri Matisse (1869-1954)

Study Questions and Problems

1. Write the expressions for the equilibrium constant K_c for the following reactions:

   a. $4NH_3(g) + 7O_2(g) \rightleftharpoons 4NO_2(g) + 6H_2O(l)$

   b. $HCN(aq) + H_2O(l) \rightleftharpoons H_3O^+(aq) + CN^-(aq)$

   c. $PCl_5(g) \rightleftharpoons PCl_3(g) + Cl_2(g)$

   d. $CaCO_3(s) \rightleftharpoons CaO(s) + CO_2(g)$

   e. $3O_2(g) \rightleftharpoons 2O_3(g)$

   f. $2H_2O(l) \rightleftharpoons H_3O^+(aq) + OH^-(aq)$

   g. $3Zn(s) + 2Fe^{3+}(aq) \rightleftharpoons 2Fe(s) + 3Zn^{2+}(aq)$

2. Write the equilibrium constant expressions for the following reactions. How are they related to one another?

   a. $2N_2O(g) + 3O_2(g) \rightleftharpoons 4NO_2(g)$

   b. $N_2O(g) + 3/2\ O_2(g) \rightleftharpoons 2NO_2(g)$

   c. $4NO_2(g) \rightleftharpoons 2N_2O(g) + 3O_2(g)$

3. Calculate the value of the equilibrium constant for the following system, given the data shown:

   $$H_2(g) + CO_2(g) \rightleftharpoons H_2O(g) + CO(g)$$

   Concentrations at equilibrium:
   $[H_2] = 1.5$ mol liter^{-1}
   $[CO_2] = 2.5$ mol liter^{-1}
   $[H_2O] = 0.5$ mol liter^{-1}
   $[CO] = 3.0$ mol liter^{-1}

4. Chlorine molecules will dissociate at high temperatures into chlorine atoms. At 3000°C, for example, K_c for the equilibrium shown is 0.55. If the partial pressure of chlorine molecules is 1.5 atm, calculate the partial pressure of the chlorine atoms:

   $$Cl_2(g) \rightleftharpoons 2Cl(g)$$

5. Suppose that 0.50 moles of hydrogen gas, 0.50 moles of iodine gas, and 0.75 moles of hydrogen iodide gas are introduced into a 2.0 liter vessel and the system is allowed to reach equilibrium.

   $$H_2(g) + I_2(g) \rightleftharpoons 2HI(g)$$

   Calculate the concentrations of all three substances at equilibrium. At the temperature of the experiment K_c equals 2.0×10^{-2}.

6. If the mechanism of a chemical equilibrium consists of two reversible elementary steps, each with its own equilibrium constant K_{c1} and K_{c2}, what expression relates the equilibrium constant K_c for the overall equilibrium to the two constants K_{c1} and K_{c2}?

7. When 2.0 mol of carbon disulfide and 4.0 mol of chlorine are placed in a 1.0 liter flask, the following equilibrium system results. At equilibrium, the flask is found to contain 0.30 mol of carbon tetrachloride. What quantities of the other components are present in this equilibrium mixture?

$$CS_2(g) \ + \ 3Cl_2(g) \ \rightleftharpoons \ S_2Cl_2(g) \ + \ CCl_4(g)$$

8. 3.0 moles each of carbon dioxide, hydrogen, and carbon are placed in a 2.0 liter vessel and allowed to come to equilibrium according to the equation:

$$CO(g) \ + \ H_2(g) \ \rightleftharpoons \ C(s) \ + \ H_2O(g)$$

If the equilibrium constant at the temperature of the experiment is 4.0, what is the equilibrium concentration of water?

9. Nitrosyl chloride NOCl decomposes to nitric oxide and chlorine when heated:

$$2NOCl(g) \ \rightleftharpoons \ 2NO(g) \ + \ Cl_2(g)$$

At 600K, the equilibrium constant K_p is 0.060. In a vessel at 600K, there is a mixture of all three gases. The partial pressure of NOCl is 675 torr, the partial pressure of NO is 43 torr and the partial pressure of chlorine is 23 torr.

a. What is the value of the reaction quotient?
b. Is the mixture at equilibrium?
c. In which direction will the system move to reach equilibrium?
d. When the system reaches equilibrium, what will be the partial pressures of the components in the system?

10. Sulfuryl chloride decomposes at high temperatures to produce sulfur dioxide and chlorine gases:

$$SO_2Cl_2(g) \ \rightleftharpoons \ SO_2(g) \ + \ Cl_2(g)$$

At 375°C, the equilibrium constant K_c is 0.045. If there are 2.0 grams of sulfuryl chloride, 0.17 gram of sulfur dioxide, and 0.19 gram of chlorine present in a 1.0 liter flask,

a. What is the value of the reaction quotient?
b. Is the system at equilibrium?
c. In which direction will the system move to reach equilibrium?

11. Ammonium chloride is placed inside a closed vessel where it comes into equilibrium at 400°C according to the equation shown. Only these three substances are present inside the vessel. If K_p for the system at 400°C is 0.640, what is the pressure inside the vessel?

$$NH_4Cl(s) \quad \rightleftharpoons \quad NH_3(g) \quad + \quad HCl(g)$$

12. Bromine and chlorine react to produce bromine monochloride according to the equation. $K_c = 36.0$ under the conditions of the experiment.

$$Br_2(g) \quad + \quad Cl_2(g) \quad \rightleftharpoons \quad 2BrCl(g)$$

If 0.180 moles of bromine gas and 0.180 moles of chlorine gas are introduced into a 3.0 liter flask and allowed to come to equilibrium, what is the equilibrium concentration of the bromine monochloride? How much BrCl is produced?

13. When ammonia is dissolved in water, the following equilibrium is established. If the equilibrium constant is 1.8×10^{-5}, calculate the hydroxide ion concentration in the solution if 0.100 mol of ammonia is dissolved in sufficient water to make 500 mL of solution.

$$NH_3(aq) \quad + \quad H_2O(l) \quad \rightleftharpoons \quad NH_4^+(aq) \quad + \quad OH^-(aq)$$

14. The following reaction is exothermic:

$$Ti(s) \quad + \quad 2Cl_2(g) \quad \rightleftharpoons \quad TiCl_4(g)$$

List all the ways the yield of the product $TiCl_4$ could be increased.

Answers to Study Questions and Problems

1. a. $K = \dfrac{[NO_2]^4}{[NH_3]^4[O_2]^7}$ $H_2O(l)$ is a liquid and therefore not included.

    b. $K = \dfrac{[H_3O^+][CN^-]}{[HCN]}$ $H_2O(l)$ is the solvent and is not included.

    c. $K = \dfrac{[PCl_3][Cl_2]}{[PCl_5]}$

    d. $K = [CO_2]$ $CaCO_3(s)$ and $CaO(s)$ are not included.

    e. $K = \dfrac{[O_3]^2}{[O_2]^3}$

    f. $K = [H_3O^+][OH^-]$ $H_2O(l)$ is the solvent and is not included.

    g. $K = \dfrac{[Zn^{2+}]^3}{[Fe^{3+}]^2}$ $Zn(s)$ and $Fe(s)$ are not included.

2.

a. $2N_2O(g) + 3O_2(g) \rightleftharpoons 4NO_2(g) \quad K_1 = \dfrac{[NO_2]^4}{[N_2O]^2[O_2]^3}$

b. $N_2O(g) + 3/2\, O_2(g) \rightleftharpoons 2NO_2(g) \quad K_2 = \dfrac{[NO_2]^2}{[N_2O][O_2]^{3/2}}$

c. $4NO_2(g) \rightleftharpoons 2N_2O(g) + 3O_2(g) \quad K_3 = \dfrac{[N_2O]^2[O_2]^3}{[NO_2]^4}$

$K_1 = (K_2)^2$ and $K_1 = (K_3)^{-1}$ and $K_2 = (K_3)^{-1/2}$

3. $K = \dfrac{[H_2O][CO]}{[H_2][CO_2]} = (0.5 \times 3.0)/(1.5 \times 2.5) = 0.40$

4. $Cl_2(g) \rightleftharpoons 2Cl(g) \qquad K_c = \dfrac{[Cl]^2}{[Cl_2]}$ and $K_p = \dfrac{P_{Cl}^2}{P_{Cl_2}}$

$K_p = K_c(RT)^{\Delta n} = 0.55 \times 0.08206 \times 3273K = 1.48 \times 10^2 = \dfrac{P_{Cl}^2}{P_{Cl_2}}$

$\Delta n = 1$
an increase of one mole of gas

and therefore $P_{Cl} = 14.9$ atm

5.

		$H_2(g)$	$+ \quad I_2(g)$	$\rightleftharpoons \quad 2HI(g)$
I	Initial	0.25	0.25	0.375
C	Change	+x	+x	−2x
E	Equilibrium	0.25+x	0.25+x	0.375−2x

Remember always to use concentrations (or partial pressures).
Convert amounts given in moles to concentrations by dividing by the volume.

Remember that the change is always in the stoichiometric ratio.

If you don't know which way the system will move, just choose a direction arbitrarily, then see if x turns out to be negative or positive.

Look out for the opportunity to take the square root of both sides—it avoids the quadratic.

$K = \dfrac{[HI]^2}{[H_2][I_2]} = \dfrac{(0.375-2x)^2}{(0.25+x)(0.25+x)} = 2.0 \times 10^{-2}$

$(0.375-2x) = (0.25+x) \times 0.1414$ so $x = 0.159$

the equilibrium concentrations are:

$[H_2] = 0.41$ moles liter^{-1}

$[I_2] = 0.41$ moles liter^{-1}

$[HI] = 0.058$ moles liter^{-1}

6. The equilibrium constant for the combination of two successive equilibria is the product of the equilibrium constants for the two steps.

$K_c = K_{c1} \times K_{c2}$

Don't make problems like these more complicated than they have to be; there's no need for any algebra in this problem.

If 0.30 moles of CCl_4 are formed, then the changes in the other concentrations can be calculated directly from the stoichiometry.

7.

		$CS_2(g)$	+	$3Cl_2(g)$	$\rightleftharpoons$	$S_2Cl_2(g)$	+	$CCl_4(g)$
I	Initial	2.0		4.0		0.0		0.0
C	Change	−0.30		−0.90		+0.30		+0.30
E	Equilibrium	1.7		3.1		0.30		0.30

Remember always to use concentrations (or partial pressures).
Convert amounts given in moles to concentrations by dividing by the volume.

8.

		$CO(g)$	+	$H_2(g)$	$\rightleftharpoons$	$C(s)$	+	$H_2O(g)$
I	Initial	1.5		1.5		3.0		0.0
C	Change	−x		−x		+x		+x
E	Equilibrium	1.5−x		1.5−x		3.0+x		x

Note that C(s) is not included in the equilibrium expression! However, it is sometimes useful to include the quantities in the table—just in case there's any likelihood of all the solid being used up.

$$K = \frac{[H_2O]}{[CO][H_2]} = \frac{(x)}{(1.5-x)(1.5-x)} = 4.0, \quad \text{therefore } x = 1.0 \text{ which is the equilibrium concentration of water}$$

9. $2NOCl(g) \quad \rightleftharpoons \quad 2NO(g) \quad + \quad Cl_2(g)$

a. $Q_p = \dfrac{P_{NO}^2 \times P_{Cl2}}{P_{NOCl}^2} = (43^2 \times 23)/\,675^2 = 0.093$

b. No, Q_p is greater than K_p; ie. there is too much product present.

c. The system will move to the left.

		$2NOCl(g)$	$\rightleftharpoons$	$2NO(g)$	+	$Cl_2(g)$
I	Initial	675		43		23
C	Change	+2x		−2x		−x
E	Equilibrium	675+2x		43−2x		23−x

Solving higher order equations like this for x using algebra is not especially easy. An iterative method using a good calculator or a spreadsheet is highly recommended. Using a spreadsheet for example, enter the expression for K in a cell, referencing another cell for x. Then put values for x in that cell until the expression yields a value of 0.060. You should be able to narrow the value for x to a sufficient number of sig. fig. in 5 or 6 tries.

d. $K_p = \dfrac{P_{NO}^2 \times P_{Cl2}}{P_{NOCl}^2} = (43-2x)^2(23-x)/(675+2x)^2 = 0.060$

x = 2.90 so the partial pressure of NOCl at equilibrium = 681 torr
and the partial pressure of NO = 37 torr
and the partial pressure of Cl_2 = 20 torr

10. $SO_2Cl_2(g) \quad \rightleftharpoons \quad SO_2(g) \quad + \quad Cl_2(g)$

SO_2Cl_2	2.0 grams	MM = 134.97 g mol−1	Moles = 0.0148
SO_2	0.17 gram	MM = 64.064 g mol−1	Moles = 0.0027
Cl_2	0.19 gram	MM = 70.91 g mol−1	Moles = 0.0027

Must convert to moles and then to moles/liter.

a. $Qp = (0.0027)^2/0.0148 = 4.93 \times 10^{-4}$.

b. No, the system is not at equilibrium. Q_p does not equal K_p.

c. Q_p is too small, the reaction will move toward product.

11. $NH_4Cl(s) \rightleftharpoons NH_3(g) + HCl(g)$

The expression for K_p is:

$K_p = P_{NH_3} \times P_{HCl} = 0.640$

Since the partial pressures of ammonia and hydrogen chloride must be equal to one another, each must equal 0.80 atm.

The total pressure is the sum of the partial pressures, so the total pressure = 1.60 atm.

The amount of NH_3 present must equal the amount of HCl present since one cannot be made without the other. Therefore their partial pressures are equal.

12.

		$Br_2(g)$ +	$Cl_2(g)$ $\rightleftharpoons$	$2BrCl(g)$
I	Initial	0.060	0.060	0.0
C	Change	–x	–x	+2x
E	Equilibrium	0.060–x	0.060–x	+2x

Remember to change amounts to concentrations.

Take the square root of both sides.

$K = \dfrac{[BrCl]^2}{[Br_2][Cl_2]} = \dfrac{(2x)^2}{(0.060-x)(0.060-x)} = 36.0$

x = 0.045

So the concentration of BrCl produced = 2x = 0.090 mol L^{-1}.

Quantity produced = 0.090 mol L^{-1} × 3 L = 0.27 mol = 31 grams.

Don't assume x is the desired answer to the problem! In this case the required quantity is the concentration of BrCl (which is 2x) and the total amount of BrCl (which is the volume of 3 liters × this concentration).

13.

		NH_3 +	H_2O $\rightleftharpoons$	NH_4^+ +	OH^-
I	Initial	0.200		0.0	0.0
C	Change	–x		+x	+x
E	Equilibrium	0.200–x		+x	+x

Remember to convert amounts into concentrations—it's easy to forget.

Can assume that 0.200–x is near enough to 0.200.

$K = \dfrac{[NH_4^+][OH^-]}{[NH_3]} = \dfrac{x^2}{(0.200-x)} = 1.8 \times 10^{-5}$

x = 1.90×10^{-3} M —this is the concentration of hydroxide ion.

14. $Ti(s) + 2Cl_2(g) \rightleftharpoons TiCl_4(g) + heat$

The yield of the product $TiCl_4$ is increased by moving the system to the right:

a.   remove heat; cool the system down.

b.   remove $TiCl_4(g)$.

c.   add chlorine gas $Cl_2(g)$.

d.   increase the pressure (or decrease the volume).

Note that:

    adding more Ti(s) has no effect.
    adding a catalyst has no effect.

Challenge 10

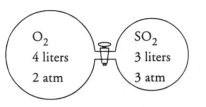

This is a more difficult variation
of the challenge encountered in
Chapter 12.

The apparatus shown consists of two glass bulbs connected by a tap which is closed. In the first bulb, volume 4 liters, there is a sample of pure oxygen at 2 atm pressure. In the second bulb, volume 3 liters, there is a sample of pure sulfur dioxide at a pressure of 3 atm.

Maintaining a constant temperature of 1000K throughout, the tap is opened, the gases are allowed to mix, and they react to produce sulfur trioxide according to the equation written below. Assume that the reaction reaches equilibrium, where $K_p = 3.00$.

$$2SO_2(g) \ + \ O_2(g) \ \rightleftharpoons \ 2SO_3(g)$$

The calculations involve solving high order equations and iteration using a spreadsheet or an advanced calculator is recommended.

x cannot be neglected in these calculations.

a. What is the partial pressure of the sulfur dioxide in the apparatus at the end of the reaction?

b. What is the partial pressure of the oxygen in the apparatus at the end of the reaction?

c. What is the partial pressure of the sulfur trioxide in the apparatus at the end of the reaction?

d. What is the total pressure in the apparatus at the end of the reaction?

CHAPTER 17

Introduction

Earlier in the text (Chapter 5) you learned to classify chemical reactions in terms of acid-base, redox, and precipitate or gas forming reactions. These reactions will be examined more closely in the next few chapters. In this chapter we will investigate acid-base reactions. Many of the acid-base reactions that you will encounter occur in aqueous solution. Like the reactions described in Chapter 16, a reaction between an acid and a base will continue until the system reaches equilibrium. We will examine various definitions of what an acid is and what a base is, and investigate how they behave in solution.

Contents

17.1   Acids, Bases, and Arrhenius

Although acids and bases had been studied for hundreds of years, it wasn't until Svante Arrhenius proposed (about 100 years ago) definitions of what acids and bases were that some systematic understanding of their behavior became possible. Arrhenius proposed that an **acid** is a substance that increases the concentration of hydrogen ions H^+ in solution. A **base** is a substance that increases the concentration of hydroxide ions OH^-. Acids and bases react with one another in a **neutralization** reaction to produce water.

The Arrhenius definitions apply to reactions in aqueous solution.

Arrhenius acid: produces H^+ in water.

Arrhenius base: produces OH^- in water.

$$H_3O^+(aq) + OH^-(aq) \rightleftharpoons 2H_2O(l)$$

17.2   The Hydronium Ion and Water Autoionization

Water is often called the universal solvent; it has many unusual properties. Most of the acid-base reactions we will examine will be in water solution so let's first examine water itself. Ions are hydrated in water—there is a strong attraction between the ion and the dipole of the water molecule. The *(aq)* notation indicates this hydration. The hydroxide ion and the hydrogen ion are hydrated

The properties perhaps do not appear unusual because they are so familiar, but compared to most other solvents, water is unusual.

Although the hydrated proton is written H_3O^+, the number of water molecules associated with the proton exceeds one.

just like any other ion. The hydrated hydrogen ion is often represented as the hydronium ion H_3O^+ to indicate that the H^+ ion does not exist alone in aqueous solution.

Even in pure water, H_3O^+ and OH^- ions exist. The water is said to **autoionize**. When two water molecules are hydrogen bonded together, there is sometimes enough energy to break the bond in a molecule instead of the bond between the molecules. Hydronium and hydroxide ions are the result:

K is however very small: 10^{-14} at 25°C.

$$H-O \overset{\text{break}}{\underset{}{\diagup}} H \quad \rightleftharpoons \quad OH^- \ + \ H_3O^+$$
$$H--O\diagdown H$$

17.3   The Brønsted Concept of Acids and Bases

Brønsted-Lowry acid: proton donor.

Brønsted-Lowry base: proton acceptor.

Johannes Brønsted and Thomas Lowry proposed a new set of definitions for acids and bases in 1923. They proposed that an acid is a substance that donates a hydrogen ion (proton) to another substance. And they proposed that a base is the substance that accepts the hydrogen ion (proton) from the acid. Consider the ionization of acetic acid:

$$CH_3CO_2H \textit{(aq)} \ + \ H_2O \textit{(l)} \ \rightleftharpoons \ H_3O^+\textit{(aq)} \ + \ CH_3CO_2^-\textit{(aq)}$$

The acetic acid molecule donates a hydrogen ion and becomes an acetate ion. Acetic acid is an acid! The water molecule accepts the hydrogen ion and becomes the hydronium ion; in this reaction water is acting as a base. In the reverse direction, the hydronium ion donates a hydrogen ion to the acetate ion, forming acetic acid and water. So the hydronium ion is an acid, and the acetate ion is a base.

Recall, from Chapter 5:

Monoprotic acids:

hydrochloric acid HCl
nitric acid HNO_3
perchloric acid $HClO_4$
sulfuric acid H_2SO_4
acetic acid CH_3CO_2H
carbonic acid H_2CO_3
hydrocyanic acid HCN
hypochlorous acid $HOCl$

Polyprotic acids:

sulfuric acid H_2SO_4
phosphoric acid H_3PO_4
carbonic acid H_2CO_3
oxalic acid $H_2C_2O_4$

Some acids are capable of donating one proton only, and are known as **monoprotic** acids. Other acids can donate more than one proton and are called **polyprotic** acids. Some bases can accept more than one proton and are called polyprotic bases. Anions of the polyprotic acids, such as SO_4^{2-} and CO_3^{2-}, are examples.

Amphiprotic substances can act as either acids or bases depending upon the conditions. Water is an example; it acts as a base when it accepts a proton from acetic acid (see reaction above), but it acts as an acid when it donates a proton to ammonia:

$$NH_3\textit{(aq)} \ + \ H_2O\textit{(l)} \ \rightleftharpoons \ NH_4^+\textit{(aq)} \ + \ OH^-\textit{(aq)}$$

Movement from one side of the equilibrium system to the other involves the transfer of the proton.

In a Brønsted-Lowry acid-base equilibrium, an acid donates a proton and a base accepts it. The reaction is a proton transfer. The reaction can go in the reverse direction; again an acid donates a proton and a base accepts it. So, on both sides of the equilibrium, there is an acid and a base. When an acid gives up its proton it forms a base; this base is called its conjugate base; together the acid and base make a **conjugate pair**. The two always differ by one proton:

$$HCN\textit{(aq)} \ + \ H_2O\textit{(l)} \ \rightleftharpoons \ H_3O^+\textit{(aq)} \ + \ CN^-\textit{(aq)}$$
$$\text{acid} \qquad\qquad \text{base} \qquad\qquad \text{acid} \qquad\qquad \text{base}$$

HCN is the conjugate acid of CN⁻.
CN⁻ is the conjugate base of HCN.
Together HCN and CN⁻ make a conjugate acid-base pair.
The two partners differ by one H^+.
Of the two partners, the acid has the H^+, the base does not.

The same applies to the H_2O and its conjugate acid H_3O^+.

Some acids are stronger than others; they are more willing to donate their protons. **Strong acids** are virtually 100% ionized; **weak acids** are only partially, sometimes only very slightly, ionized. For example, hydrochloric acid is a strong acid; one mole of hydrochloric acid yields one mole of H^+ ions in solution. Acetic acid, on the other hand, is a weak acid, one mole of acetic acid in solution yields only 0.001 mole of H^+ ions; most of the acetic acid remains as unionized molecules. The same applies to bases; there are strong bases like oxide O^{2-} and hydroxide OH^-, and weak bases like ammonia NH_3.

In a Brønsted-Lowry acid-base equilibrium, the stronger the acid, the weaker its conjugate base, and *vice versa*. This is because the more the acid wants to donate its proton, the less its conjugate base wants it back.

The stronger acid and the stronger base lie on one side of the equilibrium, and the weaker acid and the weaker base lie on the other side. The system exists predominantly on the weak side.

$$NH_3(aq) \quad + \quad H_2O(l) \quad \rightleftharpoons \quad NH_4^+(aq) \quad + \quad OH^-(aq)$$

weaker base   weaker acid   stronger acid   stronger base
equilibrium lies on this side

17.4   Strong Acids and Bases

The strongest acid that can exist in water is the hydrated proton, the hydronium ion H_3O^+. Any acid stronger than the hydronium ion ionizes completely to form the hydronium ion when placed in water. Hydrochloric acid, for example, does not exist as HCl molecules in water, it forms H_3O^+ and Cl^- ions because H_3O^+ is the strongest acid that can exist in water. All **strong acids** are **levelled** to H_3O^+ in aqueous solution. The system lies on the right (the weaker side):

$$HCl(aq) \quad + \quad H_2O(l) \quad \rightleftharpoons \quad H_3O^+(aq) \quad + \quad Cl^-(aq)$$

The strongest base that can exist in water is the hydroxide ion OH^-. Any base stronger than hydroxide is levelled to hydroxide when dissolved in water. For example, the amide ion NH_2^- is a stronger base than hydroxide:

$$NH_2^-(aq) \quad + \quad H_2O(l) \quad \rightleftharpoons \quad NH_3(aq) \quad + \quad OH^-(aq)$$

And so the amide ion cannot exist in aqueous solution; it is levelled to hydroxide. The system lies on the right (the weaker side). Other strong bases are the hydride ion H^- and the methoxide ion CH_3O^-.

Be able to identify the acids and bases in a Brønsted-Lowry acid-base equilibrium.

Be able to pair the acids and bases in conjugate pairs.

Think of a Brønsted-Lowry acid base equilibrium as a competition for the hydrogen ion. The stronger acid doesn't want it, and the stronger base gets it, so the system lies on the side of the conjugate base of the stronger acid (the weaker base) and the conjugate acid of the stronger base (the weaker acid)—i.e on the weaker side.

Table 17.3 on page 794 can be used to determine the stronger of two acids or the stronger of two bases, and therefore the side on which an equilibrium system predominantly lies.

Strong acids:

Important ones to know:

hydrochloric acid HCl
nitric acid HNO_3
perchloric acid $HClO_4$
sulfuric acid H_2SO_4

Some other ones:

hydrobromic acid HBr
hydroiodic acid HI
selenic acid H_2SeO_4

17.5   Weak Acids and Bases

Hydrated metal cations like $[Cu(H_2O)_6]^{2+}$ are often acidic. The metal withdraws electrons from the water molecule making loss of a proton easier.

Most acids and bases are weak; they ionize relatively little in aqueous solution and remain predominantly in the molecular form. The relative strength can be expressed by an equilibrium constant K_a or K_b where the larger the constant, the stronger the acid or base. For a weak acid HA:

$$HA(aq) \quad + \quad H_2O(l) \quad \rightleftharpoons \quad H_3O^+(aq) \quad + \quad A^-(aq)$$

$$K_a = \frac{[H_3O^+][A^-]}{[HA]}$$

Recall that the solvent (water) is not included in the equilibrium constant expression.

For a weak base B, a similar equilibrium equation and expression for the equilibrium constant K_b can be written:

$$B(aq) \quad + \quad H_2O(l) \quad \rightleftharpoons \quad BH^+(aq) \quad + \quad OH^-(aq)$$

$$K_b = \frac{[BH^+][OH^-]}{[B]}$$

$K_a > 1$:   strong acid
v. weak conjugate base

$K_a = 1$ to 10^{-16}:   weak acid
weak conjugate base

$K_a < 10^{-16}$:   very weak acid
strong conjugate base

Ionization constants are listed in Table 17.4 on page 799 of the text.

17.6   Water and the pH Scale

Water autoionizes, producing hydronium ions and hydroxide ions:

$$H_2O(l) \quad + \quad H_2O(l) \quad \rightleftharpoons \quad H_3O^+(aq) \quad + \quad OH^-(aq)$$

$K = \dfrac{[H_3O^+][OH^-]}{[H_2O]^2}$ but recall that the solvent need not be included so

$K_w = [H_3O^+][OH^-]$ — the ionization constant for pure water

At 25°C, $K_w = 1.0 \times 10^{-14}$, and in pure water:
$[H_3O^+] = 1.0 \times 10^{-7}$
$[OH^-] = 1.0 \times 10^{-7}$

At higher temperatures K_w is larger.

Because one hydroxide ion is produced for every hydronium ion, the two concentrations are equal in pure water. If acid is added, the $[H_3O^+]$ increases and the $[OH^-]$ decreases, but the product K_w remains the same. Likewise, if base is added, the $[OH^-]$ increases and the $[H_3O^+]$ decreases, but the product K_w remains the same.

The K_a for a weak acid is related to the K_b for its conjugate base by the expression:

$$K_w = K_a \times K_b = 1.0 \times 10^{-14} \text{ at } 25°C$$

If you know either K_a or K_b at 25°C, you can always calculate the other.

An alternative way of expressing the hydronium ion concentration is by using the pH scale. The pH of a solution is defined as the negative logarithm to the base 10 of the hydronium ion concentration:

$$pH = -\log_{10}[H_3O^+]$$

In fact:
p(anything) $= -\log_{10}$[anything]
For example,
$pK_w = -\log_{10}[K_w] = 14$ at 25°C

Similarly the pOH is defined as:

$$pOH = -\log_{10}[OH^-]$$

Read Problem solving Tips and Ideas 17.2 on page 808.

Color changes for some common indicators are shown in Figure 17.11 on page 810 of the text.

$$pH + pOH = pK_w = 14$$

The pH of a solution can be determined using a pH meter or by using indicator solutions. These indicators have different colors at different pHs.

17.7   Equilibria involving Weak Acids and Bases

The principles of the equilibrium systems encountered in Chapter 16 can be applied to aqueous solutions of weak acids and bases. The equilibrium constants K_a and K_b can be determined if the concentrations of the various species present in the solution are known. Most often these are determined by measuring the pH of the solution. If the acid or base is weak, and the initial concentration of acid (or base) is at least $100 \times K_a$ (or K_b), then the approximation that $[\text{acid}]_{\text{initial}} = [\text{acid}]_{\text{equilibrium}}$ is valid. Otherwise a quadratic equation must usually be solved.

For a weak acid:

$$HA(aq) \;+\; H_2O(l) \;\rightleftharpoons\; H_3O^+(aq) \;+\; A^-(aq)$$

$$K_a = \frac{[H_3O^+][A^-]}{[HA]} \quad \text{and because } [H_3O^+] = [A^-],$$

$$= \frac{[H_3O^+]^2}{[HA]} \quad \text{or } [H_3O^+] = \sqrt{K_a[HA]_{\text{equil}}}$$

The assumption that $[HA]_{\text{equil}} = [HA]_{\text{initial}}$ is valid if $[HA]_{\text{initial}} > 100 \times K_a$

These equilibrium problems are exactly the same as those described in Chapter 16. Try some of the study problems at the end of this study guide.

Keep the data organized in an ICE table.

Because K_a and K_b are typically quite small, the approximation that $[HA]_{eq} = [HA]_{init}$ is very often valid.

17.8   Acid-Base Properties of Salts: Hydrolysis

The conjugate bases (salts) of weak acids are indeed basic. When they dissolve in water, they produce hydroxide ions through **hydrolysis**. The general sequence of events is:

$$\text{dissociation} \;\rightarrow\; \text{hydration} \;\rightarrow\; \text{hydrolysis}$$

A salt dissociates to produce ions; the ions are solvated (hydrated) by the water molecules; and then, if the salt is a salt of a weak acid, hydrolysis by the anion of the salt occurs. Likewise, if the salt is a salt of a weak base, hydrolysis of the cation of the salt occurs.

$$A^-(aq) \;+\; H_2O(l) \;\rightleftharpoons\; HA(aq) \;+\; OH^-(aq)$$

$$K_b = \frac{[HA][OH^-]}{[A^-]}$$

Qualitatively, the pH of a salt solution can be determined by establishing the strengths of the acid and base used to make the salt:

Hydrolysis literally means the splitting of or by a water molecule; an anion of a weak acid (the conjugate base) removes a hydrogen ion from the water molecule leaving a hydroxide ion which causes the solution to be basic.

K_b is sometimes written as K_h (for hydrolysis).

Using this expression you should be able to calculate the pH of a salt solution.

Cation	Anion	pH of the salt solution
from a strong base	from a strong acid	neutral, pH = 7
from a strong base	from a weak acid	basic, pH > 7
from a weak base	from a strong acid	acidic, pH < 7
from a weak base	from a weak acid	depends upon relative strengths of acid & base

Be careful with the anions of polyprotic acids. For example, $NaHSO_4$ is acidic not neutral.

17.9   Polyprotic Acids and Bases

Some acids are capable of donating more than one proton; they are called **polyprotic**. Examples are the diprotic strong acid H_2SO_4, and the triprotic phosphoric acid H_3PO_4:

$$H_3PO_4(aq) + H_2O(l) \rightleftharpoons H_3O^+(aq) + H_2PO_4^-(aq) \quad K_{a1} = 7.5 \times 10^{-3}$$

$$H_2PO_4^-(aq) + H_2O(l) \rightleftharpoons H_3O^+(aq) + HPO_4^{2-}(aq) \quad K_{a2} = 6.2 \times 10^{-8}$$

$$HPO_4^{2-}(aq) + H_2O(l) \rightleftharpoons H_3O^+(aq) + PO_4^{3-}(aq) \quad K_{a3} = 3.6 \times 10^{-13}$$

Because the difference in the successive ionization constants is so large (about 10^{-5}), it is always reasonable to assume that there is only one conjugate acid base pair present in the solution to any appreciable extent at one time. This makes the calculation of the pH of the solution easier.

Successive K_a values for polyprotic acids usually differ by about 10^{-5}.

It gets harder and harder to remove successive hydrogen ions from a polyprotic acid.

17.10   Molecular Structure, Bonding, and Acid-Base Behavior

Some insight into the relative strengths of acids can be obtained from the strengths of the H–A bonds in the acid molecules and the affinities of the A fragment for an electron.

For example, in the series HF, HCl, HBr, HI, the bond strengths decrease and the acid strength increases. In the series $HOCl$, $HOClO$, $HOClO_2$, $HOClO_3$, the acid strength increases as the electrons in the HO bonds are more and more strongly withdrawn.

17.11   The Lewis Concept of Acids and Bases

A more general theory of acids and bases was developed by G. N. Lewis in the 1930s. He defined an acid as a substance that would accept an electron pair to form a new bond and a base as a substance that would donate the pair of electrons. An acid is therefore a substance that has an empty orbital available or can make one available. A base is a substance that has a pair of electrons available for donation. Some examples are:

A still more generalized approach than Lewis's concept is referred to as donor-acceptor theory.

The bond that is formed is called a coordinate covalent bond. The product of the Lewis acid base reaction is often called an adduct or complex.

H^+	+	H_2O	$\rightarrow$	H_3O^+
empty 1s		lone pair on O:		
BF_3	+	NH_3	$\rightarrow$	$F_3B–NH_3$
empty sp^3		lone pair on N:		
$Cu^{2+}(aq)$	+	$4NH_3(aq)$	$\rightarrow$	$[Cu(NH_3)_4]^{2+}(aq)$
empty sp^3		lone pair on N:		

adduct

complex

Review Questions

1.   Define an Arrhenius acid and an Arrhenius base. Is the Arrhenius set of definitions restricted to aqueous solutions?

2.   Define a Brønsted-Lowry acid and base.

3. What is the difference between a monoprotic acid and a polyprotic acid? Is there such a thing as a polyprotic base?

4. Define, and give an example of, an amphiprotic substance.

5. What is a conjugate acid-base pair? By what do the two components of a pair differ? How many conjugate acid-base pairs make up a simple Brønsted-Lowry acid-base equilibrium?

6. What is the difference between a strong acid and a weak acid?

7. How is the strength of an acid related to the strength of its conjugate base and *vice versa*?

8. How can you determine on what side of an equilibrium a system predominantly lies?

9. What is the levelling effect?

10. How can you tell whether an acid or base is weak or strong?

See Problem Solving Tips and Ideas 17.1 on page 803 of the text.

11. Write examples of anions acting as acids in aqueous solution.

12. Write an expression for K_w, the water ionization constant.

13. What does the numerical value of K_a or K_b tell you about the state of the aqueous system at equilibrium?

14. What is the relationship between K_a for a weak acid and K_b for its conjugate base? Prove the relationship for the weak acid acetic acid.

15. When is it reasonable to assume that change in the concentration x of a weak acid or base is negligible compared to the original concentration. In other words, when can you assume that $[acid]_{initial} = [acid]_{equilibrium}$ or $[base]_{initial} = [base]_{equilibrium}$?

16. How can you determine, qualitatively, whether a salt will be basic or acidic when dissolved in water?

17. Prove that the concentration of sulfite ion $[SO_3^{2-}]$ in a solution of sulfurous acid is independent of the concentration of sulfurous acid. What is the concentration of sulfite ion?
    $K_{a1} = 1.7 \times 10^{-2}$
    $K_{a2} = 6.4 \times 10^{-8}$

18. Define the Lewis concept of an acid and a base. How do these definitions differ from the Arrhenius and Brønsted-Lowry definitions?

Answers to Review Questions

1. Arrhenius proposed that an acid is a substance that increases the concentration of hydrogen ions H⁺ in solution. A base is a substance that increases the concentration of hydroxide ions OH⁻. The Arrhenius definitions apply only to reactions in aqueous solution.

2. Brønsted and Lowry proposed that an acid be defined as a substance that donates a hydrogen ion (proton) to another substance. And they proposed that a base be defined as the substance that accepts the hydrogen ion (proton) from the acid. The acid-base reaction is a proton transfer reaction.

3. Monoprotic acids are capable of donating one proton only (for example HCl). Polyprotic acids can donate more than one proton (for example H_2SO_4). There are such substances as polyprotic bases—they are bases that accept more than one proton. The carbonate ion (CO_3^{2-}) is an example; it accepts one proton to produce the bicarbonate ion (HCO_3^-) and then another to become carbonic acid (H_2CO_3).

4. Amphiprotic substances can act as acids or bases depending upon the conditions. The bicarbonate ion (HCO_3^-) is an example; it acts as a base when it accepts a proton to produce carbonic acid (H_2CO_3), but it acts as an acid when it donates a proton to produce the carbonate ion (CO_3^{2-}).

5. In a Brønsted-Lowry acid-base equilibrium, an acid donates a proton and a base accepts it. The reaction is a proton transfer. The reaction can go in the reverse direction; again an acid donates a proton and a base accepts it. So, on both sides of the equilibrium, there is an acid and a base. Two conjugate acid-base pairs make up a simple Brønsted-Lowry acid-base equilibrium. When an acid gives up its proton it forms a base; this base is called its conjugate base; and together the acid and base make a conjugate pair. The two always differ by one proton.

    $$HA(aq) \quad + \quad H_2O(l) \quad \rightleftharpoons \quad H_3O^+(aq) \quad + \quad A^-(aq)$$

    acid   base   acid   base

Acids stronger than H_3O^+ are called strong acids.
Acids weaker than H_3O^+ are called weak acids.

H_3O^+ is the dividing line.

6. Strong acids are virtually 100% ionized; weak acids are only partially ionized. For example, nitric acid is a strong acid; one mole of nitric acid yields one mole of H⁺ ions in solution. Nitrous acid, on the other hand, is a weak acid, one mole of nitrous acid in solution yields only 0.01 mole of H⁺ ions; most of the nitrous acid remains as un-ionized molecules. The strongest acid that can exist in water is the hydronium ion, so any acids stronger than the hydronium ion are levelled to hydronium ion in aqueous solution.

7. In a Brønsted-Lowry acid-base equilibrium, the stronger the acid, the weaker its conjugate base, and *vice versa*. This is because the more the acid wants to donate its proton, the less its conjugate base wants it back. For example consider the strong acid HCl; the conjugate base of HCl is the chloride ion. The chloride ion is very willing to give up its proton; that's what makes the acid strong. It does not want the proton back; that's what makes the base weak.

8. In a Brønsted-Lowry acid-base equilibrium, the stronger acid and the stronger base lie on one side of the equilibrium, and the weaker acid and the weaker base lie on the other side. The system exists predominantly on the weak side.

$$NH_3(aq) \ + \ H_2O(l) \ \rightleftharpoons \ NH_4^+(aq) \ + \ OH^-(aq)$$

weaker base   weaker acid   stronger acid   stronger base
equilibrium lies on this side

9. The strongest acid that can exist in water is the hydronium ion H_3O^+. Any acid stronger than the hydronium ion ionizes completely to form the hydronium ion when placed in water. Nitric acid, for example, does not exist as HNO_3 molecules in water, it forms H_3O^+ and NO_3^- ions because H_3O^+ is the strongest acid that can exist in water. All strong acids are levelled to H_3O^+ in aqueous solution. The system lies on the right (the weaker side):

$$HNO_3(aq) \ + \ H_2O(l) \ \rightleftharpoons \ H_3O^+(aq) \ + \ NO_3^-(aq)$$

Likewise, the strongest base that can exist in water is the hydroxide ion OH^-. Any base stronger than hydroxide is levelled to hydroxide when dissolved in water. For example, the hydride ion H^- is a stronger base than hydroxide:

$$H^-(aq) \ + \ H_2O(l) \ \rightleftharpoons \ H_2(aq) \ + \ OH^-(aq)$$

And so the hydride ion cannot exist in aqueous solution; it is levelled to hydroxide. The system lies on the right (the weaker side).

10. Remember the four common strong acids: perchloric $HClO_4$, nitric HNO_3, hydrochloric HCl, and sulfuric H_2SO_4. And remember the weak base ammonia NH_3; the other bases you commonly encounter are the hydroxides of Group 1 and Group 2 metals, and these are all strong.

There are other strong acids, but these are the four common ones.

11. Anions still possessing a hydrogen, in other words the anions of polyprotic acids, can act as acids in aqueous solution. Examples are the bisulfate ion HSO_4^- and the dihydrogen phosphate ion $H_2PO_4^-$:

$$HSO_4^-(aq) \ + \ H_2O(l) \ \rightleftharpoons \ H_3O^+(aq) \ + \ SO_4^{2-}(aq)$$

$$H_2PO_4^-(aq) \ + \ H_2O(l) \ \rightleftharpoons \ H_3O^+(aq) \ + \ HPO_4^{2-}(aq)$$

The hydrogens have to be acidic hydrogens. For example the hydrogens of the CH_3 group in acetic CH_3CO_2H acid are not acidic and acetic acid is only monoprotic. Likewise, the third H in H_3PO_3 is not acidic and phosphorous acid is only diprotic.

12. K_w, the water ionization constant $= [H_3O^+][OH^-]$.

13.   A large value for K_a or K_b means a proportionately large concentration of product, i.e. a high concentration of hydronium ions or hydroxide ions. The higher the value of K_a, for example, the stronger the acid. On the other hand, a smaller value for K_a means less product at equilibrium and a weaker acid.

14.   The product of the K_a for a weak acid and K_b for its conjugate base is equal to K_w, the water ionization constant.

$K_w = K_a \times K_b = 1.0 \times 10^{-14}$ at 25°C.

$$CH_3CO_2H(aq) + H_2O(l) \rightleftharpoons H_3O^+(aq) + CH_3CO_2^-(aq)$$

$$K_a = \frac{[H_3O^+][CH_3CO_2^-]}{[CH_3CO_2H]}$$

$$CH_3CO_2^-(aq) + H_2O(l) \rightleftharpoons OH^-(aq) + CH_3CO_2H(aq)$$

$$K_b = \frac{[OH^-][CH_3CO_2H]}{[CH_3CO_2^-]}$$

$$K_a \times K_b = \frac{[H_3O^+][CH_3CO_2^-]}{[CH_3CO_2H]} \times \frac{[OH^-][CH_3CO_2H]}{[CH_3CO_2^-]} = [H_3O^+][OH^-] = K_w$$

15.   The assumption that $[HA]_{equil} = [HA]_{initial}$ is valid if $[HA]_{initial} > 100 \times K_a$.

16.   Qualitatively, the pH of a salt solution can be determined by establishing the strengths of the acid and base used to make the salt: If the acid is strong, the salt will be acidic, if the base is strong the salt will be basic. If both are strong then the salt will be neutral. If both are weak, it depends upon the relative strengths of the acid and base.

Be aware of anions such as HSO_4^- and $H_2PO_4^-$ that ionize to produce acidic solutions.

17. $$HSO_3^-(aq) + H_2O(l) \rightleftharpoons H_3O^+(aq) + SO_3^{2-}(aq) \quad K_{a2} = 6.4 \times 10^{-8}$$

$$K_{a2} = \frac{[H_3O^+][SO_3^{2-}]}{[HSO_3^-]}$$

The $[H_3O^+]$ comes almost entirely from the first ionization of sulfurous acid. In this first ionization, the $[H_3O^+] = [HSO_3^-]$:

$$H_2SO_3(aq) + H_2O(l) \rightleftharpoons H_3O^+(aq) + HSO_3^-(aq) \quad K_{a1} = 1.7 \times 10^{-2}$$

So $K_{a2} = [SO_3^{2-}]$

18.   Lewis defined an acid as a substance that would accept an electron pair to form a new bond and a base as a substance that would donate the pair of electrons. The Lewis definitions are more general than the Arrhenius and Brønsted-Lowry definitions—acid-base reactions are no longer restricted to those involving a hydrogen ion transfer.

Study Questions and Problems

1. For the following aqueous equilibria, designate the Brønsted-Lowry conjugate acid-base pairs and establish the weaker side:

   a. $NH_3(aq) + H_2O(l) \rightleftharpoons NH_4^+(aq) + OH^-(aq)$

   b. $HCN(aq) + H_2O(l) \rightleftharpoons H_3O^+(aq) + CN^-(aq)$

   c. $NH_4^+(aq) + CO_3^-(aq) \rightleftharpoons NH_3(aq) + HCO_3^-(aq)$

2. Write the name and formula for the conjugate bases of the following acids:
   a. HNO_2
   b. H_2SO_4
   c. $H_2PO_4^-$
   d. HF
   e. CH_3CO_2H

3. Complete the Brønsted-Lowry equilibria, label the components acid or base, and pair up the conjugate acid-base pairs:

   a. $HSO_4^- + H_2O \rightleftharpoons$

   b. $NH_3 + H_2O \rightleftharpoons$

   c. $CN^- + H_2O \rightleftharpoons$

   d. $H^- + H_2O \rightleftharpoons$

   e. $HClO_4 + H_2O \rightleftharpoons$

4. Is the monohydrogenphosphate ion HPO_4^{2-} amphiprotic? If so, write the formulas of its conjugate acid and its conjugate base.

5. Of the following acids, determine
   a. the strongest acid.
   b. the acid that produces the lowest concentration of hydronium ions per mole of acid.
   c. the acid with the strongest conjugate base.
   d. the diprotic acid.
   e. the strong acid.
   f. the acid with the weakest conjugate base.

   $HNO_3(aq) + H_2O(l) \rightleftharpoons H_3O^+(aq) + NO_3^-(aq)$ K_a = very large

   $HSO_4^-(aq) + H_2O(l) \rightleftharpoons H_3O^+(aq) + SO_4^-(aq)$ $K_a = 1.2 \times 10^{-2}$

   $HCN(aq) + H_2O(l) \rightleftharpoons H_3O^+(aq) + CN^-(aq)$ $K_a = 4.0 \times 10^{-10}$

   $H_2CO_3(aq) + H_2O(l) \rightleftharpoons H_3O^+(aq) + CO_3^-(aq)$ $K_a = 4.2 \times 10^{-7}$

   $NH_4^+(aq) + H_2O(l) \rightleftharpoons H_3O^+(aq) + NH_3(aq)$ $K_a = 5.6 \times 10^{-10}$

   $HF(aq) + H_2O(l) \rightleftharpoons H_3O^+(aq) + F^-(aq)$ $K_a = 7.2 \times 10^{-4}$

6. Write net ionic acid-base reactions for:

    a. the reaction of acetic acid with aqueous ammonia solution
    b. the reaction of hydrofluoric acid with sodium hydroxide
    c. the reaction of ammonium chloride with potassium hydroxide
    d. the reaction of sodium bicarbonate with sulfuric acid
    e. the reaction of chlorous acid with aqueous ammonia solution
    f. the reaction of disodium hydrogen phosphate with acetic acid

7. What is the pH of

    a. 0.0010 M HCl solution?
    b. 0.15 M KOH solution?
    c. 10^{-8} M HNO_3 solution?

See Table 17.4 on page 799 of the text.

8. List the following substances in order of increasing acid strength:

    H_2O, H_2SO_3, HCN, $H_2PO_4^-$, NH_4^+, $[Cu(H_2O)_6]^{2+}$, NH_3, H_3O^+, HCO_2H, HCl.

9. Complete the table for each aqueous solution at 25°C. State whether the solutions are acidic or basic:

$[H_3O^+]$	$[OH^-]$	pH	pOH	acidic or basic
2.0×10^{-5}				
		6.25		
	5.6×10^{-2}			
			9.20	
8.7×10^{-10}				

10. What is the pH of a solution that contains 2.60 grams of NaOH in 250 mL of aqueous solution?

11. If the pH of a sample of rainwater is 4.62, what is the hydronium ion concentration $[H_3O^+]$ and the hydroxide ion concentration $[OH^-]$ in the rainwater?

12. A 0.12M solution of an unknown weak acid has a pH of 4.26 at 25°C. What is the hydronium ion concentration in the solution and what is the value of its K_a?

13. Hydroxylamine is a weak base with a $K_b = 6.6 \times 10^{-9}$. What is the pH of a 0.36 M solution of hydroxylamine in water at 25°C?

14. Suppose you dissolved benzoic acid in water to make a 0.15 M solution. What is
    a. the concentration of benzoic acid?
    b. the concentration of hydronium ion?
    c. the concentration of benzoate anion?
    d. the pH of the solution?
    K_a for benzoic acid = 6.3×10^{-5} at 25°C

15. Which of the following salts, when dissolved in water to produce 0.10 M solutions, would have the lowest pH?

    a. sodium acetate
    b. potassium chloride
    c. sodium bisulfate
    d. magnesium nitrate
    e. potassium cyanide

16. For each of the following salts, predict whether an aqueous solution would be acidic, basic, or neutral.

    a. sodium nitrate $NaNO_3$
    b. ammonium iodide NH_4I
    c. sodium bicarbonate $NaHCO_3$
    d. ammonium cyanide NH_4CN
    e. sodium hypochlorite $NaOCl$
    f. potassium acetate KCH_3CO_2

17. a. Cyanic acid HOCN has a $K_a = 3.5 \times 10^{-4}$, what is the K_b for the cyanate ion OCN^-?

    b. Phenol is a relatively weak acid, $K_a = 1.3 \times 10^{-10}$. How does the strength of its conjugate base compare with the strength of ammonia, the acetate ion, and sodium hydroxide?

18. a. What is the pH of a 0.80 M solution of sulfurous acid?
    b. What is the concentration of sulfite ion in a 0.80 M solution of sulfurous acid?
    c. What happens to the concentration of sulfite ion SO_3^{2-} if the concentration of sulfurous acid is halved?

19. Identify the Lewis acid and the Lewis base in the following reactions:

    a. Boron trichloride reacts with chloride ion to produce $[BCl_4]^-$.

    b. Nickel reacts with carbon monoxide to produce nickel tetracarbonyl $[Ni(CO)_4]$.

    c. Ammonia reacts with acetic acid to produce ammonium acetate.

    d. Sodium ions are solvated by water to produce $Na^+(aq)$.

20. Calculate the pH of a 0.35 M solution of potassium cyanide.
    K_a for HCN = 4.0×10^{-10}.

Answers to Study Questions and Problems

1. a. $NH_3(aq)$ + $H_2O(l)$ ⇌ $NH_4^+(aq)$ + $OH^-(aq)$

   base   weaker   acid   acid   stronger   base

   b. $HCN(aq)$ + $H_2O(l)$ ⇌ $H_3O^+(aq)$ + $CN^-(aq)$

   acid   weaker   base   acid   stronger   base

   c. $NH_4^+(aq)$ + $CO_3^-(aq)$ ⇌ $NH_3(aq)$ + $HCO_3^-(aq)$

   acid   stronger   base   base   weaker   acid

2. acid $\xrightarrow{\text{remove } H^+}$ conjugate base

   a. HNO_2 NO_2^- nitrite
   b. H_2SO_4 HSO_4^- bisulfate
   c. $H_2PO_4^-$ HPO_4^{2-} hydrogen phosphate
   d. HF F^- fluoride
   e. CH_3CO_2H $CH_3CO_2^-$ acetate

3. a. HSO_4^- + H_2O ⇌ H_3O^+ + SO_4^{2-}
      acid   base   acid   base

   b. NH_3 + H_2O ⇌ NH_4^+ + OH^-
      base   acid   acid   base

   c. CN^- + H_2O ⇌ HCN + OH^-
      base   acid   acid   base

   d. H^- + H_2O ⇌ H_2 + OH^-
      base   acid   acid   base

   e. $HClO_4$ + H_2O ⇌ H_3O^+ + ClO_4^-
      acid   base   acid   base

4. The HPO_4^{2-} ion is amphiprotic; it can donate or accept hydrogen ions:

   $$H_2PO_4^- \rightleftharpoons HPO_4^{2-} \rightleftharpoons PO_4^{3-}$$
   conjugate acid   conjugate base

5. $HNO_3(aq) + H_2O(l) \rightleftharpoons H_3O^+(aq) + NO_3^-(aq)$ K_a = very large

    $HSO_4^-(aq) + H_2O(l) \rightleftharpoons H_3O^+(aq) + SO_4^-(aq)$ $K_a = 1.2 \times 10^{-2}$

    $HCN(aq) + H_2O(l) \rightleftharpoons H_3O^+(aq) + CN^-(aq)$ $K_a = 4.0 \times 10^{-10}$

    $H_2CO_3(aq) + H_2O(l) \rightleftharpoons H_3O^+(aq) + CO_3^-(aq)$ $K_a = 4.2 \times 10^{-7}$

    $NH_4^+(aq) + H_2O(l) \rightleftharpoons H_3O^+(aq) + NH_3(aq)$ $K_a = 5.6 \times 10^{-10}$

    $HF(aq) + H_2O(l) \rightleftharpoons H_3O^+(aq) + F^-(aq)$ $K_a = 7.2 \times 10^{-4}$

   a. the strongest acid is HNO_3.
   b. the acid that produces the lowest concentration of hydronium ions per mole of acid is the weakest acid: HCN.
   c. the acid with the strongest conjugate base is the weakest acid: HCN.
   d. the diprotic acid is H_2CO_3.
   e. the strong acid is HNO_3.
   f. the acid with the weakest conjugate base is the strongest acid: HNO_3.

> The strongest acid is not necessarily a strong acid.

6. $CH_3CO_2H(aq) + NH_3(aq) \rightleftharpoons NH_4^+(aq) + CH_3CO_2^-(aq)$

    $HF(aq) + OH^-(aq) \rightleftharpoons F^-(aq) + H_2O(l)$

    $NH_4^+(aq) + OH^-(aq) \rightleftharpoons NH_3(aq) + H_2O(l)$

    $H_3O^+(aq) + HCO_3^-(aq) \rightleftharpoons 2H_2O(l) + CO_2(g)$

    $HClO_2(aq) + NH_3(aq) \rightleftharpoons NH_4^+(aq) + ClO_2^-(aq)$

    $HPO_4^{2-}(aq) + CH_3CO_2H(aq) \rightleftharpoons H_2PO_4^-(aq) + CH_3CO_2^-(aq)$

> Spectator ions are omitted– these are net ionic equations.
>
> Each reaction involves the transfer of a proton from the acid to the base.

7. a. HCl is a strong acid, so 0.0010 M HCl solution contains 0.0010 M H_3O^+. The pH = $-\log[H^+]$ = $-\log[0.0010]$ = 3.0.

   b. KOH is a strong base, so 0.15 M KOH solution contains 0.15 M OH^-. The pOH = $-\log[OH^-]$ = $-\log[0.15]$ = 0.82; pH = 13.2.

   c. The total quantity of H^+ in the solution is initially 10^{-8} M from the HNO_3 plus 10^{-7} M from the water. The total concentration of H^+ is approx. 1.1×10^{-7}. So the pH = 6.96.

> The solution must be acidic, so the answer is not 8!
> The solution is slightly more acidic than pure water.
> There is always some H_3O^+ and OH^- in an aqueous solution due to the autoionization of water— usually insignificant but not in this case.

8. In increasing acid strength from bottom to top:

HCl	approx 10^7
H_3O^+	55
H_2SO_3	1.2×10^{-2}
HCO_2H	1.8×10^{-4}
$[Cu(H_2O)_6]^{2+}$	1.6×10^{-7}
$H_2PO_4^-$	6.2×10^{-8}
NH_4^+	5.6×10^{-10}
HCN	4.0×10^{-10}
H_2O	1.8×10^{-16}
NH_3	very small

9.

[H_3O^+]	[OH^-]	pH	pOH	acidic or basic
2.0×10^{-5}	5.0×10^{-10}	4.70	9.30	acidic
5.6×10^{-7}	1.8×10^{-8}	6.25	7.75	acidic
1.8×10^{-13}	5.6×10^{-2}	12.75	1.25	basic
1.6×10^{-5}	6.3×10^{-10}	4.80	9.20	acidic
8.7×10^{-10}	1.1×10^{-5}	9.06	4.94	basic

For water at 25°C:

$[H_3O^+][OH^-] = K_w = 10^{-14}$

$pH + pOH = pK_w = 14$

10. Sodium hydroxide is a strong base.
Concentration of NaOH = 2.6 g/250 mL = 0.26 mol/L
The concentration of OH^- ions is therefore the same = 0.26 M
$pOH = -\log[OH] = 0.585$
$pH = 13.4$

Molar mass of NaOH
= 40 g/mol

11. $pH = -\log[H^+] = 4.62$
$[H^+] = 2.4 \times 10^{-5}$
$[OH^-] = 4.2 \times 10^{-10}$

$[OH^-]$ is much lower than $[H^+]$;
the solution is acidic.

12. $pH = 4.26$ at 25°C
$[H_3O^+] = 5.50 \times 10^{-5}$

$$K_a = \frac{[H_3O^+]^2}{0.12} = 2.52 \times 10^{-8}.$$

Learn how to take an antilog on
your calculator.

13. $$K_b = 6.6 \times 10^{-9} = \frac{[OH^-]^2}{[NH_2OH]} = \frac{[OH^-]^2}{0.36}$$

$[OH^-] = 4.87 \times 10^{-5}$
$pOH = 4.31$ and the pH = 9.69.

14.

		$C_6H_5CO_2H + H_2O \rightleftharpoons H_3O^+(aq) + C_6H_5CO_2^-(aq)$		
I	Initial	0.15	0	0
C	Change	$-x$	$+x$	$+x$
E	Equilibrium	$0.15-x$	x	x

K_a for benzoic acid $= 6.3 \times 10^{-5} = x^2/(0.15-x) = x^2/0.15$; x = 0.00307M

(0.15–x) is approximately equal
to 0.15. Actually 0.147.

Solving the quadratic yields a
value for x = 0.00304.

a. the concentration of benzoic acid at equilibrium = 0.147M
b. the concentration of hydronium ion = 0.0031 M
c. the concentration of benzoate anion = 0.0031 M
d. the pH of the solution = 2.51

Determine the acid and base
used to make the salt.
But look out for acid salts (the
anions of polyprotic acids).

No calculations are necessary.

15. a. sodium acetate K_a for acetic acid $= 1.8 \times 10^{-5}$; basic
    b. potassium chloride   strong acid : strong base : neutral solution
    c. sodium bisulfate   bisulfate ionizes ($K_a = 1.2 \times 10^{-2}$) lowest pH
    d. magnesium nitrate   strong acid : strong base : neutral solution
    e. potassium cyanide K_a for HCN $= 4.0 \times 10^{-10}$; basic

16. a. $NaNO_3$ strong acid   strong base   neutral
    b. NH_4I strong acid   weak base   acidic
    c. $NaHCO_3$ weak acid   strong base   basic
    d. NH_4CN weaker acid   weak base   slightly basic
    e. $NaOCl$ weak acid   strong base   basic
    f. KCH_3CO_2 weak acid   strong base   basic

17. a. K_a for the acid and K_b for the conjugate base are related:

$$K_a \times K_b = K_w.$$

If $K_a = 3.5 \times 10^{-4}$ for cyanic acid HOCN,
then K_b for cyanate $= 10^{-14}/3.5 \times 10^{-4} = 2.86 \times 10^{-11}$.

    b. Phenol is a relatively weak acid, $K_a = 1.3 \times 10^{-10}$.
K_b for its conjugate base $= 10^{-14}/1.3 \times 10^{-10} = 7.7 \times 10^{-5}$.

For comparison:
ammonia $K_b = 1.8 \times 10^{-5}$ (about the same)
acetate $K_b = 5.6 \times 10^{-10}$ (much weaker)
sodium hydroxide K_b = very large (much stronger).

18. a. $K_{a1} = \dfrac{[H_3O^+][HSO_3^-]}{[H_2SO_3]} = \dfrac{[H_3O^+]^2}{[H_2SO_3]} = 1.7 \times 10^{-2}$ $[H_3O^+] = 0.108$ Solve the quadratic to obtain the hydronium ion concentration.
pH = 0.97

    b. $[SO_3^{2-}] = K_{a2} = 6.4 \times 10^{-8}$

    c. Nothing; the $[SO_3^{2-}]$ is independent of $[H_2SO_3]$.   See review question 17.

19. a. Boron trichloride (acid) accepts a pair of electrons from chloride (base).   The Lewis base donates the electron pair; the Lewis acid accepts the electron pair.

    b. Nickel (acid) accepts a pair of electrons from carbon monoxide (base).

    c. Ammonia (base) donates a pair of electrons to the proton (acid) from acetic acid.

    d. Sodium ions (acid) are solvated by water (base).

20. $CN^-(aq) + H_2O(l) \rightleftharpoons HCN(aq) + OH^-(aq)$

$$K_b = \frac{[HCN][OH^-]}{[CN^-]} = \frac{[OH^-]^2}{[CN^-]} = \frac{[OH^-]^2}{0.35} = 4.0 \times 10^{-10}$$

$[OH^-] = 1.18 \times 10^{-5}$
pOH = 4.93
pH = 9.07

Equilibria

—a crossword puzzle

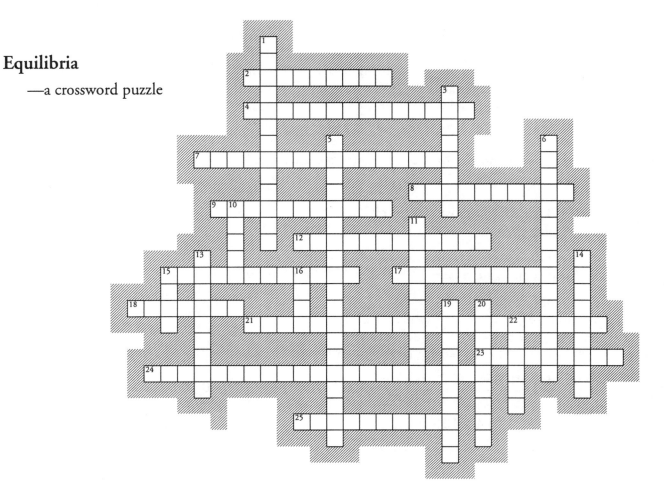

Across:

2. In a titration, it lets you know when you have reached the equivalence point.
4. When water produces *14 down* and *23 across* ions.
7. Product concentrations divided by reactant concentrations, not necessarily at equilibrium.
8. It's assumed to be complete for a strong *9 across* in solution.
9. Weak or strong, it conducts an electrical current in solution.
12. Industrial manufacture of ammonia.
15. It transfers a hydrogen ion to form its conjugate base.
17. What happens to the salts of weak acids or weak bases in solution.
18. Normally from 0 to 14, an indication of hydronium ion concentration.
21. A process for which *5 down* is large.
23. The conjugate base of water; produced by Arrhenius bases in aqueous solution.
24. The idea that a system in equilibrium will adjust to accommodate any alteration in conditions.
25. An acid with only one hydrogen atom per molecule.

Down:

1. The partner of a base.
3. Unchanging.
5. Q at equilibrium.
6. A statement of the fact that the *5 down* is *3 down* regardless of how the masses of the components present are changed—developed by Guldberg & Waage.
10. He classified an acid as an electron pair acceptor and a base as an electron pair donor.
11. Sulfuric acid, oxalic acid, phosphoric acid, carbonic acid, for example.
13. Negative this of the concentration of *14 down* is pH.
14. Ion produced by the protonation of a water molecule.
15. This increases the hydroxide ion concentration...
16. ...whereas this increases the hydronium ion concentration in aqueous solution.
19. If the reaction is reversed, then the new equilibrium constant is this of the original.
20. His was the first set of definitions describing an acid and base in aqueous solution.
22. One name for what is produced when a Lewis acid and a Lewis base react, forming a coordinate bond.

CHAPTER 18

Introduction

This chapter continues the examination of acid-base reactions. LeChatelier's principle is applied to acid-base equilibria in an examination of the common ion effect and the way in which buffer solutions work is explored. Finally acid *vs.* base titrations are investigated.

Contents

18.1   Acid-Base Reactions

An acid-base equilibrium always lies on the weaker side. Consider first, for example, the reaction between the **weak acid** acetic acid and the **weak base** ammonia to produce ammonium acetate:

$$CH_3CO_2H(aq) + NH_3(aq) \rightleftharpoons NH_4^+(aq) + CH_3CO_2^-(aq)$$

acid   base   acid   base

stronger   weaker

The strengths of acids and bases are listed in Table 17.4 on page 799. It is just coincidental that K_a for acetic acid = K_b for ammonia.

The equilibrium lies on the right. The value of the equilibrium constant can be derived from the following information:

$$CH_3CO_2H + H_2O \rightleftharpoons H_3O^+ + CH_3CO_2^- \qquad K_a = 1.8 \times 10^{-5}$$

$$NH_3 + H_2O \rightleftharpoons NH_4^+ + OH^- \qquad K_b = 1.8 \times 10^{-5}$$

$$H_3O^+ + OH^- \rightleftharpoons 2H_2O \qquad K = 1/K_w = 1.0 \times 10^{14}$$

(aq) designations have been omitted for clarity.

Note that this reaction is reversed, so that $K = 1/K_w = 10^{14}$.

Adding these equations produces the equation above, so the equilibrium constant for the neutralization reaction is the product of these equilibrium constants:

$$K_{neut} = K_a \times K_b \times 1/K_w = (1.8 \times 10^{-5})(1.8 \times 10^{-5})(1.0 \times 10^{14}) = 3.2 \times 10^4$$

The equilibrium constant is referred to here as K_{neut} just to distinguish it from the other equilibrium constants.

Because of the high value of $(1/K_w)$ (due to the fact that H_3O^+ and OH^- so readily combine), the reaction between acetic acid and ammonia is driven almost to completion.

The reaction is product favored.

Second, the reaction of any **strong acid** and any **strong base** has the net ionic equation:

$$H_3O^+(aq) + OH^-(aq) \rightleftharpoons 2H_2O(l) \qquad K = 1.0 \times 10^{14}$$

For all practical purposes, this reaction is quantitatively complete.

These equilibrium constants can be derived in the same way as K_{neut} for the weak acid–weak base neutralization discussed first.

The high value of K indicates that the system lies almost completely on the right.

A neutralization reaction does not usually result in a neutral (pH=7) solution.

Read the Problem Solving Tips and Ideas 18.1 on page 847.

The volume is doubled from 25mL to 50mL, therefore the concentration of nitrite ion is halved from 0.010M to 0.0050M

The hydroxide ion produced in the hydrolysis makes the solution basic (as predicted).

These characteristics of acid-base reactions are summarized in Table 18.1 on page 848.

To calculate the pH at the equivalence point, write the hydrolysis reaction for the conjugate partner of the weak acid or base.

Third, the reaction of any **strong acid** and a **weak base** like ammonia has the net ionic equation:

$$H_3O^+(aq) \; + \; NH_3(aq) \; \rightleftharpoons \; NH_4^+(aq) \; + \; H_2O(l) \qquad K = 1.8 \times 10^9$$

Fourth, the reaction of a **weak acid** with a **strong base**, like the previous three acid-base reactions, goes a long way toward completion. For example, the reaction between acetic acid and sodium hydroxide:

$$CH_3CO_2H(aq) + OH^-(aq) \; \rightleftharpoons \; CH_3CO_2^-(aq) \; + \; H_2O(l) \; K = 1.8 \times 10^9$$

Unless both the acid and base involved in the neutralization reaction are strong, the pH of the solution that results will not be neutral. If the concentration of the salt that is produced in the reaction is known, then the pH can be calculated. For example, consider the reaction between the weak acid nitrous acid and hydroxide ion to produce ammonium nitrite:

$$HNO_2(aq) \; + \; OH^-(aq) \; \rightleftharpoons \; H_3O^+(aq) \; + \; NO_2^-(aq)$$

K_a for nitrous acid $= 4.5 \times 10^{-4}$ at 25°C

Because the base is stronger than the acid, the final solution will be basic. Suppose 25 mL of a 0.010M solution of nitrous acid is added to 25 mL of a 0.010M solution of hydroxide, what will the pH of the final solution be?

The final volume will be 50 mL, and the concentration of the salt will be 0.0050M. The anion of the sodium nitrite salt will hydrolyze:

$$NO_2^-(aq) \; + \; H_2O(l) \quad \rightleftharpoons \quad HNO_2(aq) \; + \; OH^-(aq)$$

$$K_b = \frac{K_w}{K_a} = \frac{[HNO_2][OH^-]}{[NO_2^-]} = \frac{[OH^-]^2}{[NO_2^-]} = \frac{[OH^-]^2}{0.0050}$$

$$[OH^-]^2 = 1.0 \times 10^{-14} \times 0.0050 \, / \, 4.5 \times 10^{-4} = 1.11 \times 10^{-13}$$

$$[OH^-] = 3.33 \times 10^{-7} \text{ and the pH} = 7.52$$

After mixing equal molar amounts of an acid and a base (to reach what is called the equivalence point), a salt and water are produced. The pH of the solution at the equivalence point *depends upon the salt*, i.e. the conjugate partner of the weak acid or weak base used in the neutralization reaction.

18.2   The Common Ion Effect

According to LeChatelier's principle, if the concentration of a substance involved in an equilibrium is changed (more is added or some is removed) then the system will adjust itself to accommodate the change and maintain the value of the equilibrium constant K. This principle applies to aqueous equilibria just as it does to any other equilibrium. If the aqueous equilibrium involves the ionization of a weak acid or base, and the concentration of the conjugate partner of the weak electrolyte is changed, then the principle is referred to as the **common ion effect**.

For example, consider the weak acid acetic acid:

$$CH_3CO_2H + H_2O \rightleftharpoons H_3O^+ + CH_3CO_2^- \qquad K_a = 1.8 \times 10^{-5}$$

Suppose that some additional acetate ion is added (in the form of sodium acetate, a salt of acetic acid). Le Chatelier's principle states that the system will move from right to left in order to maintain equilibrium. The common ion suppresses the ionization of the acetic acid.

Calculation of the pH of a solution containing both a weak acid and its conjugate base (a salt of the weak acid) is in fact somewhat easier than the calculation of the pH of a solution of the weak acid alone. In general, the small change in the initial concentrations of the weak acid and its conjugate base can be safely ignored. Suppose that a solution is 0.10M in acetic acid and 0.050M in sodium acetate; what is the pH of the solution?

$$CH_3CO_2H + H_2O \rightleftharpoons H_3O^+ + CH_3CO_2^-$$

$$K_a = 1.8 \times 10^{-5} = \frac{[H_3O^+][CH_3CO_2^-]}{[CH_3CO_2H]} = \frac{[H_3O^+][0.050]}{[0.10]}$$

$$[H_3O^+] = 3.6 \times 10^{-5} \text{ and pH} = 4.44$$

18.3   Buffer Solutions

A **buffer solution** is resistant to change in pH when an acid or base is added to the solution. They are quite remarkable, particularly when you first observe one in action. Consider that when just 0.0010 mol of a strong acid (just 37 mg of HCl) is added to one liter of water, the pH drops from 7 to 3. If the same quantity of acid is added to a buffer solution, the change in pH is barely perceptible.

Buffer solutions consist of a weak acid and its conjugate base, or a weak base and its conjugate acid, in solution, preferably in nearly equimolar quantities.

Consider, for example, a solution containing ammonia and the ammonium ion in equal molar quantities:

$$NH_3\,(aq) + H_2O\,(l) \rightleftharpoons NH_4^+\,(aq) + OH^-\,(aq)$$

If acid (H_3O^+) is added to this equilibrium system, the H_3O^+ will combine with the OH^- ion and form water, this will remove OH^- from the equilibrium. According to LeChatelier's principle, the equilibrium system will shift right and restore the OH^- concentration, and the pH will therefore not change.

If base (OH^-) is added to the solution, the equilibrium will move to the left, absorbing the OH^-, and so again the pH will not change.

General expressions for the hydronium ion or hydroxide concentrations in an acidic buffer solution or basic buffer solution respectively are:

$$[H_3O^+] = \frac{[HA]}{[A^-]} K_a \qquad\qquad [OH^-] = \frac{[B]}{[BH^+]} K_b$$

$$\text{or } [H_3O^+] = \frac{[acid]}{[conjugate\ base]} K_a \qquad [OH^-] = \frac{[base]}{[conjugate\ acid]} K_b$$

For some reason the prototypical weak acid is acetic acid and it is usually the example used in discussions of aqueous equilibria. Keep in mind that there are other weak acids!

Be aware of any change in concentration due to dilution when two solutions are mixed.

Remember that the conjugate base of a weak acid is a salt of the weak acid. For example, acetic acid and sodium acetate.

Likewise, the conjugate acid of a weak base is a salt of the weak base. For example, ammonia and ammonium nitrate.

LeChatelier's principle:

A system in equilibrium will move to the right or to the left to accommodate any stress placed upon the system. At a constant temperature, the equilibrium constant is indeed constant.

Notice the inversion of the concentration terms between the two expressions; this is because $\log(A^{-1}) = -\log(A)$. Recall that the $pH = -\log[H^+]$

For the acid buffer, taking the logarithm of both sides:

$$pH = pK_a + \log(\frac{[\text{conjugate base}]}{[\text{acid}]})$$

This is known as the **Henderson-Hasselbalch equation**.

There are two requirements of a buffer solution. First the buffer should have sufficient capacity to maintain the pH after addition of anticipated amounts of acid or base. Second, it should maintain the pH at the desired value.

See the Problem Solving Tips and Ideas 18.3 on page 858 for the calculation of the pH of a buffer solution.

From a list of weak acids (or bases), one is chosen that has a pK_a close to the pH value desired. The relative amounts of the weak acid and its conjugate base are adjusted to achieve exactly the pH required. It is the relative quantities of weak acid and conjugate base that is important, the actual concentrations do not matter. If a buffer solution is diluted, both concentrations change, but the ratio of the concentrations stays the same, and the pH therefore stays the same.

An increase in concentration of the buffer components increases the buffer capacity—more acid or base can be added without change in pH.

18.4   Acid-Base Titration Curves

See Example 18.7 on pages 858 – 860 for methods to calculate the pH of the solution at various stages during a titration. The methods are summarized in the Problem Solving Tips and Ideas 18.4 on page 864 of the text.

Titration curves are illustrated in Figures 18.6, 18.7, and 18.8 on pages 862, 865, and 866 of the text.

A **titration** is the addition of a base to an acid until the **equivalence point** is reached. At the equivalence point, the quantity of base added and the quantity of acid originally present match the stoichiometry of the reaction. Titrations are a particularly valuable analytical procedure (referred to as volumetric analysis). The progress of the titration can be monitored by the use of an indicator or by the use of a pH meter.

If the titration is a strong acid–strong base titration, then the equivalence point will occur at a pH of 7. If however the acid is weak, titrated against a strong base, then the equivalence point will be at a pH>7. Likewise, if a weak base is titrated against a strong acid, the equivalence point will occur at a pH<7. If both acid and base are weak, the equivalence point will depend upon their relative strengths.

18.5   Acid-Base Indicators

Plotting the titration curve itself, or better still, the derivative or the second derivative, allows a very accurate determination of the equivalence point.

The purpose of a titration is to add an exact stoichiometric equivalent of a base to an acid, or *vice versa*. To do this it is essential to recognize when the equivalent point has been reached. The best way is to use a pH meter and plot the change in pH throughout the titration, but a very good estimate can be obtained using an **acid-base indicator**.

Some acid-base indicators and their color changes are shown in Figure 17.11 on page 810.

An acid-base indicator is itself a weak acid or base. It is usually a large organic molecule that has slightly different structures in acid and base (the weak acid has one structure and its conjugate base another structure). The two structures have different colors and by observing the change in color during the titration, you can monitor the change in pH. The idea is to choose an indicator with a K_a near that of the acid being titrated, so that the color change occurs at the correct stage during the titration.

Review Questions

1. How do you calculate the equilibrium constant for a neutralization reaction between a weak acid HA and a weak base B?

2. What is the net ionic equation for all strong acid–strong base neutralizations?

3. How can you predict whether the solution resulting from a neutralization reaction will be acidic, basic, or neutral?

4. For a generic weak acid HA, write an expression for the equilibrium constant K_a. Then write an expression for the equilibrium constant K_b for the hydrolysis of its conjugate base A^-. Then write an expression for the equilibrium constant K for the neutralization reaction of HA by a strong base OH^-. How are these equilibrium constants related to one another?

5. What is a common-ion? What is the common-ion effect?

6. What is a buffer solution? What are the two essential components of a buffer solution?

7. According to the Henderson-Hasselbalch equation, there are two terms that establish the pH of a buffer solution. What are they, and what are the relative magnitudes of their contributions to the value of the pH?

8. What are the two requirements of a buffer solution?

9. Explain how it is that the pH of a buffer solution does not change if the solution is diluted.

10. Four types of acid-base titrations are possible: strong acid–strong base, strong acid–weak base, weak acid–strong base, and weak acid–weak base. Describe how the titration curves for these four titrations differ.

11. Figure 17.11 on page 810 of the text illustrates a range of indicators for acid-base titrations. Choose appropriate indicators for the four titrations listed in the previous question.

12. Imagine titrating a diprotic acid like sulfuric acid against a strong base such as sodium hydroxide. How would the titration curve differ from that for the titration of a monoprotic strong acid like hydrochloric acid against sodium hydroxide?

13. What is significant about the point in a titration of a weak acid against a strong base at which sufficient base has been added to get exactly halfway to the equivalence point?

Answers to Review Questions

1. The reaction between the weak acid HA and the weak base B:

$$HA(aq) + B(aq) \rightleftharpoons BH^+(aq) + A^-(aq)$$

The ionization equilibria for the weak acid and the weak base are:

$$HA(aq) + H_2O(l) \rightleftharpoons H_3O^+(aq) + A^-(aq) \quad \text{equil. const.} = K_a$$

$$B(aq) + H_2O(l) \rightleftharpoons BH^+(aq) + OH^-(aq) \quad \text{equil. const.} = K_b$$

$$H_3O^+(aq) + OH^-(aq) \rightleftharpoons 2H_2O(l) \quad \text{equil. const.} = 1/K_w$$

Adding these equations produces the equation for the reaction between the weak acid HA and the weak base B, so the equilibrium constant is:

$$K_{neut} = \frac{K_a \times K_b}{K_w}$$

K_a is the acid ionization constant, and K_b is the base ionization constant, for the acid and base involved in the neutralization reaction.

The larger the K_a for the acid and the larger the K_b for the base, the larger the value for K_{neut} for the neutralization reaction.

For example, for the weak acid acetic acid and the weak base ammonia:

	K_{neut}
weak acid – weak base	10^4
weak acid – strong base	10^9
strong acid – weak base	10^9
strong acid – strong base	10^{14}

where the strong acid is a typical strong acid like HCl and the strong base is a typical strong base like NaOH.

2. Strong acids, strong bases, and salts, are all completely ionized in solution. As a result the anion of the acid, and the cation of the base are merely spectator ions and do not appear in the net ionic equation:

$$H_3O^+(aq) + OH^-(aq) \rightleftharpoons 2H_2O(l)$$

3. Recall the discussion of hydrolysis in Chapter 17. The pH of a salt solution can be estimated by establishing the strengths of the acid and base used to make the salt:

Cation	Anion	pH of the salt solution
from a strong base	from a strong acid	neutral, pH = 7
from a strong base	from a weak acid	basic, pH > 7
from a weak base	from a strong acid	acidic, pH < 7
from a weak base	from a weak acid	depends upon relative strengths of acid & base

4. For a generic weak acid HA:

$$HA(aq) + H_2O(l) \rightleftharpoons H_3O^+(aq) + A^-(aq) \quad \text{equil. const.} = K_a$$

For the hydrolysis of the conjugate base of the weak acid:

$$A^-(aq) + H_2O(l) \rightleftharpoons HA(aq) + OH^-(aq) \quad \text{equil. const.} = K_b$$

For the neutralization of the weak acid by a strong base:

$$HA(aq) + OH^-(aq) \rightleftharpoons A^-(aq) + H_2O(l) \quad \text{equil. const.} = K_{neut}$$

The K_a of an acid and the K_b for its conjugate base are related by the expression:

$$K_a = \frac{K_w}{K_b} \quad \text{or} \quad K_a \times K_b = K_w$$

The K_{neut} for the acid and the K_b for its conjugate base are the reciprocal of one another:

$$K_{neut} = \frac{1}{K_b} \text{ or } K_{neut} \times K_b = 1$$

5.   A common-ion is an ion, usually the conjugate base of a weak acid, or the conjugate acid of a weak base, that is the same as an ion already present in the solution as a participant in an equilibrium. For example suppose an equilibrium already exists in solution as a result of the ionization of hydrocyanic acid:

$$HCN\,(aq) \ + \ H_2O\,(l) \ \rightleftharpoons \ H_3O^+\,(aq) \ + \ CN^-\,(aq)$$

If additional CN^- ion were added to this equilibrium, it would be a common ion, since the ion is already present in solution as a participant in the equilibrium.

The common-ion effect refers to the effect the common ion has upon the equilibrium in solution. According to LeChatelier's principle, if the concentration of the common ion is increased (more is added) then the system will adjust itself to accommodate the change and maintain the value of the equilibrium constant K; the ionization of the weak acid (or base) will be suppressed.

6.   A buffer solution is a solution that resists a change in pH when an acid or base is added to the solution. The two essential components of a buffer solution are a weak acid and its conjugate base, or a weak base and its conjugate acid, preferably in nearly equimolar quantities.

7.   The Henderson-Hasselbalch equation is:

$$pH = pK_a + \log\left(\frac{[\text{conjugate base}]}{[\text{acid}]}\right)$$

The two terms are first pK_a and second $\log\left(\dfrac{[\text{conjugate base}]}{[\text{acid}]}\right)$

The first term establishes the order of magnitude of the $[H_3O^+]$ of the buffer solution, i.e. the approximate pH. The relative concentrations of the weak acid and its conjugate base in the second term are adjusted to obtain the exact pH required. In order to maintain a reasonable buffer capacity in both directions (addition of acid and addition of base) the concentrations of the weak acid and its conjugate base should not differ by more than 10:1 or 1:10.

The pK_a term is the coarse adjustment and the second term is the fine adjustment of the pH of the buffer solution.

8.   There are two requirements of a buffer solution. First the buffer should have sufficient capacity to maintain the pH after addition of anticipated amounts of acid or base. Second, it should maintain the pH at the desired value.

9.   The second term in the Henderson-Hasselbalch equation contains the ratio of the concentration of the conjugate base of a weak acid to the concentration of the weak acid. If the buffer solution is diluted, then both concentrations change by the same factor, and the ratio of their concentrations remains unchanged. Therefore the value of the second term in the Henderson-Hasselbalch equation remains the same and the pH stays the same. This property of buffer solutions is of considerable practical importance during experiments in which the volume of the solution changes.

10.   Examine the titration curves illustrated in the text (Figures 18.6, 18.7, and 18.8). The most obvious difference is the shape of the titration curve. When a weak acid or weak base is used, the pH at the starting point is further from the bottom (or top) of the diagram (depending upon how weak the acid or base is). Therefore the length of the vertical portion of the curve is shortened for a titration involving a weak acid or base. Another difference in the appearance is the slope of the curve before the equivalence point is reached. Finally, the equivalence point itself occurs at a pH of 7 only if the titration is a strong acid–strong base titration. If the acid is weak, titrated against a strong base, then the equivalence point will be at a pH>7. If a weak base is titrated against a strong acid, the equivalence point will occur at a pH<7. And if both acid and base are weak, the equivalence point will depend upon their relative strengths.

11.   The actual indicators chosen depend upon the strengths of the acids and bases. The idea is to choose an indicator with a K_a near that of the acid being titrated, so that the color change occurs at the correct stage during the titration. For a strong acid–strong base titration, bromthymol blue would be a good choice. For a weak acid like acetic acid titrated against a strong base, (equivalence point 8.72), phenolphthalein is a common choice. For a weak base–strong acid titration, an indicator in the range of thymol blue or methyl orange would be used. Titrations involving both a weak acid and a weak base are not often done; the equivalence points are generally less satisfactory and there is usually no good reason to choose a weak acid to titrate a weak base, or *vice versa*.

Two fairly distinct steps are evident provided that the ionization constants for the diprotic acid differ by more than about 10^4.

12.   The neutralization of a diprotic acid such as sulfuric acid occurs in two distinct stages; the curve has the stepwise appearance of one titration curve followed by another, with two vertical portions at the two equivalence points.

13.   At a point exactly halfway to the equivalence point, half the acid has been neutralized and the concentration of the remaining acid equals the concentration of its conjugate base that has been produced. The second term in the Henderson-Hasselbalch equation therefore equals zero and the pH = pK_a for the acid.

log 1 = zero

Study Questions and Problems

1. Calculate the equilibrium constant for the neutralization of hydrocyanic acid by ammonia:

    $$HCN\,(aq) \; + \; NH_3\,(aq) \; \rightleftharpoons \; NH_4^+\,(aq) \; + \; CN^-\,(aq)$$

    K_a for hydrocyanic acid = 4.0×10^{-10} at 25°C

    K_b for ammonia = 1.8×10^{-5} at 25°C

2. Is the solution that results from the neutralization of hydrocyanic acid by ammonia basic or acidic?

3. If exactly 50 mL of a 0.050M solution of hydrochloric acid is added to exactly 50 mL of 0.050M ammonia, what is the pH of the resulting solution?

4. a. Calculate the pH of a 0.20M solution of formic acid HCO_2H.
    b. Now suppose sufficient sodium formate is added to make the solution 0.10M in formate ion (without changing the total volume). Would you expect the pH to increase or decrease?
    c. Calculate the pH of the new solution.
    d. What would the pH be if the concentration of formate ion was increased to 0.20M?
    e. What do you notice about the pH of this solution?

5. a. What is the pH of 100 mL of pure water at 25°C?
    b. What would the pH of this 100mL water sample be if 0.10mL of 12M HCl was added to it? (Assume the volume doesn't change.)
    c. Calculate the pH of a buffer solution composed of 0.20M ammonia and 0.20M ammonium chloride.
    d. Calculate the pH of 100 mL of this buffer solution if 0.10mL of 12M hydrochloric acid is added to it. (Assume the volume doesn't change).

6. From the list of weak acids shown in Table 17.4 on page 799, choose an appropriate acid for the preparation of a buffer with a pH equal to 7.25. Calculate the relative quantities of the acid and its conjugate base required for the buffer solution.

7. Repeat question 6 for a buffer requiring a pH = 9.25.

8. Benzoic acid is a weak monoprotic acid ($Ka = 6.3 \times 10^{-5}$). Calculate the pH of the solution at the equivalence point when 25.0 mL of a 0.100 M solution of benzoic acid is titrated against 0.050 M sodium hydroxide.

9. A solution contains KH_2PO_4 and K_2HPO_4 and has a pH of 7.10. What is the mole ratio of K_2HPO_4 to KH_2PO_4?

Answers to Study Questions and Problems

Notice that the stronger the base, and the stronger the acid, the more the neutralization reaction is product favored (the higher the value of K_{neut}).

1. The equilibrium constant for the neutralization of any weak acid by any weak base is given by the expression:

$$K_{neut} = \frac{K_a \times K_b}{K_w}$$

which in this case = $(4.0 \times 10^{-10})(1.8 \times 10^{-5})(1 \times 10^{14}) = 0.72$

2. K_a for hydrocyanic acid = 4.0×10^{-10} at 25°C

   K_b for ammonia = 1.8×10^{-5} at 25°C

   The base is stronger than the acid, so the equivalence point will be slightly basic.

3. $$HCl(aq) + NH_3(aq) \rightleftharpoons NH_4^+(aq) + Cl^-(aq)$$

   The concentration of the salt is 0.025M (the volume has doubled from 50 mL to 100 mL). The ammonium ion will hydrolyze because NH_3 is a weak electrolyte (K_b for ammonia = 1.8×10^{-5}).

   $$NH_4^+(aq) + H_2O(l) \rightleftharpoons NH_3(aq) + H_3O^+(aq) \quad K = K_w/K_b = 5.6 \times 10^{-10}$$

   $$K = \frac{[NH_3][H_3O^+]}{[NH_4^+]} = \frac{[H_3O^+]^2}{[0.025]} \quad \text{so } [H_3O^+] = 3.74 \times 10^{-6}M$$
   $$pH = 5.43$$

We will assume that the change in the concentration of the formic acid is negligible. Solving the quadratic yields a value for the hydronium ion concentration equal to 5.91 × 10⁻³ and a pH = 2.23.

4. a. The pH of a 0.20M solution of formic acid HCO_2H:

   $$HCO_2H + H_2O \rightleftharpoons H_3O^+ + HCO_2^- \quad K_a = 1.8 \times 10^{-4}$$

   $$K_a = \frac{[H_3O^+][HCO_2^-]}{[HCO_2H]} = \frac{[H_3O^+]^2}{[0.20]} \quad \text{so } [H_3O^+] = 6.0 \times 10^{-3}M$$
   $$pH = 2.22$$

   b. Addition of formate ion will suppress the ionization of the formic acid and reduce the hydronium ion concentration. The pH will increase.

Solving the quadratic yields a pH = 3.45.

   c. $$K_a = \frac{[H_3O^+][HCO_2^-]}{[HCO_2H]} = \frac{[0.10][H_3O^+]}{[0.20]} \quad \text{so } [H_3O^+] = 4.0 \times 10^{-4}M$$
   $$pH = 3.44$$

   d. If the concentration of formate ion is increased to 0.20M, then the $[H_3O^+]$ will equal the K_a for the acid = 1.8×10^{-4}.

   e. The pH of this solution = the pK_a for the acid = 3.74.

Notice a further increase in pH; a greater suppression of the ionization of formic acid.

5. a. The pH of 100 mL of pure water at 25°C = 7.0.

   b. HCl is a strong acid, so the concentration of hydronium ion will be (0.10 mL × 1L/1000mL × 12 mol/L = 0.0012 moles in 100 mL = 0.012 M). So the pH = 1.92.

The pH falls from 7 to 1.92–a drop of 5 pH units.

c. A buffer solution composed of equimolar (0.20M) quantities of ammonia and ammonium chloride has a pOH equal to the pK_b for the base.
pOH = pKb = 4.74, so pH = 9.26

d. The hydronium ion reacts with the ammonia to produce ammonium ion:

$$H_3O^+(aq) + NH_3(aq) \rightleftharpoons NH_4^+(aq) + H_2O(l) \qquad K = 1.8 \times 10^9$$

K is very large, the reaction is virtually complete.

The concentration of HCl added is 0.012 M so the concentration of NH_4^+ initially increases by that amount and the concentration of ammonia decreases by the same amount. Summarize in a table:

		$H_3O^+(aq) +$	$NH_3(aq) \rightleftharpoons$	$NH_4^+(aq) +$	$H_2O(l)$
I	Initial	0	0.20	0.20	
C	Change	+0.012			
N	New concn:	0	0.188	0.212	

Assume that the reaction goes to completion, and then adjusts backward slightly to reach equilibrium. For the equilibrium, use either the ionization of the weak electrolyte, or the hydrolysis of the conjugate partner—the result is the same.

Now let the system reach equilibrium; the system moves back to the left:

		$H_3O^+(aq) +$	$NH_3(aq) \rightleftharpoons$	$NH_4^+(aq) +$	$H_2O(l)$
I	Initial		0.188	0.212	
C	Change	+x	+x	–x	
E	Equilibrium	x	0.188+x	0.212–x	

$\xrightarrow{\text{assume completion}}$

$\xleftarrow{\text{then adjust}}$ backward

Solve for x, $x = 6.265 \times 10^{-10}$ so $[H^+] = 6.265 \times 10^{-10}$
pH = 9.20 —the pH changes very little!

Alternatively, using the ionization of ammonia:

		$NH_3(aq) +$	$H_2O \rightleftharpoons$	$NH_4^+(aq) +$	$OH^-(aq)$
I	Initial	0.188		0.212	0
C	Change	–x		+x	+x
E	Equilibrium	0.188–x		0.212+x	x

Compared to adding 0.10 mL of 12M HCl to water with a decrease in pH from 7.00 to 1.92, this decrease from 9.26 to 9.20 is negligible.

Solve for x, $x = 1.596 \times 10^{-5}$ so $[OH^-] = 1.596 \times 10^{-5}$
$[H^+] = 6.265 \times 10^{-10}$ and pH = 9.20 —the result is the same.

6. Desired pH = 7.25:
possible choices (look for a pK_a near 7.25):

H_2S	$K_a = 1 \times 10^{-7}$	$pK_a = 7.00$
$H_2PO_4^-$	$K_a = 6.2 \times 10^{-8}$	$pK_a = 7.21$
HSO_3^-	$K_a = 6.2 \times 10^{-8}$	$pK_a = 7.21$
$HClO$	$K_a = 3.5 \times 10^{-8}$	$pK_a = 7.46$

For $H_2PO_4^-$ or HSO_3^- use the Henderson-Hasselbalch equation to determine the ratio of acid and its conjugate base required:

$$pH = pK_a + \log(\frac{[\text{conjugate base}]}{[\text{acid}]})$$

$$\log(\frac{[\text{conjugate base}]}{[\text{acid}]}) = 7.25 - 7.21 = 0.04(24)$$

$$\frac{[\text{conjugate base}]}{[\text{acid}]} = 1.10$$

The desired pH is slightly more basic than the pK_a, so slightly more base than acid is required.

Maintain a sufficient number of significant figures. Recall that only the mantissa of a logarithm is counted in determining significant figures.

[The *mantissa* of a logarithm is the part of the number after the decimal point. The part before the decimal point is called the *characteristic* of the logarithm.]

7.   Desired pH = 9.25:
possible choices (look for a pK_a near 9.25):

$[Co(H_2O)_6]^{2+}$ $K_a = 1.3 \times 10^{-9}$ $pK_a = 8.89$
$B(OH)_3$ $K_a = 7.3 \times 10^{-10}$ $pK_a = 9.14$
NH_4^+ $K_a = 5.6 \times 10^{-10}$ $pK_a = 9.25$
HCN $K_a = 4.0 \times 10^{-10}$ $pK_a = 9.40$

The ammonia/ammonium ion buffer appears to be perfect in an equimolar ratio (equal number of moles of ammonia and ammonium ion).
For an equimolar buffer, $pH = pK_a$ or pK_b

8.   The salt sodium benzoate is formed in the reaction. However, although the number of moles of sodium benzoate formed equals the number of moles of acid (or base) used, its concentration is decreased by dilution:

Volume of benzoic acid = 25 mL
Volume of sodium hydroxide = 50 mL   (i.e. 25 mL × 0.100 M/0.050 M)
Total volume = 75 mL
Concentration of sodium benzoate = 0.100M × 25mL/75mL = 0.033 M

At the equivalence point:

$Bz^-(aq)$ + $H_2O(l)$ $\rightleftharpoons$ $HBz(aq)$ + $OH^-(aq)$

$K = K_w/K_a = 1.59 \times 10^{-10} = [OH^-]^2/0.033$

$[OH^-] = 2.29 \times 10^{-6}$ and   pOH = 5.64;   pH = 8.36

9.   The two salts KH_2PO_4 and K_2HPO_4 are conjugate partners; they form a buffer solution. You can use the Henderson-Hasselbalch equation to determine the ratio:

$$pH = pK_a + \log(\frac{[\text{conjugate base}]}{[\text{acid}]})$$

$$7.10 = 7.21 + \log(\frac{[\text{conjugate base}]}{[\text{acid}]})$$

$$\frac{[\text{conjugate base}]}{[\text{acid}]} = 0.776$$

The mole ratio of K_2HPO_4 (base) to KH_2PO_4 (acid) is 0.776 to 1.

CHAPTER 19

Introduction

Another type of chemical reaction you learned to recognize in Chapter 5 was a precipitation reaction. A precipitation reaction is an exchange reaction in which one of the products is an insoluble compound. In this chapter these reactions, in which a precipitate is formed, are examined in more detail. A precipitation reaction is just another example of a system in equilibrium and the same principles that have been discussed in the previous three chapters apply to these reactions as well.

Contents

19.1   The Solubility of Salts

An insoluble substance is defined as one that has a solubility less than about 0.01 mol per liter of water. General solubility rules are listed in Figure 5.4 on page 184. Many minerals are insoluble sulfides, oxides, and carbonates and methods had to be developed for extracting the various metals from these ores. Very often a sequence of solution and precipitation reactions is involved.

Some common minerals are listed in Table 19.1 on page 878 in the text.

19.2   The Solubility Product Constant K_{sp}

When a salt dissolves to the greatest extent possible, the solution is said to be **saturated**. For a salt such as sodium bromide, which is only very sparingly soluble in water, the concentration of a saturated solution is very small. At 25°C, the concentration of silver ions and bromide ions in a saturated solution is only 5.7×10^{-7} M.

For practical purposes, silver bromide is classified as insoluble. However, all insoluble salts are in reality very slightly soluble, some more so than others.

$$AgBr(s) \quad \rightleftharpoons \quad Ag^+(aq) \; + \; Br^-(aq)$$

Recall that substances in a different phase are not included in the equilibrium constant expression.

There is often considerable association between ions in solution, particularly when the charges are +2, or −2, or greater. Simple K_{sp} calculations can be inaccurate.

Do not confuse *solubility* and *solubility product.*

Read the Problem Solving Tips and Ideas 19.1 on page 883 and try some of the problems at the end of this study guide.

The K_{sp} for AgCl is greater but its solubility is smaller.

Read the Problem Solving Tips and Ideas 19.2 on page 889 regarding the question whether or not a precipitate will form.

When $Q_{sp} = K_{sp}$ the salt in solution is in equilibrium with excess solid salt not in solution.

The equilibrium constant for the system is:

$$K_{sp} = [Ag^+(aq)][Br^-(aq)] = (5.7 \times 10^{-7})(5.7 \times 10^{-7})$$
$$= 3.3 \times 10^{-13}$$

The equilibrium constant is called the solubility product constant, or more simply, the solubility product, symbol K_{sp}. If the solubility product is known, then the concentrations of ions in solution can be calculated.

19.3   Determining K_{sp} from Experimental Measurements

Values for the K_{sp} constants for various salts are determined experimentally. If the solubility can be determined, then the K_{sp} can be calculated.

19.4   Estimating Salt Solubility from K_{sp}

Just as K_{sp} can be calculated from the solubility of a salt, so the solubility can be calculated from the K_{sp}. Comparison of solubility products does not necessarily indicate relative solubilities—it depends upon the stoichiometries of the salts. For example, compare silver chloride and silver chromate:

AgCl: $K_{sp} = 1.8 \times 10^{-10}$ but its solubility = 1.3×10^{-5} mol/L

Ag_2CrO_4: $K_{sp} = 9.0 \times 10^{-12}$ but its solubility = 1.3×10^{-4} mol/L

Comparison of solubilities by examining the K_{sp} values is useful only when the stoichiometries of the salts are the same.

19.5   Precipitation of Insoluble Salts

A salt will precipitate from solution if the solubility product is exceeded. To determine whether or not a salt will precipitate it is useful to calculate the reaction quotient Q_{sp}.

If $Q_{sp} > K_{sp}$ then the salt will precipitate.
If $Q_{sp} < K_{sp}$ then the solution is unsaturated and no precipitate will occur.
If $Q_{sp} = K_{sp}$ then the solution is saturated.

The quotient Q_{sp} is the product of the actual concentrations of the ions in solution. For example,

for silver chloride, $Q_{sp} = [Ag^+][Cl^-]$
for nickel sulfide, $Q_{sp} = [Ni^{2+}][S^{2-}]$
for silver chromate, $Q_{sp} = [Ag^+]^2[CrO_4^{2-}]$

Suppose, for example, that a solution contains 3.0×10^{-4} M Ag^+ ions and sufficient potassium chromate is added to make the solution 5.0×10^{-3} M in chromate ions. Will precipitation of silver chromate occur?

$$Q_{sp} = [Ag^+]^2[CrO_4^{2-}] = (3.0 \times 10^{-4})^2(5.0 \times 10^{-3}) = 4.5 \times 10^{-10}$$

This value exceeds the K_{sp} (9.0×10^{-12}), so precipitation will occur.

19.6   Solubility and the Common Ion Effect

The common ion effect in the last chapter described the effect on the ionization of a weak acid or base of the addition of a common ion (the conjugate partner of the weak electrolyte). The effect was the suppression of the ionization. The effect is exactly the same in the case of a sparingly soluble salt. Addition of a common ion suppresses the solubility of the salt.

In calculations involving a common ion, you can usually make the assumption that the concentration of the common ion is derived entirely from the added soluble salt, and not at all from the small amount of the sparingly soluble salt actually in solution.

> LeChatelier's principle:
>
> A system in equilibrium will move to the right or to the left to accommodate any stress placed upon the system.

> Read Problem Solving Tips and Ideas 19.3 on page 891 for precipitation in the presence of a common ion.

19.7   Solubility, Ion Separations, and Qualitative Analysis

Qualitative analysis is the determination of which substances exist in a sample, not how much of the substance exists in the sample. In the qualitative analysis of aqueous solutions, the task is often to determine what cations and what anions are present. The solubility of various salts can be employed to good effect to establish which ions are present. For example, if a solution contains silver and copper salts, and sodium sulfide is added, both will precipitate. If sodium chloride is added, only silver will precipitate. Such experiments allow the identification of both metal cations and anions in a solution.

19.8   Simultaneous Equilibria

Sometimes more than one reaction occurs in solution at the same time; such systems are called **simultaneous equilibria**. Suppose, for example, that sodium bromide (a source of bromide ion) is added to a saturated solution of silver chloride. What happens?

For silver chloride:

$$AgCl(s) \rightleftharpoons Ag^+(aq) + Cl^-(aq) \qquad K_{sp1} = [Ag^+][Cl^-] = 1.8 \times 10^{-10}$$

For silver bromide:

$$AgBr(s) \rightleftharpoons Ag^+(aq) + Br^-(aq) \qquad K_{sp2} = [Ag^+][Br^-] = 3.3 \times 10^{-13}$$

Reverse this equation and add to the first equation:

$$AgCl(s) + Br^-(aq) \rightleftharpoons AgBr(s) + Cl^-(aq) \quad K = K_{sp1} \times 1/K_{sp2} = 545$$

So the reaction is product favored; the bromide ions cause precipitation of silver bromide and chloride ions replace the bromide ions in solution.

> Recall that for successive (simultaneous) equilibria, the equilibrium constant for the overall reaction (the sum of the individual equations) is the product of the equilibrium constants for the individual steps.

19.9   Solubility and pH

Any salt that contains an anion that is the conjugate base of a weak acid dissolves in water to a greater extent than indicated by the value of K_{sp}. This is because these anions are hydrolyzed in aqueous solution:

$$X^-(aq) + H_2O(l) \rightleftharpoons HX(aq) + OH^-(aq)$$

Addition of a strong acid to this equilibrium removes hydroxide OH^- and shifts the equilibrium further to the right. As a result, more salt dissolves.

In general, the solubility of a salt containing the conjugate base of a weak acid is increased by addition of a strong acid (a decrease in pH). For example, consider the solution of magnesium hydroxide in water:

(aq) designations have been omitted for clarity.

$$Mg(OH)_2(s) \rightleftharpoons Mg^{2+} + 2OH^- \qquad K_{sp} = [Mg^{2+}][OH^-]^2 = 1.5 \times 10^{-11}$$

This hydroxide is essentially insoluble. If, however, a strong acid is added the equilibrium is pulled over to the right, the acid removes the hydroxide OH^-

$$OH^- \text{ }(aq) + H_3O^+(aq) \rightleftharpoons 2H_2O(l) \qquad K = 1/K_w = 1.0 \times 10^{14}$$

Doubling this equation and adding it to the first equation produces the overall reaction. The equilibrium constant for the second equation must be squared and multiplied by the equilibrium constant for the first equation:

This reaction is essentially complete, K = 1.5 × 10¹⁷!

Metal carbonates and sulfides can be treated in the same way.

$$Mg(OH)_2(s) + 2H_3O^+ \rightleftharpoons Mg^{2+} + 4H_2O(l) \qquad \begin{aligned} K &= (1/K_w)^2 \times K_{sp} \\ &= 1.5 \times 10^{17} \end{aligned}$$

19.10   Solubility and Complex Ions

Metal ions are solvated in aqueous solution; each metal ion is surrounded by water molecules. The solvated ion is called a complex ion and is often written as $[M(H_2O)_6]^{n+}$. The formation of the complex ion is a Lewis acid–base reaction. Other Lewis bases, such as ammonia, also complex with metal ions and such reactions can influence the solubility of metal ions. For example, consider the solution of silver chloride:

$$AgCl(s) \rightleftharpoons Ag^+(aq) + Cl^-(aq) \qquad K_{sp} = [Ag^+][Cl^-] = 1.8 \times 10^{-10}$$

Addition of ammonia complexes the metal ion:

This equilibrium constant for the formation of a metal complex (ion) is often referred to as a formation constant.

Silver chloride is much more soluble in the presence of ammonia.

See Figure 19.13 on page 899 of the text for a sequence of reactions involving silver ions in solution.

$$Ag^+(aq) + 2NH_3(aq) \rightleftharpoons [Ag(NH_3)_2]^+ \qquad K = 1.6 \times 10^7$$

Adding the two equations, and multiplying the equilibrium constants:

$$AgCl(s) + 2NH_3(aq) \rightleftharpoons [Ag(NH_3)_2]^+ + Cl^-(aq) \qquad K = 2.9 \times 10^{-3}$$

Because metal ions invariably form complex ions in solution, K_{sp} calculations can be quite inaccurate when these reactions are ignored.

19.11   Equilibria in the Environment

The carbon dioxide cycle on this planet is chemistry on an immense scale. The atmosphere contains 700 billion tons of carbon dioxide (0.0325% of the atmosphere). This CO_2 is in equilibrium with CO_2 in the oceans (concentration about $10^{-5}M$). The oceans contain about 40 trillion tons of CO_2 in the form of dissolved gas, as the bicarbonate or carbonate ions, as calcium or magnesium carbonates, or as organic matter. Understanding the various equilibria involved in the carbon dioxide cycle allows a better understanding of the greenhouse effect, global warming, and the capability of the ocean to absorb increasing amounts of carbon dioxide.

19.12 Chemical Equilibria: an Epilogue

Although the past three chapters have dealt with chemical systems in equilibrium, many chemical reactions occur under nonequilibrium conditions. Movement toward equilibrium is the driving force for many processes.

Review Questions

1.   Review the solubility rules listed in Table 5.4 on page 184 of the text.

2.   What is the difference between the solubility and the solubility product?

3.   If the K_{sp} for one salt is higher than for another salt, does that mean its solubility is greater?

4.   For a sparingly soluble salt such as MX where M is a metal and X is an anion and the solubility of the salt is x mol L^{-1}, write a general expression for the solubility product constant K_{sp}. Do the same thing for a salt having the stoichiometry MX_2, where again M is a metal and X is an anion and the solubility of the salt is x mol L^{-1}.

5.   What is Q_{sp}?

6.   Summarize the possible relationships between Q_{sp} and K_{sp} and indicate what each of the relationships mean.

7.   How does the presence of a common ion influence the solubility of a salt? Does this effect apply only to sparingly soluble salts?

8.   If two or more equilibria occur simultaneously in a system, how can you obtain the value of the equilibrium constant for the overall reaction?

9.   Explain how the solubility of a salt, containing as an anion the conjugate base of a weak acid, can be increased by the addition of a strong acid. Then explain how the solubility of a salt, containing as an anion the conjugate base of a strong acid, cannot usually be increased by the addition of a strong acid.

10.   When does the K_{sp} not reflect the solubility of a salt?

11.   Explain how the solubility of a salt can be increased by complex ion formation.

Answers to Review Questions

1.   It is useful to get to know the solubility rules listed in Table 5.4 on page 184 of the text. You may find you know most of them already.

2. The solubility of a salt is the concentration of the salt in a saturated solution. In other words it is a measure of how much salt will dissolve at a particular temperature. The solubility is usually expressed in units such as grams per 100mL or moles/L. The solubility product (constant) is an equilibrium constant (K_{sp}); it is a measure of the state of a system involving the salt and its saturated solution at equilibrium. The form of the solubility product (constant) depends upon the stoichiometry of the salt.

3. No, it depends upon the stoichiometry of the salt. If the stoichiometries of the salts are all the same (e.g. AgCl, AgBr, AgI), then the K_{sp} indicates the trend in solubility (the higher the K_{sp} the higher the solubility). If however, the stoichiometries are different (e.g. AgCl and Ag_2CrO_4), then no such relationship exists.

4. For a sparingly soluble salt such as MX:
   Solubility = x mol L^{-1}, so in a saturated solution $[M]$ = x mol L^{-1}
                                                          and $[X]$ = x mol L^{-1}

   $K_{sp} = [M][X] = x^2$

   For a sparingly soluble salt such as MX_2:
   Solubility = x mol L^{-1}, so in a saturated solution $[M]$ = x mol L^{-1}
                                                          and $[X]$ = 2x mol L^{-1}

   $K_{sp} = [M][X]^2 = (x) \times (2x)^2 = 4x^3$

   The stoichiometry of the salt MX_2 determines that the concentration of X is twice the concentration of M.

5. Q_{sp} is the reactant quotient for the state of a system involving the solution of a sparingly soluble salt. Comparison of Q_{sp} and K_{sp} determines whether more salt will dissolve, whether precipitation will occur, or whether the system is at equilibrium.

6. The reaction quotient Q_{sp} is compared with the K_{sp} and

If $Q_{sp} > K_{sp}$	then the salt will precipitate; the K_{sp} is exceeded.
If $Q_{sp} < K_{sp}$	then the solution is unsaturated and no precipitate will occur—more salt will dissolve at that temperature.
If $Q_{sp} = K_{sp}$	then the solution is at equilibrium (saturated solution).

7. The presence of a common ion suppresses the solubility of the salt. This is an application of LeChatelier's principle. Addition of a common ion moves the equilibrium to the side of the solid salt.
   The effect applies to any saturated solution; for example, addition of HCl to a saturated solution of NaCl causes NaCl to precipitate.

   Often the addition of an uncommon ion increases the solubility of a salt because it increases the overall ionic concentration and ionic interactions become more important. This is sometimes referred to as the salt effect.

8. The equilibrium constant for the overall reaction is the product of the equilibrium constants for the individual reactions that add up to the overall reaction. Remember that if a reaction is reversed, the equilibrium constant is the reciprocal; if the stoichiometric coefficients are doubled, the equilibrium constant is squared.

9. An anion that is the conjugate base of a weak acid is hydrolyzed in solution:

$$A^-(aq) + H_2O(l) \rightleftharpoons HA(aq) + OH^-(aq) \qquad K_b = \frac{[HA][OH^-]}{[A^-]}$$

Addition of a strong acid removes hydroxide ion OH^- from this equilibrium which pulls the system to the right, removing A^- ion. This then causes more salt to dissolve. An overall equilibrium constant can be derived from the K_{sp} of the solution process and the K_b for the hydrolysis. When a strong acid is added to the solution of a salt containing as an anion the conjugate base of a strong acid, nothing happens because there is no hydrolysis of the anion (but see comments in margin at right).

> If the strong acid happens to be the conjugate acid of the anion of the salt, then the common ion effect will suppress the solubility.
>
> Addition of Cl^- and the formation of complex ions such as $[AgCl_2]^-$ can increase the solubility.

10. Solubility equilibria are represented by K_{sp} only when nothing further happens to the ions in solution. If hydrolysis occurs (see previous question), then this will increase the solubility of the salt. If ions form ion-pairs in solution this will increase the solubility. Likewise, if a complex ion is formed (see next question), this will increase the solubility of the salt.

> Increasing the ionic strength of the solution can also increase the solubility.

11. If a complex ion is formed the solubility of the salt will increase. Again, the overall equilibrium constant can be obtained by multiplying the K_{sp} for the solution process and the K_{form} for the complex ion formation. For example consider how the solubility of cobalt(II) sulfide can be increased by addition of ammonia:

$$CoS(s) \rightleftharpoons Co^{2+}(aq) + S^{2-}(aq) \qquad K_{sp} = [Co^{2+}][S^{2-}] = 5.9 \times 10^{-21}$$

Addition of ammonia complexes the metal ion:

$$Co^{2+}(aq) + 6NH_3(aq) \rightleftharpoons [Co(NH_3)_6]^{2+} \qquad K_{form} = 7.7 \times 10^4$$

Adding the two equations, and multiplying the equilibrium constants:

$$CoS(s) + 6NH_3(aq) \rightleftharpoons [Co(NH_3)_6]^{2+} + S^{2-}(aq) \qquad K = 4.5 \times 10^{-16}$$

So the K_{sp} for cobalt(II) sulfide is 100,000 times greater in the presence of ammonia.

Study Questions and Problems

> If you're not part of the solution, you're part of the problem.
>
> Eldridge Cleaver
> (1935-1998)

1. Predict, on the basis of the solubility rules, which of the following salts are soluble, and which are insoluble. For those that are "insoluble" look up their solubility products.

   a. AgI c. $BaSO_4$ e. $NiCO_3$
   b. Na_3PO_4 d. $(NH_4)_2SO_4$ f. $Cu(OH)_2$

2. Write solubility product expressions for the following salts:

   a. $PbSO_4$ c. CuS
   b. $Ca_3(PO_4)_2$ d. CaF_2

3.   If the molar concentration of lead bromide $PbBr_2$ in an aqueous solution is 1.6×10^{-6}, what is the concentration of Pb^{2+} ions and what is the concentration of Br^- ions?

4.   If the molar solubility of silver iodide is 1.22×10^{-8} mol L^{-1}, what is the solubility product constant for silver iodide?

5.   What is the molar solubility of cadmium sulfide CdS if its solubility product constant is 3.6×10^{-29}? What is the solubility of cadmium sulfide in mg L^{-1} (or μg L^{-1})?

6.   What is the concentration of strontium ions in a saturated solution of strontium fluoride SrF_2 if its solubility product constant is 2.5×10^{-9}? What is the concentration of fluoride ions, and what is the molar solubility of the salt?

7.   What is the concentration of magnesium ions in a saturated solution of magnesium fluoride MgF_2 if its solubility product constant is 6.4×10^{-9}? What is the concentration of magnesium ions if the solution also contains 0.30M sodium fluoride?

8.   The solubility product constant for silver sulfite is 1.5×10^{-14}. From which of the following mixtures of silver nitrate and sodium sulfite would silver sulfite precipitate?

     a.   50 mL of 1.0×10^{-4} M Ag^+ and 50 mL of 1.0×10^{-4} M $SO_3{}^{2-}$
     b.   25 mL of 1.0×10^{-3} M Ag^+ and 25 mL of 1.0×10^{-5} M $SO_3{}^{2-}$
     c.   50 mL of 1.0×10^{-5} M Ag^+ and 100 mL of 1.0×10^{-3} M $SO_3{}^{2-}$

9.   Calculate the solubility in moles per liter of cobalt(II) sulfide in a solution that contains 0.030 M cobalt(II) chloride.
     K_{sp} for cobalt sulfide CoS = 5.9×10^{-21}.

10. Addition of a strong acid would increase the solubility of which of the following salts?

     AgCl $CaSO_4$
     CdS $CaCO_3$
     $PbBr_2$ $CaHPO_4$
     $Cd(OH)_2$ AuCl

11. Calculate the concentration of zinc ions $[Zn^{2+}]$ in a saturated 0.10M H_2S solution that contains 0.40M hydrochloric acid HCl.

     $ZnS(s) + H_2O(l) \rightleftharpoons Zn^{2+}(aq) + HS^-(aq) + OH^-(aq)$ K = 2.0×10^{-25}

Answers to Study Questions and Problems

1.   a.   AgI   insoluble $K_{sp} = [Ag^+][I^-]$ $= 1.5 \times 10^{-16}$
      b. Na_3PO_4 soluble
      c. $BaSO_4$ insoluble $K_{sp} = [Ba^{2+}][SO_4^{2-}]$ $= 1.1 \times 10^{-10}$
      d. $(NH_4)_2SO_4$ soluble
      e. $NiCO_3$ insoluble $K_{sp} = [Ni^{2+}][CO_3^{2-}]$ $= 6.6 \times 10^{-9}$
      f. $Cu(OH)_2$ insoluble $K_{sp} = [Cu^{2+}][OH^-]^2$ $= 1.6 \times 10^{-19}$

2.   a. $K_{sp} = [Pb^{2+}][SO_4^{2-}]$ $= 1.8 \times 10^{-8}$
      b. $K_{sp} = [Ca^{2+}]^3[PO_4^{2-}]^2$ $= 1.0 \times 10^{-25}$
      c. $K_{sp} = [Cu^{2+}][S^{2-}]$ $= 8.7 \times 10^{-36}$
      d. $K_{sp} = [Ca^{2+}][F^-]^2$ $= 3.9 \times 10^{-11}$

3.   Molar concentration of lead bromide "$PbBr_2$" $= 1.6 \times 10^{-6}$
      Molar concentration of Pb^{2+} is the same $= 1.6 \times 10^{-6}$
      Molar concentration of Br^- is twice as great $= 3.2 \times 10^{-6}$

> There is one Pb^{2+} ion per formula unit.
>
> There are two Br^- ions per formula unit.

4.   Silver iodide AgI has a 1:1 stoichiometry, the concentration of silver ions
      is 1.22×10^{-8} mol L^{-1} and the concentration of iodide ions is the same.
      The solubility product constant
      $K_{sp} = [Ag^+][I^-] = (1.22 \times 10^{-8})(1.22 \times 10^{-8}) = 1.5 \times 10^{-16}$

5. $K_{sp} = [Cd^{2+}][S^{2-}] = 3.6 \times 10^{-29}$.
      If x = the concentration of both the cadmium and sulfide ions, then
      $x^2 = 3.6 \times 10^{-29}$, and $x = 6.0 \times 10^{-15}$—this is the molar solubility.
      The molar mass of cadmium sulfide is 144.5 g mol^{-1},
      so the concentration $= 8.67 \times 10^{-13}$g/L $= 8.67 \times 10^{-10}$mg/L
                            $= 8.67 \times 10^{-7}$ μg/L —not very soluble!

6. $K_{sp} = [Sr^{2+}][F^-]^2 = 2.5 \times 10^{-9}$.
      If x = the concentration of the strontium ions, then
      the concentration of the fluoride ions = 2x.
      $(x)(2x)^2 = 2.5 \times 10^{-9}$ so $x = 8.6 \times 10^{-4}$
      The concentration of strontium ions $= 8.6 \times 10^{-4}$ M
      The concentration of fluoride ions $= 1.7 \times 10^{-3}$ M
      The molar solubility of strontium fluoride $SrF_2 = 8.6 \times 10^{-4}$ M.

7. $K_{sp} = [Mg^{2+}][F^-]^2 = 6.4 \times 10^{-9}$.
      If x = the concentration of the magnesium ions, then
      the concentration of the fluoride ions = 2x.
      $(x)(2x)^2 = 6.4 \times 10^{-9}$ so $x = 1.2 \times 10^{-3}$
      The concentration of magnesium ions $= 1.2 \times 10^{-3}$ M

      In 0.30M sodium fluoride, the fluoride ion concentration is 0.30M,
      $[Mg^{2+}][F^-]^2 = [Mg^{2+}][0.30]^2 = 6.4 \times 10^{-9}$
      $[Mg^{2+}]$ = concentration of magnesium ions $= 7.1 \times 10^{-8}$ M

> In the presence of NaF, the amount of fluoride derived from the solution of MgF_2 is infinitesimal and can be ignored.
>
> Note that the solubility is reduced in the presence of the common ion.

8. $K_{sp} = [Ag^+]^2[SO_3^{2-}] = 1.5 \times 10^{-14}$.

a. $Q_{sp} = [5.0 \times 10^{-5}]^2[5.0 \times 10^{-5}] = 1.3 \times 10^{-13}$.

b. $Q_{sp} = [5.0 \times 10^{-4}]^2[5.0 \times 10^{-6}] = 1.3 \times 10^{-12}$.

c. $Q_{sp} = [3.3 \times 10^{-6}]^2[6.7 \times 10^{-4}] = 7.4 \times 10^{-15}$.

In the first two cases the value of the reaction quotient Q_{sp} exceeds the solubility product K_{sp}, and so in both cases a precipitate of silver sulfite will occur. In the last example, the value of the reaction quotient Q_{sp} is less than the solubility product K_{sp}, and so the solution is unsaturated and no precipitate occurs. Note that it is important to consider the dilution of the solutions when they are mixed.

9. K_{sp} for cobalt sulfide $= [Co^{2+}][S^{2-}] = 5.9 \times 10^{-21}$.
Virtually all the cobalt ion in the solution is from the cobalt(II) chloride, so the concentration of the cobalt ion $[Co^{2+}] = 0.030$M.
Therefore the sulfide concentration $[S^{2-}] = 5.9 \times 10^{-21}/0.030$
$$= 2.0 \times 10^{-19} \text{ M}$$

This is the concentration of the sulfide in the solution and therefore also the concentration of the cobalt sulfide salt in solution.

10. Any anion that is a conjugate base of a weak acid will hydrolyze in aqueous solution to form hydroxide ions. Addition of a strong acid will remove these hydroxide ions and pull the equilibrium to the right, increasing the solubility of the salt. Of the salts listed, CdS, $CaCO_3$, $CaHPO_4$, and $Cd(OH)_2$ contain the anion that is a conjugate base of a weak acid.

11. $[H_2S] = 0.10$ M
$[HCl] = 0.40$ M

$ZnS(s) \rightleftharpoons Zn^{2+}(aq) + S^{2-}(aq)$

$S^{2-}(aq) + H_2O(l) \rightleftharpoons HS^-(aq) + OH^-(aq) \qquad K_b = 1.0 \times 10^5$

Adding these two equations:

$ZnS(s) + H_2O(l) \rightleftharpoons Zn^{2+}(aq) + HS^-(aq) + OH^-(aq) \quad K = 2.0 \times 10^{-25}$

In acidic solution, most of the HS$^-$ ions and OH$^-$ ions are protonated:

$HS^-(aq) + H_3O^+(aq) \rightleftharpoons H_2S(aq) + H_2O(l) \qquad K = 1.0 \times 10^7$

$H_3O^+(aq) + OH^-(aq) \rightleftharpoons 2H_2O(l) \qquad K = 1.0 \times 10^{14}$

Adding these three equations (and multiplying the constants):

$ZnS(s) + H_3O^+(aq) \rightleftharpoons Zn^{2+}(aq) + H_2S(aq) + 2H_2O(l) \quad K = 2.0 \times 10^{-4}$

$$K_{spa} = \frac{[Zn^{2+}][H_2S]}{[H_3O^+]^2} = \frac{[Zn^{2+}][0.10]}{[0.40]^2} = 2.0 \times 10^{-4}$$

$[Zn^{2+}] = 3.2 \times 10^{-4}$ M

CHAPTER 20

Introduction

Many chemical reactions and physical processes take place by themselves. Such reactions and processes are said to be spontaneous or product–favored—they may happen quite slowly, or they may happen at explosive speed, but they will happen. In this chapter we will investigate why some reactions are spontaneous while others are not. We will explore ways to predict when a reaction is spontaneous. The second law of thermodynamics is the law that determines spontaneity.

Contents

20.1   Spontaneous Reactions and Speed: Thermodynamics *vs.* Kinetics

All systems eventually reach equilibrium; the move toward equilibrium is the driving force behind a reaction. If the equilibrium constant is very small, the reaction is said to be reactant–favored; equilibrium is reached after only very few reactant molecules have formed product molecules. If the equilibrium constant is large, the reaction is product–favored; equilibrium is reached only when most of the reactant molecules have formed products.

Even though a reaction is driven toward equilibrium, it may not progress there very quickly. The speed of a reaction is determined by the kinetics not the thermodynamics of the reaction.

Chemical kinetics is the subject of Chapter 15.

Think of chemical thermo-dynamics as the difference between the beginning and the end and chemical kinetics as how you get from one to the other.

20.2   Directionality of Reactions: Entropy

One of the reasons why a reaction may be product–favored is that it is exothermic. In an exothermic reaction, chemical potential energy is released and the kinetic energies of the molecules in the vicinity of the reaction increase—the temperature rises. This dispersal of energy is one of two driving forces for chemical reactions and physical processes.

The second driving force is the dispersal of matter. For example, when a gas is placed in a flask, it disperses to fill the entire flask; when two gases are placed in a flask, they mix. This tendency for randomness is a result of probability. If you were to examine all the possible ways in which all the molecules in two gases A and B could be arranged, you would find that the chance of having all

Try problem #1 in the Study Questions and Problems.

the molecules of A on one side of the flask, and all the molecules of B on the other side of the flask, is infinitesimally small—there is a far greater probability of arrangements in which the gas molecules are mixed together.

The dispersal of energy is the more important driving force at lower temperatures whereas the dispersal of matter is the more important driving force at higher temperatures.

If both matter and energy are dispersed, the reaction is spontaneous.

If only one of the two is dispersed, then the spontaneity depends upon the relative magnitude of the two and the temperature. If neither are dispersed, then the reaction is not spontaneous (it is reactant-favored).

The name for the disorder of matter is **entropy**; symbol S. A change in entropy can be determined by measuring the heat absorbed by matter divided by the temperature (in K) at which it is absorbed. Since the entropy of all substances in a perfect crystalline state at a temperature of 0K is zero, the absolute entropy of any substance at any temperature can be determined by adding up all the entropy changes required to get to that temperature.

General observations on the entropies of substances are:

> Entropy increases from solid to liquid to gas.
> Entropies of complex molecules are higher than entropies of simpler molecules.
> Entropies increase in ionic solids as the interionic attraction decreases.
> Entropy usually increases when solutions are prepared from pure solute and solvent.
> Entropy increases when a dissolved gas escapes from a solution.

The **second law of thermodynamics** states that a reaction is spontaneous when the entropy of the universe increases. This means, in effect, that the total disorder in the universe is continually increasing and there's absolutely nothing you can do about it. If order is created in some small part of the universe, it is always at the expense of even greater disorder somewhere else.

The entropy change for a system when a reaction occurs can be calculated by adding the entropies of the products and subtracting the entropies of the reactants:

$$\Delta S^\circ_{system} = \Sigma(S^\circ \text{(products)} - S^\circ \text{(reactants)})$$

The entropy change for the surroundings occurs because heat leaves the system and enters the surroundings (or *vice versa*). Provided that the energy transfer is slow and it occurs at a constant temperature, the entropy change in the surroundings can be calculated by:

$$\Delta S^\circ_{surroundings} = \frac{q_{surr}}{T} = -\frac{q_{sys}}{T} = -\frac{\Delta H_{sys}}{T}$$

The change in the entropy of the entire universe when a reaction occurs is the sum of the entropy changes in the system and in the surroundings:

$$\Delta S^\circ_{universe} = \Delta S^\circ_{system} + \Delta S^\circ_{surroundings}$$

Side notes (left margin):

This is the third law of thermodynamics.

Determining entropies in this way requires integration to obtain the area under an entropy-temperature graph.

Some absolute molar entropies of a variety of substances at 298 K are listed in Table 20.1 on page 917 in the text.

The solution of gases is an exception; see the next statement in this box.

$\Delta S_{universe} >$ zero.

Disorder is created (entropy is increased) by
    the dispersal of energy,
    the dispersal of matter,
    or both.

This should remind you of a similar equation for the enthalpy change $\Delta H^\circ_{reaction}$ encountered in Chapter 6.

A reaction is product-favored (spontaneous) when there is an increase in the entropy of the system or when the reaction is exothermic ($\Delta H^\circ_{reaction}$ is negative), or both.

20.3   Gibbs Free Energy

Establishing the entropy change for the entire universe is not easy, and so a new thermodynamic function was proposed by J. Willard Gibbs that is a function of the system alone. This function is now called **Gibbs free energy** and is given the symbol G. Establishing the change in Gibbs free energy ΔG allows you to establish whether or not the reaction is spontaneous.

$$\Delta S^\circ_{universe} = \Delta S^\circ_{system} + \Delta S^\circ_{surroundings} = \Delta S^\circ_{sys} - \frac{\Delta H}{T}_{sys}$$

> The superscript ° denotes standard conditions (1 bar pressure).
>
> T is in degrees K.

Multiplying through by –T:

$$-T\Delta S^\circ_{universe} = -T\Delta S^\circ_{sys} + \Delta H^\circ_{sys} \text{ and this is defined as } \Delta G^\circ_{sys}$$

So, $\Delta G^\circ_{sys} = \Delta H^\circ_{sys} - T\Delta S^\circ_{sys}$ or more simply $\Delta G^\circ = \Delta H^\circ - T\Delta S^\circ$

> Usually the subscripts $_{sys}$ are omitted because this is an equation in which all terms apply to the system.

The free energy of the system must decrease if the entropy of the universe is to increase and the reaction is to be spontaneous.

ΔH°_{sys}	ΔS°_{sys}	ΔG°_{sys}	result
–	+	must be –	reaction is spontaneous
–	–	depends upon the temperature	
+	+	depends upon the temperature	
+	–	must be +	reaction cannot be spontaneous

It is useful to define a standard free energy of formation ΔG°_f analogous to the standard enthalpy of formation first encountered in Chapter 6. The free energy change for a reaction can then be calculated in the same way that the enthalpy of reaction was calculated:

> Note that the ΔG°_f for an element in its standard state is zero. Recall that ΔH°_f for an element in its standard state is zero.

$$\Delta H^\circ_{reaction} = \Sigma(\Delta H^\circ_f (products) - \Delta H^\circ_f (reactants)) \quad \text{for the enthalpy change}$$

$$\Delta G^\circ_{reaction} = \Sigma(\Delta G^\circ_f (products) - \Delta G^\circ_f (reactants)) \quad \text{for the free energy change}$$

> Note however that standard entropies of elements are *not* zero!
>
> Check Appendix L in the text.

The value of the free energy change indicates the maximum amount of useful work that can be obtained from a chemical reaction. Some energy is always required to ensure an increase in entropy during the process, and therefore not all is available for work.

The expression for Gibbs free energy change ΔG°_{sys} contains T, the temperature in degrees K. This means that if both ΔH°_{sys} and ΔS°_{sys} are negative, or both are positive, the sign of the free energy change will change if the value of T is changed enough. As a result, a reaction that is spontaneous in one direction at one temperature will become spontaneous in the reverse direction at another temperature. Consider for example, the melting of ice. For this process, both ΔH°_{sys} and ΔS°_{sys} are positive (an endothermic process and an increase in disorder).

> Even though ΔH°_{sys} and ΔS°_{sys} may not change very much when the temperature changes, ΔG°_{sys} does change because its expression contains T.

If ice is removed from the freezer and placed on the kitchen table, it melts. The process ice → water is spontaneous. ΔG° is negative because $-T\Delta S^\circ$ is greater than $+ \Delta H^\circ$—the reaction is said to be entropy–driven. The T is sufficiently high. However, if the water is placed back in the freezer, the reaction becomes spontaneous in the reverse direction. The water freezes because the T is now low enough to make $\Delta H°$ greater than $-T\Delta S°$. The free energy change $\Delta G°$ is now positive for the melting process and negative for the freezing process. Therefore the water freezes—this is an enthalpy–driven process.

20.4 Thermodynamics and the Equilibrium Constant

$\Delta G°$ is the difference between pure reactants and pure products.
ΔG is a differential quantity—it is the slope of the graph measuring how the free energy of the system changes as the reaction proceeds.
The two are related by the expression:

$$\Delta G = \Delta G° + RT\ln Q$$

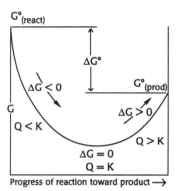

The standard free energy of reaction $\Delta G°$ is the difference between the free energy of the pure reactants (in their standard states) and the free energy of the pure products (in their standard states). In reality of course the reaction does not progress instantaneously from pure reactants to pure products. Usually some equilibrium is reached containing a mixture of reactants and products. The movement toward equilibrium is measured by ΔG.

ΔG and $\Delta G°$ are related by the expression:

$$\Delta G = \Delta G° + RT\ln Q \qquad \text{where Q is the reaction quotient}$$

The reaction quotient quantitatively describes the composition of the system. Recall that if $Q < K$, the system moves to the right, if $Q > K$, the system moves to the left, and if $Q = K$, the system has reached equilibrium. If the system is at equilibrium, $\Delta G = 0$, the movement away from equilibrium in either direction means that ΔG = positive and therefore is nonspontaneous (it doesn't happen).

$$0 = \Delta G° + RT\ln K \qquad \text{or} \qquad \Delta G° = -RT\ln K \qquad \text{at equilibrium}$$

20.5 Thermodynamics and Time

The three laws of thermodynamics are:

First law: The energy of the universe is constant: $\Delta E_{universe} = 0$
Second law: The entropy of the universe is increasing: $\Delta S_{universe} > 0$
Third law: The entropy of a perfect crystal at 0K is zero.

All natural processes result in chaos. Matter, energy, or both, are dispersed.

Review Questions

1. Describe the differences between the kinetics and the thermodynamics of a chemical reaction.

2. What two components drive a naturally occurring process?

3. What do you understand by the word "entropy"?

4. What types of processes or reactions lead to an increase in entropy?

5. How are entropy, enthalpy, and the spontaneity of a reaction related?

6.   What is Gibbs free energy? Why is the energy referred to as free?

7.   What is the Gibbs free energy of formation? For what substances is the Gibbs free energy of formation zero?

8.   Explain why, if the entropy change is positive, and the enthalpy change is negative, the reaction has to be spontaneous.

9.   How is the change in Gibbs free energy related to the spontaneity of a physical process or chemical reaction?

10.   What is the difference between the standard free energy change $\Delta G°$ and the free energy change ΔG?

11.   What is the relationship between $\Delta G°$ and ΔG?

12.   What does ΔG equal at equilibrium? What does $\Delta G°$ equal at equilibrium?

13.   How is the change in the entropy of the surroundings related to the quantity of heat released (or absorbed) by the system at constant temperature?

14.   From your answer to the previous question it would appear that a certain quantity of heat has a greater effect on the entropy of the surroundings at a lower temperature than it does at a higher temperature. Does this make sense?

15.   What is the relationship between ΔG and the equilibrium constant K? What is the relationship between $\Delta G°$ and the equilibrium constant K?

16.   Why is the temperature important in determining the value of the free energy change?

17.   For two simultaneous equilibria, what is the relationship between the equilibrium constants for the two reactions and the equilibrium constant for the overall reaction? What therefore is the relationship between the standard free energy changes for the two individual reactions and the standard free energy change for the overall reaction?

18.   Of what significance is your answer to the previous question to the coupling of chemical reactions?

19.   State briefly the three laws of thermodynamics.

Answers to Review Questions

1. All systems eventually reach equilibrium; the move toward equilibrium is determined by the thermodynamic properties of the system (the value of the free energy change ΔG and the magnitude of equilibrium constant K). The speed at which the system moves toward its final state is determined by the kinetic properties of the system. Even though a reaction is spontaneous and is driven toward equilibrium, it may not get there very quickly.

2. The dispersal of energy and the dispersal of matter are the two driving forces for chemical reactions and physical processes. For every spontaneous reaction, one of these driving forces (or both) must operate.

3. The word "entropy" means disorder. The higher the disorder, the higher the entropy.

4. Typical processes or reactions that lead to an increase in entropy are:

   Changes from solid to liquid to gas.
   Breakdown of a large molecule into smaller molecules.
   Breakdown of an ionic solid.
   Reactions in which the number of gas molecules increase.
   Making a solution or a mixture.
   The expansion of a gas.

5. The two contributing driving forces for a spontaneous reaction are an increase in entropy and a decrease in enthalpy of the system. Both result in an increase in the entropy of the universe.

6. Gibbs free energy is a thermodynamic function proposed by J. Willard Gibbs that is a function of the system alone. Establishing the change in Gibbs free energy ΔG allows you to establish whether or not the reaction is spontaneous, without worrying about the rest of the universe.

$$\Delta G° = \Delta H° - T\Delta S°$$

   The free energy of the system must decrease if the entropy of the universe is to increase and the reaction is spontaneous. Gibbs free energy is referred to as free because only this quantity of energy is available to do work. The remainder is responsible for ensuring that the entropy of the universe increases.

7. The standard free energy of formation $\Delta G_f°$ of a substance is the standard free energy change when one mole of the substance is prepared from its constituent elements in their standard states. The free energy change for a reaction can then be calculated in the same way that the enthalpy of reaction was calculated:

$$\Delta G^\circ_{reaction} = \Sigma(\Delta G^\circ_f \text{ (products)} - \Delta G^\circ_f \text{ (reactants)})$$

The Gibbs free energy of formation is zero for all elements in their standard states.

8. If $\Delta G^\circ = \Delta H^\circ - T\Delta S^\circ$ and the entropy change ΔS° is positive, and the enthalpy change ΔH° is negative, then ΔG° must be negative and the reaction has to be spontaneous regardless of the temperature.

9. The Gibbs free energy change ΔG° equals $-T\Delta S^\circ$ for the universe, so a negative change in ΔG° means that the entropy of the universe is increasing and the reaction or process is spontaneous.

10. ΔG° is the difference between the Gibbs free energy of the pure products and the Gibbs free energy of the pure products. ΔG is a differential quantity—it is the slope of the graph measuring how the free energy of the system changes as the reaction proceeds. Even though ΔG° is negative for a reaction, the final state of the system can be reached from *both* ends, the reactant end and the product end. The free energy change ΔG is negative down the reactant slope forwards, and down the product slope backwards, to the point of equilibrium where ΔG equals zero.

11. The two are related by the expression: $\Delta G = \Delta G^\circ + RT\ln Q$.

12. ΔG equals zero at equilibrium. ΔG° equals the same quantity regardless of the state of the system. It doesn't matter whether the system is at equilibrium or not. It is the difference between the free energies of the pure products and the pure reactants.

13. $\Delta S^\circ_{surroundings} = -\dfrac{\Delta H^\circ_{sys}}{T}$

Heat entering the surroundings increases the disorder (entropy) of the surroundings.

14. Yes; it makes sense. A certain quantity of heat can cause a comparatively large amount of chaos in an ordered system at a low temperature whereas at a high temperature the same amount of heat would cause an imperceptible change in the high degree of disorder already present.

15. There is no relationship between ΔG and the equilibrium constant K; at equilibrium the value of ΔG is zero but that has no influence on the value of K.
ΔG and ΔG° are related by the expression $\Delta G = \Delta G^\circ + RT\ln Q$; so when ΔG equals zero at equilibrium, the reaction quotient Q = the equilibrium constant K, and $\Delta G^\circ = -RT\ln K$. The position of equilibrium is determined by the value of ΔG° for the reaction.

16. Even though $\Delta H°$ and $\Delta S°$ may not change very much when the temperature changes, $\Delta G°$ does change because its expression contains T. For this reason you will often see $\Delta G°$ written as $\Delta G°_T$ where the T specifies the temperature:

$$\Delta G°_T = \Delta H° - T\Delta S°$$

17. Suppose the two equilibria are:

$$A + B \rightleftharpoons C \qquad K_1 \qquad \Delta G°_1 \qquad \Delta G°_1 = -RT\ln K_1$$
$$C + D \rightleftharpoons E \qquad K_2 \qquad \Delta G°_2 \qquad \Delta G°_2 = -RT\ln K_2$$

Recall that adding logarithms (exponents) is equivalent to multiplying numbers.

For the overall reaction:

$$A + B + D \rightleftharpoons E \quad K = K_1 \times K_2 \text{ and } \Delta G° = \Delta G°_1 + \Delta G°_2$$

Recall some of the examples in aqueous and heterogeneous equilibria examined in earlier chapters. For example, the neutralization of a weak acid, or the solution of a sparingly soluble salt of a weak acid by adding a strong acid.

18. Coupled reactions often occur in which one reaction is nonspontaneous on its own ($\Delta G°_1$ is positive) but is coupled with a second reaction for which the change in free energy is sufficient ($\Delta G°_2$ is negative) to drive the first reaction in addition to itself. The overall $\Delta G°$ is negative even though one of the individual free energy changes ($\Delta G°_1$) is positive.

19. The three laws of thermodynamics are:
    First law: The energy of the universe is constant: $\Delta E_{universe} = 0$
    Second law: The entropy of the universe is increasing: $\Delta S_{universe} > 0$
    Third law: The entropy of a perfect crystal at 0K is zero.

That is what learning is. You suddenly understand something you've understood all your life, but in a new way.

Doris Lessing
(1919-)

Study Questions and Problems

1. Imagine tossing two coins in the air.

   a. Predict the distribution of various combinations of heads and tails.
   b. What is the probability of the result being two heads?
   c. What is the most probable result?

   Now imagine tossing three coins in the air.

   d. What is the probability of a three heads result?
   e. Which system has the highest entropy, the two-coin system or the three-coin system?

2. Which one of the following pairs of samples have the higher entropy?

   a. $Br_2(l)$ or $Br_2(g)$

   b. $C_2H_6(g)$ or $C_3H_8(g)$

   c. $MgO(s)$ or $NaCl(s)$

   d. $KOH(s)$ or $KOH(aq)$

3.   Predict the entropy change for the following processes:

a. $O_2(g) \rightarrow 2O(g)$

b. $2O_3(g) \rightarrow 3O_2(g)$

c. $CH_4(g) + 2O_2(g) \rightarrow CO_2(g) + 2H_2O(g)$

d. $NaCl(s) \rightarrow Na^+(aq) + Cl^-(aq)$

e. $C_2H_5OH(l) \rightarrow C_2H_5OH(g)$

f. $Ag^+(aq) + Cl^-(aq) \rightarrow AgCl(s)$

4.   Calculate the standard entropy changes for the reactions listed in question 3 and verify your predictions.

5.   The standard entropy of carbon tetrachloride *(l)* is 216.40 $JK^{-1}mol^{-1}$
     The standard entropy of carbon tetrachloride *(g)* is 309.85 $JK^{-1}mol^{-1}$
     Calculate the standard molar entropy of vaporization for CCl_4.

     The standard enthalpy of formation of CCl_4 *(l)* is –135.44 $kJmol^{-1}$
     The standard enthalpy of formation of CCl_4 *(g)* is –102.90 $kJmol^{-1}$
     Calculate the standard molar enthalpy of vaporization for CCl_4.

     The standard free energy of formation of CCl_4 *(l)* is –65.21 $kJmol^{-1}$
     The standard free energy of formation of CCl_4 *(g)* is –60.59 $kJmol^{-1}$
     Calculate the standard molar free energy of vaporization for carbon tetrachloride at 298.15K.

$$CCl_4(l) \rightarrow CCl_4(g)$$

     Is the entropy change and the enthalpy change what you expected?
     Are the data consistent with one another?

6.   The heat necessary to melt pure acetic acid at its melting point (16.6°C) is 11.50 $kJmol^{-1}$. Calculate the entropy change when acetic acid melts at its normal melting point.

7.   Consider the decomposition of calcium carbonate:

$CaCO_3(s) \rightleftharpoons CaO(s) + CO_2(g)$ $\Delta G_f^\circ(CaCO_3(s)) = -1128.79\ kJmol^{-1}$
                                         $\Delta G_f^\circ(CaO(s)) = -604.03\ kJmol^{-1}$
                                         $\Delta G_f^\circ(CO_2(g)) = -394.359\ kJmol^{-1}$

a.   What is the standard free energy change at 298.15°C for the reaction shown?

b.   Calculate the equilibrium constant for the reaction.

c.   Calculate the vapor pressure of carbon dioxide above calcium carbonate at 298.15°C.

8. When benzene is burned:

$$C_6H_6(l) + 15/2\ O_2(g) \rightarrow 6CO_2(g) + 3H_2O(g)$$

if
$$\Delta G_f^\circ(C_6H_6(l)) = +124.5\ kJmol^{-1}$$
$$\Delta G_f^\circ(H_2O(g)) = -228.6\ kJmol^{-1}$$
$$\Delta G_f^\circ(CO_2(g)) = -394.4\ kJmol^{-1}$$

and
$$\Delta H_f^\circ(C_6H_6(l)) = +49.0\ kJmol^{-1}$$
$$\Delta H_f^\circ(H_2O(g)) = -241.8\ kJmol^{-1}$$
$$\Delta H_f^\circ(CO_2(g)) = -393.5\ kJmol^{-1}\ \text{at } 298.15K\ (25°C)$$

a. How much heat is released in this reaction at 25°C ?

b. Calculate the standard entropy change for the reaction at 25°C.

9. Of the following reactions, which are spontaneous at any temperature, which are never spontaneous regardless of the temperature, which may be nonspontaneous at a low temperature but may well become spontaneous at a higher temperature, and which are spontaneous only at low temperature?

		ΔH	ΔS
a.	$C_8H_{18}(l) + 25/2\ O_2(g) \rightarrow 8CO_2(g) + 9H_2O(g)$	–	+
b.	$N_2(g) + 2F_2(g) \rightarrow N_2F_4(g)$	–	–
c.	$Cl_2(g) \rightarrow 2Cl(g)$	+	+
d.	$2O_3(g) \rightarrow 3O_2(g)$	–	+
e.	$2C(s) + 2H_2(g) \rightarrow C_2H_4(g)$	+	–

Answers to Study Questions and Problems

1. a. The distribution of various combinations of heads (h) and tails (t) is:

    hh, ht, th, tt   or   two heads, 2 × one of each, two tails (1:2:1)

    b. The probability of the result being two heads is 1 in 4.

    c. The most probable result is one head and one tail.

    d. The distribution for three coins is:

    hhh, hht, hth, thh, htt, tht, tth, ttt   (or a 1:3:3:1 distribution)

    The probability of a three heads result is 1 in 8.

    e. The three-coin system has the highest entropy because they are more possible arrangements for the coins in this system.

This is a statistical definition of entropy.

2. The higher entropy:

    a. $Br_2(g)$; gases have a greater disorder (175 *vs.* 152 $JK^{-1}mol^{-1}$ at 298K)

    b. $C_3H_8(g)$; the larger molecule has a higher entropy (270 *vs.* 230 $JK^{-1}mol^{-1}$ at 298K)

Tabulated values for the entropies of many common substances are listed in Appendix L of the text.

c. NaCl(s); weaker interionic attraction (72 *vs.* 27 JK^{-1}mol^{-1} at 298K)

d. KOH(aq); in solution (92 vs. 79 JK^{-1}mol^{-1} at 298K)

3. Predicted entropy changes:

a. $O_2(g) \rightarrow 2O(g)$
An increase in entropy; twice as many particles and therefore more disorder. (Even though the molar entropy of $O_2(g)$ is larger than $O(g)$ —there are twice as many moles of $O(g)$ compared to $O_2(g)$).

b. $2O_3(g) \rightarrow 3O_2(g)$
An increase in entropy; as in (a) there is an increase in the number of particles.

c. $CH_4(g) + 2O_2(g) \rightarrow CO_2(g) + 2H_2O(g)$
Very difficult to predict; number of particles in the gas state remains unchanged; anticipate very little change.

d. $NaCl(s) \rightarrow Na^+(aq) + Cl^-(aq)$
All solution processes except the dissolving of a gas result in an increase in disorder.

e. $C_2H_5OH(l) \rightarrow C_2H_5OH(g)$
A change in state from liquid to gas results in an increase in disorder.

f. $Ag^+(aq) + Cl^-(aq) \rightarrow AgCl(s)$
Anticipate a decrease in entropy; formation of a solid precipitate from ions in solution.

4. Use the expression: $\Delta S^\circ_{system} = \Sigma(S^\circ \text{(products)} - S^\circ \text{(reactants)})$

a. $O_2(g) \rightarrow 2O(g)$
$\Delta S^\circ_{system} = \Sigma(2 \times 161 - 1 \times 205) = +117$ JK^{-1}mol^{-1}

b. $2O_3(g) \rightarrow 3O_2(g)$
$\Delta S^\circ_{system} = \Sigma(3 \times 205 - 2 \times 239) = +137$ JK^{-1}mol^{-1}

c. $CH_4(g) + 2O_2(g) \rightarrow CO_2(g) + 2H_2O(g)$
$\Delta S^\circ_{system} = \Sigma((2 \times 189 + 214) - (2 \times 205 + 186) = -4$ JK^{-1}mol^{-1}
(a very small change as anticipated)

d. $NaCl(s) \rightarrow Na^+(aq) + Cl^-(aq)$
$\Delta S^\circ_{system} = \Sigma(116 - 72) = +44$ JK^{-1}mol^{-1}

e. $C_2H_5OH(l) \rightarrow C_2H_5OH(g)$
$\Delta S^\circ_{system} = \Sigma(283 - 161) = +122$ JK^{-1}mol^{-1}

f. $Ag^+(aq) + Cl^-(aq) \rightarrow AgCl(s)$
$\Delta S^\circ_{system} = \Sigma(92 - (73 + 57) = -38$ JK^{-1}mol^{-1}

5. The standard entropy of carbon tetrachloride(l) is 216.40 $JK^{-1}mol^{-1}$
The standard entropy of carbon tetrachloride(g) is 309.85 $JK^{-1}mol^{-1}$
The standard molar entropy of vaporization for CCl_4:

$$\Delta S^\circ_{vap} = \Sigma(309.85 - 216.40) = +93.45 \ JK^{-1}mol^{-1}$$

Many thermodynamic quantities
are listed for a temperature of
298.15K.

The standard enthalpy of formation of $CCl_4(l)$ is -135.44 $kJmol^{-1}$
The standard enthalpy of formation of $CCl_4(g)$ is -102.90 $kJmol^{-1}$
The standard molar enthalpy of vaporization for CCl_4:

$$\Delta H^\circ_{vap} = \Sigma(-102.90 - (-135.44)) = +32.54 \ kJmol^{-1}$$

The standard free energy of formation of $CCl_4(l)$ is -65.21 $kJmol^{-1}$
The standard free energy of formation of $CCl_4(g)$ is -60.59 $kJmol^{-1}$
The standard molar free energy of vaporization for carbon tetrachloride at 298.15K:

$$\Delta G^\circ_{vap} = \Sigma(-60.59 - (-65.21)) = +4.62 \ kJmol^{-1}$$

The entropy change is positive as expected for a change in state from liquid to vapor. The enthalpy change is positive as expected for a process in which bonds have to be broken.

The standard free energy change for the vaporization process can be calculated from the entropy and enthalpy changes:

Remember to change kJ into J
(or vice versa).

$$
\begin{aligned}
\Delta G^\circ_{vap} &= \Delta H^\circ_{vap} - T\Delta S^\circ_{vap} \\
&= +32.54 - (298.15 \times 93.45/1000) \ kJmol^{-1} \\
&= +4.68 \ kJmol^{-1} \text{ so the data are consistent.}
\end{aligned}
$$

6. At the melting point the system is in equilibrium and ΔG is zero. Therefore $\Delta H_{fus} - T\Delta S_{fus}$ must equal zero or, in other words, ΔH_{fus} must equal $T\Delta S_{fus}$.

Remember to use degrees K
and to change kJ to J to obtain
the entropy change in its usual
units of $JK^{-1}mol^{-1}$.

$$\Delta S_{fus} = \Delta H_{fus}/T = 11,500/289.75 = +39.7 \ JK^{-1}mol^{-1}$$

7. The decomposition of calcium carbonate:

$$CaCO_3(s) \rightleftharpoons CaO(s) + CO_2(g) \quad
\begin{aligned}
&\Delta G^\circ_f(CaCO_3(s)) = -1128.79 \ kJmol^{-1} \\
&\Delta G^\circ_f(CaO(s)) = -604.03 \ kJmol^{-1} \\
&\Delta G^\circ_f(CO_2(g)) = -394.359 \ kJmol^{-1}
\end{aligned}$$

a. The standard free energy change at 298.15°C:

$$
\begin{aligned}
\Delta G^\circ_{reaction} &= \Sigma(\Delta G^\circ_f(\text{products}) - \Delta G^\circ_f(\text{reactants})) \\
&= \Sigma((-604.03 - 394.359) - (-1128.79)) \\
&= +130.40 \ kJ
\end{aligned}
$$

b. The equilibrium constant for the reaction:

$$\Delta G° = -RT\ln K$$

$$+130.40 \text{ kJ} \times 1000 \text{ J/kJ} = -8.314 \text{ JK}^{-1}\text{mol}^{-1} \times 298.15 \text{ K} \times \ln K_p$$

$$\ln K_p = -52.61$$

$$K_p = 1.4 \times 10^{-23}$$

c. Vapor pressure of carbon dioxide above calcium carbonate at 298.15°C:

$$K_p = P_{CO2} = 1.4 \times 10^{-23} \text{ atm.}$$

Recall that components in phases other than the gas phase are omitted from the equilibrium expression.

8. When benzene is burned:

$$C_6H_6(l) + 15/2 \, O_2(g) \rightarrow 6CO_2(g) + 3H_2O(g)$$

if $\Delta G_f°(C_6H_6(l)) = +124.5 \text{ kJmol}^{-1}$
        $\Delta G_f°(H_2O(g)) = -228.6 \text{ kJmol}^{-1}$
        $\Delta G_f°(CO_2(g)) = -394.4 \text{ kJmol}^{-1}$

and $\Delta H_f°(C_6H_6(l)) = +49.0 \text{ kJmol}^{-1}$
        $\Delta H_f°(H_2O(g)) = -241.8 \text{ kJmol}^{-1}$
        $\Delta H_f°(CO_2(g)) = -393.5 \text{ kJmol}^{-1}$ at 298.15K (25°C)

$$\Delta H°_{reaction} = \Sigma(\Delta H_f°(\text{products}) - \Delta H_f°(\text{reactants})) = -3135.4 \text{ kJ}$$

$$\Delta G°_{reaction} = \Sigma(\Delta G_f°(\text{products}) - \Delta G_f°(\text{reactants})) = -3176.7 \text{ kJ}$$

a. Heat is released in this reaction at 25°C = 3135.4 kJ/mole C_6H_6

b. Standard entropy change for the reaction at 298.15K (25°C):

$$\Delta G° = \Delta H° - T\Delta S° \quad \text{so } T\Delta S° = +41.3 \text{ kJ} = 41,300 \text{ J}$$

$$\Delta S° = +138.5 \text{ JK}^{-1}$$

$S°(H_2O) = 188.8 \text{ JK}^{-1}\text{mol}^{-1}$
$S°(CO_2) = 213.7 \text{ JK}^{-1}\text{mol}^{-1}$
$S°(C_6H_6) = 172.8 \text{ JK}^{-1}\text{mol}^{-1}$
$S°(O_2) = 205.1 \text{ JK}^{-1}\text{mol}^{-1}$

$\Delta S°_{system} = \Sigma(S°_{(prod)} - S°_{(react)})$

$\Delta S°_{system} = 137.6 \text{ JK}^{-1}$

An increase in disorder.

9.   ΔH   ΔS

a. $C_8H_{18}(l) + 25/2 \, O_2(g) \rightarrow 8CO_2(g) + 9H_2O(g)$ –   +

ΔG must be negative regardless of temperature

b. $N_2(g) + 2F_2(g) \rightarrow N_2F_4(g)$ –   –

ΔG depends on temperature, will be favored by low temperature

c. $Cl_2(g) \rightarrow 2Cl(g)$ +   +

ΔG depends on temperature, will be favored by high temperature

d. $2O_3(g) \rightarrow 3O_2(g)$ –   +

ΔG must be negative regardless of temperature

e. $2C(s) + 2H_2(g) \rightarrow C_2H_4(g)$ +   –

ΔG always positive regardless of temperature. Reaction never happens by itself.

Across:

3.   *and 54 across:* q and w—two ways to increase the internal energy of a system.
4.   A process in which heat is absorbed.
6.   Prefix for 1000 ×.
10. *see 59 down.*
14. The state of the system at the end.
15. At its freezing point, its solid state is less dense than its liquid state.
16. Entropy.
17. A constant volume calorimeter.
18. The smallest particle of an element.
20. SI unit for energy.
21. The unit for *48 down* was named after him.
23. The part of the *53 down* under examination.
25. An older unit of energy.
26. The study of heat and work.
28. A system to (or from) which neither energy nor matter can be added (or removed).
30. Newtons per square meter.
32. In its standard state has an enthalpy of formation equal to zero.
33. A statement of regularity in natural occurrence or phenomena.
35. The state of the system at the beginning.
38. Sometimes described as the "heat within."
39. Energy due to position.
40. Where the needle points.
41. A process in which heat is released.
42. The enthalpy change is independent of the route.
45. To decrease the entropy.
47. A common laboratory constant pressure calorimeter.
50. A perfect one has zero entropy at zero K.
52. A property of a system that is independent of its previous history.
54. *see 3 across.*
55. S = k logW is engraved on his tombstone in Vienna.
56. A bond resulting from the transfer of electron(s).
58. Changing state from liquid to solid.
60. The state of lowest entropy and...
61. The first law.
63. ...the state of highest entropy.
64. Δ.
65. *see 12 down.*
66. A system to (or from) which both energy and matter can be added (or removed).
67. The sign of ΔH for an endothermic process.
68. The energy required to break up a *50 across.*

Down:

1.   Regarded by many as the greatest physicist between Newton and Einstein.
2.   Heat travels from this...
5.   ...to this.
6.   The rate and mechanism of reactions.
7.   *see 29 down.*
8.   The measure of how hot an object is.
9.   Conversion from liquid to gas.
11. The ability to absorb heat per degree rise in temperature.
12. *and 43 down, 65 across:* the three laws of *26 across.*
13. Middle name of *20 across.*
19. SI unit for an amount of a substance.
22. Energy in the form of heat.
24. A process for which the entropy of the universe increases.
27. In magnitude a degree the same as a K.
29. *and 7 down:* His is *46 down.*
31. A device for determining heat change by measuring temperature change.
32. S.
34. The heat required to raise the temperature of one gram by one degree K.
36. His were the laws of multiple proportions and partial pressures.
37. The change from vapor to liquid.
38. It's conserved so the first law says.
43. *see 12 down.*
44. Everything but the system.
46. Available for useful work.
48. Energy divided by time.
49. He derived an expression for the efficiency of heat engines.
51. Conversion from solid to vapor.
52. This state of a substance is its most stable state at 1 bar pressure and a specified temperature (usually 25°C).
53. *23 across plus 44 down.*
57. The sign of the free energy change for a spontaneous process.
58. Its heat melts the solid state.
59. *and 10 across:* The first to distinguish temperature and heat capacity.
62. Like *49 down's*, or Born-Haber, a sequence of processes that start and end at the same state.

Thermodynamics

—a crossword puzzle

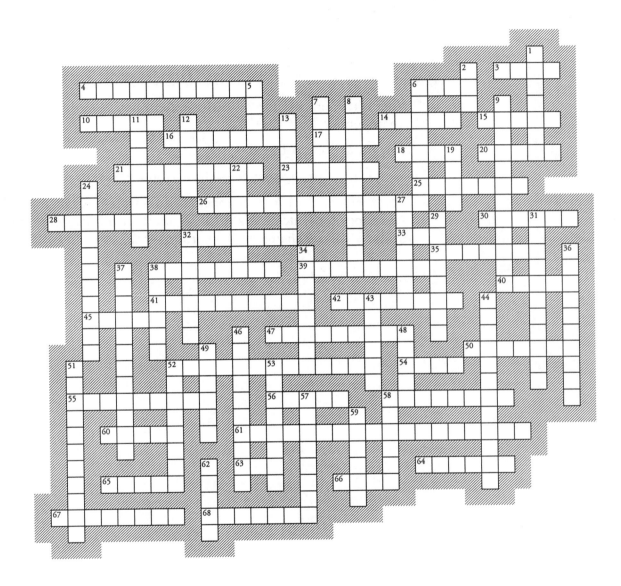

Challenge 11:

It is possible, with considerable care, to supercool liquid water to a temperature of −10°C and maintain it in this thermodynamically unstable state for some length of time. Suppose you did this, and had 100 grams of the supercooled water in a perfectly isolated container. Now suppose that at some later time it suddenly and spontaneously reverts to its thermodynamically stable state and changes to a mixture of ice and water. Assume the pressure (1 atm) and the volume do not change.

a.   What is the temperature of the ice-water mixture at the end of the experiment?

b.   How much ice is formed?

c.   What is the entropy change in the surroundings?

d.   How then can the process happen? Liquid water is changing to ice —a change for which ΔS is *negative*. Spontaneous processes *always* involve a *positive* change in the entropy of the universe.

ΔH_{fusion} (ice) = 333 Jg^{-1}
Heat capacity of water = 4.184 JK^{-1}g^{-1}

CHAPTER 21

Introduction

One of the ways in which energy can be supplied or is liberated in a chemical reaction is in the form of an electrical current. This chapter examines the ways in which electrical energy can be extracted from chemical reactions, and the ways in which the application of an electrical potential can cause reactions to happen. You first encountered redox reactions in Chapter 5; here we will look at them in more detail.

Contents

21.1   Oxidation–Reduction Reactions

Oxidation and reduction occurs because of the transfer of one or more electrons from the substance being oxidized to the substance being reduced. Oxidation-reduction (**redox**) reactions constitute a major class of reactions.

A **battery** is a collection of **electrochemical cells** (**voltaic cells**) in which a chemical (electron–transfer) reaction produces an electrical current at a particular voltage. Conversely, the application of electricity is used in a process called **electrolysis**, where electrical energy causes a chemical reaction to take place—as in the commercial production or purification of metals.

Balancing oxidation–reduction equations is necessary if quantitative calculations are to be done regarding the amount of energy released or absorbed in a chemical reaction. Often a redox reaction is divided into two halves; one half representing the oxidation process and the second half representing the reduction process. Balancing the reaction involves balancing these two halves. For example consider the reaction:

$$Cu^{2+}(aq) \ + \ Zn(s) \ \rightarrow \ Cu(s) \ + \ Zn^{2+}(aq)$$

The copper ions are reduced and the zinc metal is oxidized. So the reaction can be divided into the two half reactions:

The other major class is acid-base reactions.

Voltaic cell:
Chemical → Electrical

Electrolysis cell:
Electrical → Chemical

Voltaic cells are sometimes called galvanic cells.

There are a variety of systematic ways to balance a redox equation. Adopt a method that you understand and use it. Remember that redox reactions are balanced both in terms of mass and in terms of charge.

Reduction: $Cu^{2+}(aq) + 2e^- \rightarrow Cu(s)$ requires two electrons

Oxidation: $Zn(s) \rightarrow Zn^{2+}(aq) + 2e^-$ provides two electrons

In this case the equation is already balanced, both by mass and by charge. The oxidation process supplies 2 electrons and the reduction process requires 2 electrons.

21.2   Chemical Change Leading to Electric Current

When zinc metal reacts with aqueous copper(II) solution, the reaction is spontaneous; electrons are transferred from the zinc to the copper(II) ion. If these two reactants (the zinc metal and the copper(II) solution) are separated, then the electrons can be made to travel through an external wire as they go from the zinc to the copper(II). On their way, the electrons can do work (drive a motor, light a lamp, etc.).

Two half-cells, one in which the zinc is oxidized, and another in which the copper(II) is reduced, are connected by an external wire through which the electrons move, and by a **salt-bridge** through which ions move to complete the circuit. The salt bridge contains a solution of a salt such as sodium sulfate. As copper(II) ions are removed by reduction at the **cathode**, they are balanced by the addition of sodium ions or the removal of sulfate ions. Similarly, the addition of zinc ions by oxidation at the **anode** is balanced by the removal of sodium ions or the addition of sulfate ions. So cations move one way through the bridge (anode to cathode) and anions move the other way (cathode to anode).

> **Oxidation** always occurs at the **anode**.
> **Reduction** always occurs at the **cathode**.

21.3   Electrochemical Cells and Potentials

Electromotive force (emf) is the force that makes electrons move. The force is due to the difference in the electric potential of an electron at the anode compared to the cathode. Just as water runs downhill to a lower potential, so electrons run downhill from the anode (higher potential) to the cathode (lower potential). The amount of work that can be done by the electrons as they move from anode to cathode depends upon two things: the potential difference (V) and the number of electrons—the total charge (C).

> Energy (or work) J = Potential difference V × Number of electrons (charge) C

One coulomb is the quantity of electrical charge that passes a point in an electrical circuit when a current of one ampere flows for one second. Because the charge on an electron is very small, 6.242×10^{18} electrons are required to carry one coulomb of charge. One mole of electrons carries a charge equal to one Faraday where one Faraday = 96485 coulombs.

The cell potential depends upon the identity of the substances used and the concentrations of the solutes in solution or the pressures of the gases at the

---

Sidebar (left column):

General procedure for balancing redox reactions:

1. Recognize the reaction as a redox reaction.

2. Separate into half-reactions.

3. Balance each half-reaction for mass. Balance O by adding H_2O. Balance H by adding H^+.

4. Balance each half-reaction for charge by adding electrons.

5. Multiply each half reaction by a factor to ensure that the electrons gained equal the electrons lost.

6. Add, and cancel common reactants and products.

Try some of the examples in the Study Questions and Problems.

Read the Problem Solving Tips & Ideas 21.1 on page 957, particularly for what to do for redox reactions in basic solution.

On a flashlight cell, the anode is marked − and the cathode +. Electrons travel from the − to the + (anode to cathode).

Flashlight cells are commonly called batteries although technically a battery is a collection of cells in series (as in a car battery).

Joule			
Volt	×	Amp	× Sec
Volt	×	Coulomb	
Watt			× Sec

$6.242 \times 10^{18} \times 96485.3$
$= 6.022 \times 10^{23}$

electrodes. Because the potential varies if the conditions are changed, standard conditions have been defined: Standard solutions are 1.0 M and standard gas pressures are 1 bar. A standard cell potential has the symbol E°.

By definition, cell potentials for product-favored (spontaneous) reactions are positive. The relation between the cell potential E° and the value of the standard free energy change ΔG°_{rxn} for the reaction is:

$$\Delta G^{\circ}_{rxn} = -nFE^{\circ}$$

The reaction mentioned earlier has a standard cell potential E° of +1.10V:

$$Cu^{2+}(aq) \; + \; Zn(s) \; \rightarrow \; Cu(s) \; + \; Zn^{2+}(aq) \qquad E^{\circ} = +1.10V$$

$$\Delta G^{\circ}_{rxn} \; = \; -nFE^{\circ} \; = \; -2.00 \text{ mol} \times 96485 \text{ C} \times 1.10 \text{ V} \times 1 \text{ kJ}/1000\text{J}$$

$$= \; -212 \text{ kJ (a product-favored or spontaneous reaction)}$$

There are two contributing components to a cell potential, the reaction at the anode (oxidation) and the reaction at the cathode (reduction). One cannot occur without the other and it is impossible therefore to measure just a half-cell potential. However, such half-cell potentials would be very useful because then you could calculate the potential of any cell simply by adding two together. So an arbitrary standard has been chosen to which all others are compared—the standard hydrogen electrode (SHE) given a half-cell potential = zero.

Half-cell potentials are, by convention, always listed as standard reduction potentials. This is the potential for the half-cell as a reduction half-cell. If placed in a cell as the oxidation half-cell (anode), then the potential is reversed in sign.

> oxidized form $\; + \; ne^{-} \; \xrightarrow{\text{reduction}} \;$ reduced form
>
> The more positive E°_{red}, the better the oxidizing power and the easier the reduction. The more negative E°_{red}, the better the reducing power and the easier the oxidation.

21.4 Using Standard Potentials

The standard reduction potential for $Ni^{2+}|Ni$ is listed as –0.25V. What does this mean? First, the fact that it is negative means that nickel is oxidized more easily than hydrogen. That means that it is a better reducing agent than hydrogen. Second, the value –0.25V is the contribution a nickel electrode would make to the cell potential in which it is the cathode.

The standard electrode reduction potential for $Zn^{2+}|Zn$ is listed as –0.76V. This means that zinc is even better than nickel as a reducing agent. It is even more easily oxidized. So what happens when a cell is constructed from a $Ni^{2+}|Ni$ electrode and a $Zn^{2+}|Zn$ electrode? The more easily oxidized element will indeed be oxidized at the anode:

$$Zn(s) \, | \, Zn^{2+}(aq) \, \| \, Ni^{2+}(aq) \, | \, Ni(s)$$
+0.76V   –0.25V   E° = +0.51V (E° is positive—
oxidation   reduction   therefore spontaneous)

Suppose that the nickel electrode is used with a copper electrode in a cell. The standard reduction potentials are $Ni^{2+}|Ni$ –0.25V and $Cu^{2+}|Cu$ +0.34V.

The superscript ° denotes standard conditions (1 bar pressure and 1M solution).

Recall J = V × C

Standard reduction potentials are listed in Table 21.1 on page 970 and in Appendix M. For example, the reduction:

$Cu^{2+}(aq) + 2e^{-} \rightarrow Cu(s)$ E° = 0.34V

In the table of standard reduction potentials, each element is shown with the oxidized form on the left and the reduced form on the right.

The more negative the reduction potential (the lower it is in the table), the better the reducing power of the reduced form of the element, the more easily the reduced form is oxidized.

$H_2|H^+\|Ni^{2+}|Ni$ E° = –0.25V
$Ni|Ni^{2+}\|H^+|H_2$ E° = +0.25V
Therefore nickel is oxidized and H^+ is reduced. Ni is a better reducing agent than H_2.

In general get into the habit of writing the anode on the left and the cathode on the right. Oxidation then occurs on the left and reduction on the right; and the reaction proceeds from left to right.

If the cell potential turns out to be negative, then the cell is written the wrong way around.

Note that the sign of the potential for the $Zn^{2+}|Zn$ electrode is changed for oxidation.

Nickel is now the more easily oxidized of the two:

Ni*(s)* | Ni²⁺*(aq)* ‖ Cu²⁺*(aq)* | Cu*(s)*
+0.25V   +0.34V   E° = +0.59V (E° is positive—
oxidation   reduction   therefore spontaneous)

In the table of standard electrode reduction potentials, with the most negative at the bottom of the table, the element lower in the table will reduce solutions of ions of elements higher in the table . For example, as in the previous examples, zinc metal will reduce Ni²⁺, and nickel will reduce Cu²⁺. Zinc, therefore, will reduce Cu²⁺:

Zn*(s)* | Zn²⁺*(aq)* ‖ Cu²⁺*(aq)* | Cu*(s)*
+0.76V   +0.34V   E° = +1.10V

the reaction proceeds from left to right i.e. Zn*(s)* → Zn²⁺*(aq)* and Cu²⁺*(aq)* → Cu*(s)*

21.5   Electrochemical Cells at Nonstandard Conditions

In reality, oxidation-reduction reactions rarely occur under standard conditions. Concentrations, even if 1M initially, do not remain 1M for long. The **Nernst equation** relates a nonstandard cell potential with the standard cell potential:

$$E = E° - (RT/nF)\ln Q \quad \text{where Q is the reaction quotient}$$

At 298K (25°C): $E = E° - (0.0257/n)\ln Q$

If the concentration of product is low, then Q is less than 1, and $\ln Q$ is therefore negative. This results in a cell potential greater (more positive) than the standard potential. As product is formed, Q becomes larger and the cell potential decreases. Eventually, the system reaches equilibrium, when Q = K, and the cell potential becomes zero:

$$\text{Zero} = E° - (RT/nF)\ln K \quad \text{or} \quad \ln K = nE°/0.0257V \text{ at } 25°C$$

Measuring the standard cell potential E° is one very useful way to determine the value of an equilibrium constant K.

21.6   Batteries and Fuel Cells

A **primary battery** (cell) uses an oxidation-reduction reaction that is irreversible. Once the battery has reached equilibrium it is of no further use. A **secondary battery** (cell) is rechargeable. The chemical reactions in secondary batteries are reversible.

The common **dry cell** was invented by Leclanché in 1866 and involves the oxidation of zinc and the reduction of ammonium ion. The **alkaline** battery again uses the oxidation of zinc, but in alkaline conditions. The cell involves the reduction of manganese dioxide MnO_2. The alkaline cell does not produce gases at the cathode like the Leclanché cell. The **mercury cell** again involves the oxidation of zinc under alkaline conditions at the anode, but the cathode involves the reduction of mercury(II) oxide to mercury. **Lithium cells** are used

---

Sidebar notes (left column):

In this cell notation, a single vertical bar represents a phase change and a double vertical bar represents a salt-bridge. For example, Ni*(s)*|Ni²⁺*(aq)* represents a half-cell in which a nickel electrode is immersed in a solution of Ni²⁺.

Note that the sign of the potential for the Ni²⁺|Ni electrode is changed for oxidation.

Sometimes a table of standard reduction potentials will have the most negative potential at the top.

R is the gas constant in SI units = 8.314510 JK⁻¹mol⁻¹.

F = 9.6485309 × 10⁴ JV⁻¹mol⁻¹

n = moles of electrons transferred in the equation

All systems eventually reach equilibrium.

The disposal of mercury cells and nickel-cadmium cells is of considerable environmental concern.

when weight is of some importance. Lithium is a strong reducing agent at the bottom of the table, and lithium cells have high cell potentials (3V).

An automobile lead storage battery is one of the most prevalent secondary batteries. One electrode is lead and the other is lead(IV) oxide. The battery can supply an electrical current when the lead is oxidized to lead sulfate and the lead(IV) oxide is reduced to lead sulfate. The process is reversed when the car's alternator reverses the potential.

In a **fuel cell**, the reactants for the electrochemical process are supplied on a continuous basis, as needed, and the system never reaches equilibrium. Very often the fuel used is hydrogen (it is oxidized at the anode) and the oxidizing agent is oxygen (it is reduced at the cathode).

21.7   Corrosion: Redox Reactions in the Environment

Corrosion is the deterioration of metals in an oxidation-reduction reaction. The common example is the rusting of iron. For corrosion to occur there must be anodic areas where the metal is oxidized, cathodic areas where some substance is reduced (often water), an electrical connection between the two areas, and an electrolytic solution also connecting the two areas. These requirements are easily fulfilled.

Usually the cathodic reduction is the rate determining step, slower than the oxidation of the metal at the anode. Reduction of water is a slow process, slower than the reduction of oxygen, so in the absence of oxygen, corrosion is slow. Corrosion may also be inhibited by the formation of insoluble oxides or hydroxides on the metal.

Protection against corrosion involves inhibiting the anodic oxidation, inhibiting the cathodic reduction, or doing both. Anodic protection can be achieved by painting the metal. Protection can also be achieved by forcing the metal to become the cathode by coating it with a more readily oxidized metal as in the galvanizing of iron. In this case the zinc forms a **sacrificial anode**.

21.8   Electrolysis: Chemical Change from Electric Energy

Chemical change can produce electrical energy in a voltaic cell. Equally important, an electrical current at a sufficient potential can cause a chemical change to occur. This process is called **electrolysis**.

In an electrolytic cell, electrons are forced onto the cathode by an external potential. Reduction occurs at the cathode (always). The cathode is negative in an electrolytic cell because that electrode is the (external) source of electrons for the reduction process.

In general, if an external potential is applied to an electrolytic cell, the reaction that occurs is the one with the lowest cell potential. However, the conditions are often nonstandard (concentrations are not 1M), and kinetic surface effects at the electrodes (overpotentials) often influence the products formed.

In the electrolysis of aqueous sodium iodide for example, two oxidation process could happen at the anode:

It is the anode that is negative in a voltaic cell because the anode is the source of electrons from the oxidation process.

See Problem Solving Tips and Ideas 21.2 on page 986 of the text.

Change the sign of the reduction potential for this oxidation.

| iodide: | $2I^-(aq) \rightarrow I_2(aq) + 2e^-$ | $E° = -0.535V$ |
| water: | $6H_2O(l) \rightarrow O_2(g) + 4H_3O^+(aq) + 4e^-$ | $E° = -1.23V$ |

And three reduction processes could occur at the cathode:

The concentration of hydronium ion is very small, and the reduction of water is more likely to occur.

sodium:	$2Na^+(aq) + e^- \rightarrow Na(s)$	$E° = -2.71V$
water:	$2H_2O(l) + 2e^- \rightarrow H_2(aq) + 2OH^-(aq)$	$E° = -0.83V$
hydronium:	$2H_3O^+(aq) + 2e^- \rightarrow 2H_2O(l) + 2H_2(g)$	$E° = 0.00V$

Taking into account the concentrations of the species in solution, the overall reaction will be the one requiring the lowest applied potential:

Note that the potential is negative–this is not a spontaneous reaction! This potential must be applied to make the reaction happen.

iodide:	$2I^-(aq) \rightarrow I_2(aq) + 2e^-$	$E° = -0.535V$
water:	$2H_2O(l) + 2e^- \rightarrow H_2(aq) + 2OH^-(aq)$	$E° = -0.83V$
	Applied potential required:	$E° = -1.37V$

In order for a species to be reduced in preference to water, it must have a reduction potential less negative than –0.83V. For example, tin, copper, silver, and gold can all be plated out from aqueous solution.

21.9   Counting Electrons

In order to reduce one silver ion Ag^+, one electron is required. To reduce one mole of silver ions, one mole of electrons is required. One mole of electrons is one Faraday, equal to 96485 coulombs. A coulomb is an amount of charge; in fact it is the charge passed when one ampere of current flows for one second.

The quantity of mass of metal deposited at a cathode, or removed from an anode, can be determined easily by mass difference. The current and time passing through an electrolytic cell are also easily measured. The relationship between these quantities is the molar mass of the metal and the stoichiometry of the oxidation or reduction (how many electrons are gained or lost; i.e. the charge on the ions).

Time (sec) × current (amps) → charge (coulombs) → moles electrons
Moles electrons → moles of metal → mass of metal

21.10   The Commercial Production of Chemicals by Electrochemical Methods

Charles Hall in the US, and Paul Héroult in France, developed the same method for producing aluminum simultaneously in 1886. Both were born in 1863 and both died in 1914.

The production of aluminum by the Hall–Heroult method, chlorine and sodium hydroxide by the electrolysis of brine, sodium metal from fused sodium chloride in a Downs cell, are all examples of the commercial production of vast quantities of materials by electrolysis. Many metals are refined by electrochemical methods.

The Washington Monument, built in 1884, was capped with aluminum, then a very precious metal.

Review Questions

1.   Define the words cell and battery.

2.   Which electrode is the anode and which electrode is the cathode in an electrochemical cell?

3.　What is a half-reaction? What is half-cell?

4.　What is the standard electrode? What is a standard electrode reduction potential? What is its symbol?

5.　How are the standard cell potential and the standard free energy change of a reaction related?

6.　Describe the general procedure for balancing the equation for an oxidation-reduction reaction. What do you do if the reaction takes place in basic solution?

7.　What is a salt-bridge? Why is such a device necessary?

8.　What is the relationship between charge, potential, and energy?

9.　What is the Faraday constant? What is its significance?

10.　How is the value for the standard electrode reduction potential related to the ease with which the element is oxidized or reduced, or the oxidizing or reducing ability of the element?

11.　Describe the Nernst equation, what is its use?

12.　How is the standard cell potential related to the equilibrium constant?

13.　What is a fuel cell; how does it differ from an ordinary alkaline cell?

14.　What is the difference between a primary cell and a secondary cell?

15.　What is corrosion? What are the requirements for corrosion to occur?

16.　What is the difference between an electrolysis cell and a voltaic cell? Are the anodes, and the cathodes, the same in both? Do the terminals have the same sign in both?

17.　How can you determine which reactions will occur at the electrodes in an electrolytic cell?

18.　Describe how an electrolysis experiment could be used to determine the molar mass of a metal.

19.　Sometimes you will see the Nernst equation incorporating the number 0.0257, sometimes 2.303, sometimes 0.0592. From where do these numbers come and what is the relationship between them?

Answers to Review Questions

1. An electrochemical cell can be a voltaic cell that produces electricity as a result of an electron transfer reaction or it can be an electrolytic cell in which electrical energy is applied to cause a chemical reaction to occur. A battery is a collection of cells in series.

2. The anode is the electrode at which oxidation occurs. The cathode is the electrode at which reduction occurs. This is true of both voltaic and electrolytic cells.

3. A redox reaction is conveniently divided into two halves; one half representing the oxidation process and the second half representing the reduction process. The half-cells are the two halves of an electrochemical cell in which the two half-reactions take place.

4. The standard half-reaction is the reaction at the standard hydrogen electrode (1M acid H^+ and 1 bar pressure H_2). The SHE is assigned a half-cell potential = zero.

$$2H_3O^+ (aq) + 2e^- \rightarrow 2H_2O(l) + 2H_2(g) \quad E° = 0.00V$$

By convention, half-cell potentials are always listed as standard reduction potentials. The symbol is $E°$. This is the potential for the half-cell as a reduction half-cell (at the cathode). The potential is reversed in sign if the half-cell placed in a cell as the oxidation half-cell (at the anode). Half-cell potentials are useful because you can calculate the potential of any cell simply by adding two half-cell potentials together.

$$E°_{cell} = E°_{oxid} + E°_{red}$$

5. $\Delta G°_{rxn} = -nFE°$

6. General procedure for balancing the equation for an oxidation-reduction reaction:

> 1. Recognize the reaction as a redox reaction.
> 2. Separate into half-reactions.
> 3. Balance each half-reaction for mass. Balance elements other than H and O, then balance O by adding H_2O and balance H by adding H^+.
> 4. Balance each half-reaction for charge by adding electrons.
> 5. Multiply each half reaction by a factor to ensure that the electrons gained equal the electrons lost.
> 6. Add, and cancel common reactants and products.

For reactions in basic solution, balance the equation for acidic solution first and then add sufficient hydroxide ion OH^- to cancel all the hydronium ions H_3O^-, and then cancel any water molecules present on both sides.

7. A salt-bridge provides a means by which the electrical circuit through an electrochemical cell can be completed without allowing the solutions in the two half-cells to mix. The salt bridge contains a solution of an electrolyte—a salt. As metal ions are removed from the reduction half-cell, they are balanced by the addition of cations or the removal of anions via the salt-bridge. Cations move through the bridge from anode to cathode and anions move through the bridge from cathode to anode.

8. Energy (J) = potential (V) × charge (C).

9. The Faraday constant = 96485 coulombs. It is the charge carried by one mole (Avogadro's number) of electrons.

10. In the table of standard reduction potentials, each element is shown with the oxidized form on the left and the reduced form on the right.
    The more positive the reduction potential (the higher it is in the table), the better the oxidizing power of the oxidized form of the element, the more easily the oxidized form is reduced.
    The more negative the reduction potential (the lower it is in the table), the better the reducing power of the reduced form of the element, the more easily the reduced form is oxidized.

Oxidized form	Reduced form	$E°$
$F_2(g) + 2e^- \rightarrow 2F^-(aq)$		+2.87V
F_2 is a good oxidizing agent		
F_2 is easily reduced		
$Li^+(aq) + e^- \rightarrow Li(s)$		−3.045V
Li is a good reducing agent		
Li is easily oxidized		

11. The Nernst equation relates a nonstandard cell potential with the standard cell potential:

    $$E = E° - (RT/nF)\ln Q \quad \text{where Q is the reaction quotient}$$

    At 298K (25°C): $E = E° - (0.0257/n)\ln Q$

    The Nernst equation is used when the electrochemical cell is run under nonstandard conditions. The concentrations may not be 1M, or the gas pressures may not be 1 bar.

12. At equilibrium, when Q = K, the cell potential becomes zero, so:

    $$\text{Zero} = E° - (RT/nF)\ln K \quad \text{or} \quad \ln K = nE°/0.0257V \text{ at } 25°C$$

    Measuring the standard cell potential $E°$ is a useful way to determine the value of an equilibrium constant K.

13. A fuel cell is an electrochemical cell in which the reactants for the reaction taking place are continually replenished and the products of the reaction are continually removed. Unlike an alkaline cell, the fuel cell never reaches equilibrium (it never dies).

14. A primary battery uses an oxidation-reduction that is irreversible. Once the battery has reached equilibrium it is of no further use. A secondary battery (cell) is rechargeable. The chemical reactions in secondary batteries are reversible. Automobile lead storage batteries and nickel-cadmium rechargeable batteries are the most common secondary batteries.

15. Corrosion is the deterioration of a metal in an oxidation-reduction reaction. For corrosion to occur there must be anodic areas where the metal is oxidized, cathodic areas where some substance is reduced (often water), an electrical connection between the two areas, and an electrolytic solution also connecting the two areas.

16. In an electrolysis cell the application of electricity is used to produce a chemical reaction. In a voltaic cell a chemical electron transfer reaction produces an electrical current.

    In both cells, oxidation always occurs at the anode and reduction always occurs at the cathode. However, in a voltaic cell, the anode is labelled negative because the anode is the source of electrons. In an electrolysis cell, the cathode is labelled negative because that is the source of electrons (where the electrons are pumped into the cell).

17. In general, if an external potential is applied to an electrolytic cell, the reaction that occurs is the one with the lowest cell potential (the smallest sum of half-cell potentials).

    However, the conditions are usually nonstandard (the concentrations are not 1M), and kinetic surface effects at the electrodes (overpotentials) often influence the products actually formed. If the concentration of one species is very much smaller than another, then even though it may require a lower potential under standard conditions, it may not be the species reduced or oxidized.

time (s) × current (A)
↓
charge (coulombs)
↓
moles electrons
↓
moles of metal
↓
mass of metal

18. Faraday's laws of electrolysis relate the charge passed through the cell to the mass lost at the anode or mass gained at the cathode. The charge can be calculated from the current and time. The charge is related to the moles of electrons, which in turn is related to the moles of the element deposited or removed. The relation between mass and moles is molar mass.

19. The Nernst equation: $E = E° - (RT/nF)\ln Q$ where Q = reaction quotient

    At 298K (25°C): $E = E° - (0.0257/n)\ln Q$

    or using $\log_{10}$, $E = E° - (0.0592/n)\log Q$

    The ratio between 0.0257 and 0.0592 is 2.303, i.e. $\ln 10 = 2.303$

Imagination is more important than knowledge.

*Albert Einstein
(1879-1959)*

Study Questions and Problems

1. Write equations for the reaction between iron and a solution of silver nitrate to produce Fe(II) ions and silver metal:

    a. write the balanced half-cell reactions
    b. write the overall balanced equation for the reaction.
    c. draw a diagram of the cell and calculate the standard cell potential

2. Write equations for the reaction between manganese(IV) dioxide and the hypochlorite ion to produce the permanganate ion and the chloride ion in basic solution:

   a. write the balanced half-cell reactions
   b. write the overall balanced equation for the reaction.
   c. draw a diagram of the cell and calculate the standard cell potential

3. The following reaction occurs in acidic solution. Balance the equation according to the general procedure outlined in the margin, describing each step of the process in turn:

   $$Fe^{3+}(aq) + NH_3OH^+(aq) \rightarrow N_2O(g) + Fe^{2+}(aq)$$

   General procedure for balancing redox reactions:

   1. Recognize the reaction as a redox reaction.

   2. Separate into half-reactions.

   3. Balance each half-reaction for mass. Balance O by adding H_2O. Balance H by adding H^+.

   4. Balance each half-reaction for charge by adding electrons.

   5. Multiply each half reaction by a factor to ensure that the electrons gained equal the electrons lost.

   6. Add, and cancel common reactants and products.

4. Balance the following reactions in acidic solution:

   a. $Al(s) + Ag^+(aq) \rightarrow Al^{3+}(aq) + Ag(s)$

   b. $Fe^{2+}(aq) + Cr_2O_7^{2-}(aq) \rightarrow Cr^{3+}(aq) + Fe^{3+}(aq)$

   c. $MnO_4^-(aq) + H_2SO_3(aq) \rightarrow Mn^{2+}(aq) + SO_4^{2-}(aq)$

5. Balance the following reactions in basic solution:

   a. $AsO_2^-(aq) + ClO^-(aq) \rightarrow AsO_3^-(aq) + Cl^-(aq)$

   b. $MnO_4^-(aq) + C_2O_4^{2-}(aq) \rightarrow MnO_2(s) + CO_3^{2-}(aq)$

   c. $N_2H_4(aq) + O_2(g) \rightarrow N_2(g) + H_2O_2(aq)$

6. Consider the following pairs of half-reactions, decide which of the two half-reactions will occur at the anode and which will occur at the cathode, draw diagrams for the cells, and calculate the standard cell potentials:

   a. $Co^{2+}(aq) + 2e^- \rightarrow Co(s)$
      $Ag^+(aq) + e^- \rightarrow Ag(s)$

   b. $Ni^{2+}(aq) + 2e^- \rightarrow Ni(s)$
      $Cu^{2+}(aq) + 2e^- \rightarrow Cu(s)$

   c. $Sn^{2+}(aq) + 2e^- \rightarrow Sn(s)$
      $Mg^{2+}(aq) + 2e^- \rightarrow Mg(s)$

7. Examine a table of standard electrode reduction potentials and determine whether the following statements are true or false.

   a. Magnesium will react with water to produce hydrogen gas.
   b. A piece of nickel immersed in a solution of silver nitrate will become coated with silver.
   c. In basic solution, manganese dioxide will oxidize mercury to mercury(II) oxide.
   d. The iodate ion in acidic solution will oxidize copper metal.
   e. Nitric acid will oxidize tin to Sn^{2+}, producing nitric oxide NO.

8. The reaction of copper metal with silver ions in a solution of silver nitrate is spontaneous. Calculate the standard cell potential to show that this is so. From the cell potential calculate the value of the equilibrium constant for the reaction at 25°C. From the equilibrium constant, or from the cell constant, calculate the standard free energy change for the reaction. Indicate clearly how these three quantities are related.

9. A copper-zinc voltaic cell is constructed using 100 mL solutions of 1M solutions of copper sulfate and zinc sulfate with a sodium sulfate salt bridge. After some time t had passed at 25°C, the concentration of the Zn^{2+} ions in the anode half cell had increased to 1.50 M and the concentration of the Cu^{2+} ions in the cathode half-cell had decreased to 0.50 M.

    a. Calculate the initial cell potential.
    b. Calculate the cell potential at time t.
    c. Calculate the total charge provided by the cell.
    d. Calculate (approximately) the energy provided by the cell.

10. Possible anode reactions in the electrolysis of sea water are:

    $$2Cl^-(aq) \rightarrow Cl_2(g) + 2e^- \qquad E° = -1.36V$$
    $$2H_2O \rightarrow 4H^+(aq) + O_2(g) + 4e^- \qquad E° = -1.229V$$

    Possible cathode reactions in the electrolysis of sea water are:

    $$Na^+(aq) + e^- \rightarrow Na(s) \qquad E° = -2.714V$$
    $$2H_2O + 2e^- \rightarrow 2OH^-(aq) + H_2(g) \qquad E° = -0.8277V$$

    Which two half-reactions will occur when sea water is electrolyzed?

11. If the solution of the salt in the cell described in question 10 is saturated, chlorine is produced at the anode instead of oxygen. Using an applied potential of 5.0V to overcome kinetic effects, and a current of a quarter-million amps,
    a. how long would it take to produce 1000 kg of chlorine gas?
    b. how much sodium hydroxide would also be produced?
    c. how much energy would be used in the process?

Answers to Study Questions and Problems

1. a. oxidation: $Fe(s) \rightarrow Fe^{2+}(aq) + 2e^-$
        reduction: $Ag^+(aq) + e^- \rightarrow Ag(s)$

    b. the half-cell reactions are already balanced; the reduction reaction must be multiplied by two so that the number of electrons lost equals the number of electrons gained:
        $Fe(s) + 2Ag^+(aq) \rightarrow 2Ag(s) + Fe^{2+}(aq)$

    c. $Fe(s)|Fe^{2+}(aq) \parallel Ag^+(aq)|Ag(s)$
        +0.44V   +0.80V $E° = +1.24V$ ($E°$ is positive)

2. a. reduction: ClO^- *(aq)* $\rightarrow$ Cl^- *(aq)*
         ClO^- *(aq)* + $2H^+$ *(aq)* + $2e^-$ $\rightarrow$ Cl^- *(aq)* + H_2O

    oxidation: MnO_2 *(s)* $\rightarrow$ MnO_4^- *(aq)*
         MnO_2 *(s)* + $2H_2O$ $\rightarrow$ MnO_4^- *(aq)* + $4H^+$ *(aq)* + $3e^-$

    b. the half-cell reactions are already balanced (for acidic solution); the reduction reaction must be multiplied by three and the oxidation reaction by two so that the number of electrons lost equals the number gained; then the acidic conditions must be changed to basic conditions by adding hydroxide ($2OH^-$) to both sides:

    $2MnO_2$ *(s)* + $3ClO^-$ + H_2O $\rightarrow$ $3Cl^-$ + $2MnO_4^-$ + $2H^+$

    $2MnO_2$ *(s)* + $3ClO^-$ + $2OH^-$ $\rightarrow$ $3Cl^-$ + $2MnO_4^-$ + H_2O (aq) notation omitted.

    c. MnO_2 *(s)*|MnO_4^- *(aq)* || ClO^-, Cl^- *(aq)*
        $-0.588V$ $+0.89V$ $E° = +0.30V$ ($E°$ is positive)

3. Fe^{3+} *(aq)* + NH_3OH^+ *(aq)* $\rightarrow$ N_2O *(g)* + Fe^{2+} *(aq)*

    Step 1: Recognize the reaction:

    There is an obvious change in the oxidation number of iron, from +3 to +2. Sometimes a change in oxidation number isn't quite so clear.

    Step 2: Separate into half-reactions:

    The Fe^{3+} *(aq)* is reduced to Fe^{2+} *(aq)*; this means that the other component in the reaction must be oxidized.

    Reduction: Fe^{3+} *(aq)* $\rightarrow$ Fe^{2+} *(aq)*

    Oxidation: NH_3OH^+ *(aq)* $\rightarrow$ N_2O *(g)*

    Step 3: Balance the half-reactions for mass:

    Reduction: Fe^{3+} *(aq)* $\rightarrow$ Fe^{2+} *(aq)*

    Oxidation: $2NH_3OH^+$ *(aq)* $\rightarrow$ N_2O *(g)* + H_2O + $6H^+$ *(aq)*

    Step 4: Balance the half-reactions for charge:

    Reduction: Fe^{3+} *(aq)* + e^- $\rightarrow$ Fe^{2+} *(aq)*

    Oxidation: $2NH_3OH^+$ *(aq)* $\rightarrow$ N_2O *(g)* + H_2O + $6H^+$ *(aq)* + $4e^-$

    Step 5: Balance the electrons lost and gained:

    Reduction: $4Fe^{3+}$ *(aq)* + $4e^-$ $\rightarrow$ $4Fe^{2+}$ *(aq)*

    Oxidation: $2NH_3OH^+$ *(aq)* $\rightarrow$ N_2O *(g)* + H_2O + $6H^+$ *(aq)* + $4e^-$

    Step 6: Combine the two half-reactions:

    $4Fe^{3+}$ *(aq)* + $2NH_3OH^+$ *(aq)* $\rightarrow$ $4Fe^{2+}$ *(aq)* + N_2O *(g)* + H_2O + $6H^+$ *(aq)*

4.   In acidic solution:

    a. $Al(s) + Ag^+(aq) \rightarrow Al^{3+}(aq) + Ag(s)$

        oxidation: $Al(s) \rightarrow Al^{3+}(aq) + 3e^-$

  ×3 reduction: $Ag^+(aq) + e^- \rightarrow Ag(s)$

        overall:

        $Al(s) + 3Ag^+(aq) \rightarrow Al^{3+}(aq) + 3Ag(s)$

    b. $Fe^{2+}(aq) + Cr_2O_7{}^{2-}(aq) \rightarrow Cr^{3+}(aq) + Fe^{3+}(aq)$

  ×6 oxidation: $Fe^{2+}(aq) \rightarrow Fe^{3+}(aq) + e^-$

        reduction: $Cr_2O_7{}^{2-}(aq) + 14H^+(aq) + 6e^- \rightarrow 2Cr^{3+}(aq) + 7H_2O$

        overall:

        $6Fe^{2+}(aq) + Cr_2O_7{}^{2-}(aq) + 14H^+(aq) \rightarrow 2Cr^{3+}(aq) + 6Fe^{3+}(aq) + 7H_2O$

    c. $MnO_4{}^-(aq) + H_2SO_3(aq) \rightarrow Mn^{2+}(aq) + SO_4{}^{2-}(aq)$

  ×5 oxidation: $H_2SO_3(aq) + H_2O \rightarrow SO_4{}^{2-}(aq) + 4H^+ + 2e^-$

  ×2 reduction: $MnO_4{}^-(aq) + 8H^+(aq) + 5e^- \rightarrow Mn^{2+}(aq) + 4H_2O$

        overall:

        $2MnO_4{}^-(aq) + 5H_2SO_3(aq) \rightarrow 2Mn^{2+}(aq) + 4H^+(aq) + 3H_2O + 5SO_4{}^{2-}(aq)$

5.   In basic solution:

    a. $AsO_2{}^-(aq) + ClO^-(aq) \rightarrow AsO_3{}^-(aq) + Cl^-(aq)$

        oxidation: $AsO_2{}^-(aq) + H_2O \rightarrow AsO_3{}^-(aq) + 2H^+ + 2e^-$

        reduction: $ClO^-(aq) + 2H^+ + 2e^- \rightarrow Cl^-(aq) + H_2O$

        overall:

        $AsO_2{}^-(aq) + ClO^-(aq) \rightarrow AsO_3{}^-(aq) + Cl^-(aq)$

> If you look at this equation carefully, you'll notice that it is already balanced.

    b. $MnO_4{}^-(aq) + C_2O_4{}^{2-}(aq) \rightarrow MnO_2(s) + CO_3{}^{2-}(aq)$

  ×3 oxidation: $C_2O_4{}^{2-}(aq) + 2H_2O \rightarrow 2CO_3{}^{2-}(aq) + 4H^+ + 2e^-$

  ×2 reduction: $MnO_4{}^-(aq) + 4H^+(aq) + 3e^- \rightarrow MnO_2(s) + 2H_2O$

        overall:

        $3C_2O_4{}^{2-}(aq) + 2MnO_4{}^-(aq) + 2H_2O \rightarrow 2MnO_2 + 6CO_3{}^{2-}(aq) + 4H^+(aq)$

        convert (add 4 OH⁻ to both sides and cancel H_2O):

        $3C_2O_4{}^{2-}(aq) + 2MnO_4{}^-(aq) + 4OH^-(aq) \rightarrow 2MnO_2 + 6CO_3{}^{2-}(aq) + 2H_2O$

> You will probably find it easier to balance the equations assuming acidic conditions, and then convert to basic conditions at the end—even though the half-reactions are unrealistic.

    c. $N_2H_4(aq) + O_2(g) \rightarrow N_2(g) + H_2O_2(aq)$

        oxidation: $N_2H_4(aq) \rightarrow N_2(g) + 4H^+ + 4e^-$

  ×2 reduction: $O_2(g) + 2H^+ + 2e^- \rightarrow H_2O_2(aq)$

        overall:

        $N_2H_4(aq) + 2O_2(g) \rightarrow N_2(g) + 2H_2O_2(aq)$

6. Look up the half-cell standard reduction potentials:

The half-cell with the more negative potential will be the anode.

Change the sign of the reduction potential for oxidation at the anode.

a. $Co^{2+}(aq) + 2e^- \rightarrow Co(s)$ $E° = -0.28V$ anode
$Ag^+(aq) + e^- \rightarrow Ag(s)$ $E° = +0.80V$
$Co(s)|Co^{2+}(aq) \parallel Ag^+(aq)|Ag(s)$ $E° = +0.28 + 0.80 = +1.08V$

b. $Ni^{2+}(aq) + 2e^- \rightarrow Ni(s)$ $E° = -0.25V$ anode
$Cu^{2+}(aq) + 2e^- \rightarrow Cu(s)$ $E° = +0.337V$
$Ni(s)|Ni^{2+}(aq) \parallel Cu^{2+}(aq)|Cu(s)$ $E° = +0.25 + 0.337 = +0.59V$

c. $Sn^{2+}(aq) + 2e^- \rightarrow Sn(s)$ $E° = -0.14V$
$Mg^{2+}(aq) + 2e^- \rightarrow Mg(s)$ $E° = -2.37V$ anode
$Mg(s)|Mg^{2+}(aq) \parallel Sn^{2+}(aq)|Sn(s)$ $E° = -0.14 + 2.37 = +2.23V$

7. a. Magnesium and water:
$Mg^{2+}(aq) + 2e^- \rightarrow Mg(s)$ $E° = -2.37V$
$2H_2O(l) + 2e^- \rightarrow H_2(aq) + 2OH^-(aq)$ $E° = -0.83V$

$Mg(s) + 2H_2O(l) \rightarrow Mg^{2+}(aq) + 2OH^-(aq)$ $E° = +1.54V$

The reaction is spontaneous.

b. Nickel and silver nitrate:
$Ni^{2+}(aq) + 2e^- \rightarrow Ni(s)$ $E° = -0.25V$
$Ag^+(aq) + e^- \rightarrow Ag(s)$ $E° = +0.80V$

$Ni(s) + 2Ag^+(aq) \rightarrow Ni^{2+}(aq) + 2Ag(s)$ $E° = +1.05V$

The reaction is spontaneous.

c. Manganese dioxide and mercury:
$MnO_2(s) + 2H_2O + 2e^- \rightarrow Mn(OH)_2(s) + 2OH^-$ $E° = -0.05V$
$HgO(s) + H_2O + 2e^- \rightarrow Hg(l) + 2OH^-$ $E° = +0.0984V$

$MnO_2(s) + Hg(l) + H_2O \rightarrow Mn(OH)_2(s) + HgO(s)$ $E° = -0.148V$

This reaction is *not* spontaneous; it will not happen.

d. Iodate ion and copper in acidic solution:
$Cu^{2+}(aq) + 2e^- \rightarrow Cu(s)$ $E° = +0.337V$
$IO_3^-(aq) + 6H^+ + 5e^- \rightarrow 1/2 I_2(aq) + 3H_2O$ $E° = +1.195V$

$2IO_3^-(aq) + 12H^+ + 5Cu(s) \rightarrow 5Cu^{2+}(aq) + I_2(aq) + 6H_2O$
                                       $E° = +0.852V$

The reaction is spontaneous.

e. Nitric acid will oxidize tin to Sn^{2+}, producing nitric oxide NO?
$Sn^{2+}(aq) + 2e^- \rightarrow Sn(s)$ $E° = +0.14V$
$NO_3^-(aq) + 4H^+(aq) + 3e^- \rightarrow NO(g) + 2H_2O$ $E° = +0.96V$

$2NO_3^-(aq) + 8H^+(aq) + 3Sn(s) \rightarrow 2NO(g) + 4H_2O + 3Sn^{2+}(aq)$
                                       $E° = +0.82V$

The reaction is spontaneous.

All three quantities $E°_{cell}$, $\Delta G°_{rxn}$, and K, are related:

$$\Delta G°_{rxn} = -nFE°_{cell} = -RTlnK$$

8. The standard cell potential is
$E°_{cell} = E°_{oxid} + E°_{red} = -0.337 + 0.7994 = +0.46V.$
The reaction is spontaneous.

At equilibrium at 25°C, $E° = (RT/nF)lnK$, and $lnK = nE°/0.0257$
$K = 3.52 \times 10^{15}$

$\Delta G°_{rxn} = -nFE° = -2 \times 96485 \times 0.46 = -88.8$ kJ
and $\Delta G°_{rxn} = -RTlnK = -8.314 \times 298.15K \times 35.8 = -88.8$ kJ

9. $Cu^{2+}(aq) + Zn(s) \rightarrow Cu(s) + Zn^{2+}(aq) \qquad E° = +1.100V$

   a. The initial potential $E°_{cell} = E°_{oxid} + E°_{red} = +0.763 + 0.337 = 1.100V$

   b. The cell potential at time t:
   Use the Nernst equation:
   $E = E° - (0.0257/n)lnQ = +1.100 - (0.0257/2)ln(1.50/0.50)$
   $= +1.100 - 0.014 = 1.086V$

Not much of a decrease in the cell potential.

   c. Quantity of copper deposited = 0.50 M × 0.100L = 0.05 mol Cu
   Quantity of electrons = 0.10 mol
   Charge = 0.10 faraday = 9648.5 C

   d. Approximate energy provided = average potential × charge
   = 1.093V × 9648.5C
   = 10.5 kJ

10. In general choose the lowest combination of half-cell potentials (the lowest sum) but be aware that the relative concentrations do matter.

    $2H_2O \rightarrow 4H^+(aq) + O_2(g) + 4e^- \qquad E° = -1.229V$

    $2H_2O + 2e^- \rightarrow 2OH^-(aq) + H_2(g) \qquad \underline{E° = -0.8277V}$

    $\qquad\qquad\qquad\qquad\qquad\qquad\qquad E° = -2.06V$

11. $2Cl^-(aq) \rightarrow Cl_2(g) + 2e^- \qquad E° = -1.36V$

    a. 1000 kg = 1,000,000 g = 1,000,000 / 70.905 = 14103 moles Cl_2

    number of electrons = 28,207 moles (2 electrons for each Cl_2)
    = 28,207 faradays
    = 28,207 × 96,485 C = 2.72×10^9 C

    2.72×10^9 C / 2.5×10^5A = 1.09×10^4 seconds
    = 3.02 hours

    $2H_2O + 2e^- \rightarrow 2OH^-(aq) + H_2(g) \quad E° = -0.8277V$

    b. number of OH^- = 28,207 moles (1 electron for each)
    mass of sodium hydroxide = 1.13 million grams

J = VC or J = VAs

    c. Energy = Potential × Charge = 5.0V × 2.72×10^9 C = 1.36×10^{10} J
    = 3780 kWh

EXAMINATION 4

Introduction

This examination tests your knowledge and understanding of the chemistry in Chapters 15 through 21 of Kotz & Treichel—kinetics, equilibria, acids and bases, precipitation reactions, the second law of thermodynamics, and redox reactions—a considerable amount of material. The first set of questions are true–false questions and the second set of questions are multiple choice questions. Try the questions before looking at the answers provided at the end of this study guide.

Many of the multiple choice questions involve calculations. However, before you start using your calculator, think about and formulate the problem. Decide first how you are going to solve the problem! Calculating the answer should then be routine.

True–false questions

1. Exothermic reactions occur faster than endothermic reactions.

2. The activation energy for the reverse reaction is always greater than the activation energy for the forward reaction.

3. The exponents in a rate equation for an elementary step equal the stoichiometric coefficients in the equation for that step.

4. The rate constant k is constant under all conditions.

5. A homogeneous catalyst is one that has a constant composition throughout, right down to the molecular level.

6. The half-life of a substance undergoing first-order decomposition is independent of the amount initially present.

7. The rate law for a chemical reaction can be derived from the stoichiometry of the equation representing the reaction.

8. If the temperature is increased, the value of the equilibrium constant increases.

9. At equilibrium the rate of the reaction in the forward direction equals the rate of the reaction in the reverse direction.

10. Liquids and solids are not included in an equilibrium constant expression.

11. When there is no difference in the number of moles of gases on the reactant side and the number on the product side of the equation representing the reaction, $K_p = K_c$.

12.   When the value of K is numerically large, then the products will predominate at equilibrium.

13.   The equilibrium state can be reached from the product side as well as from the reactant side.

14.   If $\Delta G°$ is positive, then the reaction does not occur.

15.   If K does not equal Q, then ΔG does not equal zero.

16.   According to LeChatelier, increasing the temperature will always increase the quantity of product.

17.   When a reaction and its equation are reversed, the K for the new reaction is the negative of the K for the original reaction.

18.   When an acid is described as stronger, this means that it is more concentrated.

19.   Autoionization is another name for neutralization.

20.   The higher the pH, the stronger the acid.

21.   All strong acids are levelled to H_3O^+ in aqueous solution.

22.   A Brønsted acid is always a Lewis acid, but a Lewis acid is not always a Brønsted acid.

23.   Conjugate acids and bases in aqueous solution always differ by only one hydrogen ion.

24.   The solution of the salt of a weak acid is basic.

25. K_a for an acid × K_b for its conjugate base = K_w.

26.   The pH of pure water is always 7.0 regardless of the temperature.

27.   If a strong acid is added to water, not only does the hydronium ion concentration increase but the hydroxide ion concentration must decrease.

28.   Equilibrium constants for neutralization reactions are generally quite large.

29.   At a point halfway to the equivalence point in the titration of a weak acid *vs.* a strong base, the pH = pK_a.

30.   When a salt solution is saturated, the concentration of the salt equals its solubility.

31.   The solubility product constant is always smaller if the solubility of the salt is lower.

32.   If $Q_{sp} < K_{sp}$, the salt will precipitate.

33.   The overall equilibrium constant for a reaction that is the sum of two simultaneous equilibria is the sum of the two constants for the two individual equilibria: $K_{sum} = K_1 + K_2$.

34.   The presence of a common ion suppresses the solubility of a salt.

35.   Sodium phosphate is insoluble.

36.   Spontaneous reactions happen very quickly.

37.   Entropy increases as the temperature increases.

38.   The entropy of the universe is constant.

39.   If $\Delta H_{sys} < 0$, and $\Delta S_{sys} > 0$, then the reaction must be spontaneous.

40.   At equilibrium, $\Delta G° = -RT\ln K$.

41.   Vaporization is an entropy-driven process.

42. $\Delta G°$ depends upon the temperature, even though $\Delta H°$ and $\Delta S°$ do not vary a great deal with change in temperature.

43.   The easiest way to increase E° is to increase the size of the cell.

44.   An anode is the negative terminal in a voltaic cell.

45.   A faraday is the charge on one mole of electrons.

46.   A battery is a voltaic cell that can be recharged.

47.   Reduction always occurs at the cathode.

48.   One coulomb is the quantity of charge carried by one electron.

49.   One watt.second = one volt.coulomb = one volt.amp.sec.

50.   If the standard reduction potential for a metal is more negative the metal is a better reducing agent.

Multiple choice questions

1.   The oxidation of hydrazine to nitrogen dioxide is represented by the equation:

$$N_2H_4(g) + 3O_2(g) \rightarrow 2NO_2(g) + 2H_2O(g)$$

If water is formed in this reaction at a rate of 42 mol $L^{-1}s^{-1}$, at what rate is oxygen used up?

a.   14 mol $L^{-1}s^{-1}$ c.   28 mol $L^{-1}s^{-1}$ e.   63 mol $L^{-1}s^{-1}$
b.   21 mol $L^{-1}s^{-1}$ d.   42 mol $L^{-1}s^{-1}$ f.   84 mol $L^{-1}s^{-1}$

2. The half-life $t_{1/2}$ of a radioactive isotope is ten days. How many days does it take for the radioactive isotope to decay to one-eighth of its original activity?

   a. 20 days   b. 30 days   c. 32 days   d. 48 days   e. 80 days

3. The reduction of nitric oxide by hydrogen is represented by the equation:

   $$2NO + 2H_2 \rightarrow N_2 + 2H_2O$$

   The following data were recorded in a series of experiments in order to determine the rate equation:

	Initial [NO] mol liter^{-1}	Initial [H$_2$] mol liter^{-1}	Initial rate mol liter^{-1} s^{-1}
Experiment 1:	0.5×10^{-3}	0.5×10^{-3}	2.5×10^{-5}
Experiment 2:	0.5×10^{-3}	1.0×10^{-3}	5.0×10^{-5}
Experiment 3:	1.0×10^{-3}	1.0×10^{-3}	2.0×10^{-4}

   The rate equation for the reaction is

   a. Rate = k [NO]
   b. Rate = k [H$_2$]
   c. Rate = k [NO] [H$_2$]
   d. Rate = k [NO]^2 [H$_2$]^2
   e. Rate = k [NO]^2 [H$_2$]
   f. Rate = k [NO]^2

4. The decomposition of dinitrogen tetroxide to nitrogen dioxide has a rate constant of 1.5×10^7 s^{-1} at 48°C. The activation energy for the reaction is 55 kJmol^{-1}. Calculate the rate constant at 24°C.

   a. 1.4×10^7 s^{-1}
   b. 1.8×10^7 s^{-1}
   c. 2.8×10^6 s^{-1}
   d. 3.4×10^6 s^{-1}
   e. 1.4×10^5 s^{-1}
   f. 3.6×10^5 s^{-1}

5. Consider the following equilibrium system:

   $$N_2 + 3H_2 \rightleftharpoons 2NH_3$$

   If 1.0 mol of N$_2$, 3.0 mol of H$_2$, and 4.0 mol of NH$_3$ are present in a 1.0 liter vessel, what is the value of the equilibrium constant, K_c?

   a. 4.0   c. 1.35   e. 1.8   g. 1.7
   b. 0.59   d. 0.79   f. 1.33   h. 5.3

6. Carbon disulfide and chlorine react as follows:

   $$CS_2(g) + 3Cl_2(g) \rightleftharpoons S_2Cl_2(g) + CCl_4(g)$$

   When 2.0 mol of CS$_2$ and 4.0 mol of Cl$_2$ are placed in a one-liter flask and the system allowed to come to equilibrium, the flask is found to contain 0.30 mol of CCl$_4$. How much Cl$_2$ is present at equilibrium?

   a. 0.30 mol   c. 1.2 mol   e. 2.0 mol   g. 3.4 mol
   b. 0.90 mol   d. 1.7 mol   f. 3.1 mol   h. 3.7 mol

7. A mixture of 3.00 mol of NO and 2.00 mol of O_2 were placed in a 1.00 liter flask. When equilibrium had been established according to the equation shown, it was discovered that 2.00 mol of NO_2 had been formed. Calculate the value of the equilibrium constant K_c at this temperature.

$$2NO + O_2 \rightleftharpoons 2NO_2$$

a. 0.17   c. 1.00   e. 2.50   g. 3.50
b. 0.50   d. 2.00   f. 3.00   h. 4.00

8. Carbon dioxide can be produced according to the equation:

$$CaCO_3(s) \rightleftharpoons CaO(s) + CO_2(g) \qquad \Delta H = \text{positive}$$

How could the amount of CO_2 present at equilibrium be increased?

a. heating the system
b. decreasing the volume of the system
c. adding more $CaCO_3$

d. cooling the system
e. increasing the pressure
f. adding more CaO

9. In the reaction $HSO_4^- + OH^- \rightleftharpoons H_2O + SO_4^{2-}$, the HSO_4^- ion is the

a. conjugate acid of H_2O
b. conjugate base of H_2O
c. conjugate acid of SO_4^{2-}

d. conjugate base of SO_4^-
e. conjugate acid of OH^-
f. conjugate base of OH^-

10. What is the strongest acid that can exist in liquid ammonia?

a. H_3O^+ c. NH_3 e. HCl   g. HNO_3
b. OH^- d. NH_2^- f. NH_4^+ h. $HClO_4$

11. What is the pH of an aqueous solution in which the hydrogen ion concentration is 2.0×10^{-2} mol/liter?

a. 0.85   c. 1.70   e. 1.05   g. 2.70
b. 2.0   d. 12.0   f. 12.30   h. 11.30

12. A solution made by dissolving NaCN in water will be

a. basic because Na^+ hydrolyzes
b. basic because CN^- hydrolyzes
c. neutral because neither ion hydrolyzes
d. acidic because Na^+ hydrolyzes
e. acidic because CN^- hydrolyzes

13. What is the pH of a 0.50 M NH_4NO_3 solution? K_b for $NH_3 = 1.8 \times 10^{-5}$.

a. 4.78   c. 9.22   e. 9.56
b. 10.05   d. 3.96   f. 13.7

14. A solution is made by dissolving 0.50 mol of formic acid (HCO_2H) and 0.10 mol of sodium formate (HCO_2Na) in enough water to make 1.00 liter of solution. What is the pH of this solution? K_a for formic acid is 2.0×10^{-4}.

    a. 4.00   c. 3.00   e. 0.30   g. 4.40
    b. 2.00   d. 10.0   f. 11.0   h. 4.60

15. At a certain temperature, the molar solubility of bismuth sulfide, Bi_2S_3, is 2.0×10^{-12} mole per liter. Calculate the solubility product constant, K_{sp}, for bismuth sulfide at this temperature.

    a. 2.4×10^{-23} c. 4.0×10^{-24} e. 3.2×10^{-59}
    b. 2.3×10^{-57} d. 3.5×10^{-57} f. 2.0×10^{-12}

16. The solubility product K_{sp} for $PbBr_2$ is 9.0×10^{-6}. What is the solubility (in moles per liter) of $PbBr_2$ in a 0.30 M NaBr solution?

    a. 1.0×10^{-4} c. 1.0×10^{-5} e. 2.5×10^{-5}
    b. 3.0×10^{-5} d. 3.0×10^{-3} f. 2.7×10^{-7}

17. From which of the following solutions would strontium fluoride precipitate? The solubility product for SrF_2 is 8.0×10^{-10}.

    a. a solution 1.0×10^{-2} M in Sr^{2+} and 1.0×10^{-3} M in F⁻
    b. a solution 1.0×10^{-5} M in Sr^{2+} and 1.0×10^{-4} M in F⁻
    c. a solution 1.0×10^{-6} M in Sr^{2+} and 1.0×10^{-6} M in F⁻

18. When copper or nickel is electroplated with silver, the silver solution contains cyanide ion to reduce the effective concentration of Ag^+ in the solution. Estimate the concentration of Ag^+ in a 0.10 M solution of silver nitrate to which excess potassium cyanide (1.0 M) has been added. K_f for $[Ag(CN)_2]^- = 1.0 \times 10^{21}$

    a. 1.6×10^{-20} M   c. 1.6×10^{-22} M   e. 3.2×10^{-11} M
    b. 1.6×10^{-21} M   d. 3.2×10^{10} M   f. 0.16 M

19. The enthalpy change $\Delta H°$ for the combustion of carbon monoxide is -283.0 kJ/mol. The free energy change $\Delta G°$ is -257.2 kJ/mol. If 2.0 moles of carbon monoxide are burned and the maximum amount of useful work is extracted from the system, how much heat is liberated?

    a. 25.8 kJ   c. 257.2 kJ   e. 514.4 kJ
    b. 51.6 kJ   d. 283.0 kJ   f. 566 kJ

20. What is the free energy change $\Delta G°$ at 25°C and 1 atm for the following reaction?

$$H_2(g) \ + \ F_2(g) \ \rightleftharpoons \ 2HF(g) \qquad \Delta G°_f(HF)(g) \ = -273.2 \text{ kJ/mol}$$

    a. +136.6 kJ   c. −546.4 kJ   e. +546.4 kJ   g. −273.2 kJ
    b. −136.6 kJ   d. +1093 kJ   f. +273.2 kJ   h. −1093 kJ

21. Calculate the equilibrium constant K_p for the reaction at 25°C:

$$2NO(g) \;+\; O_2(g) \;\rightleftharpoons\; 2NO_2(g) \qquad \begin{aligned} \Delta G_f^\circ(NO)(g) &= +86.55 \text{ kJ/mol} \\ \Delta G_f^\circ(NO_2)(g) &= +51.31 \text{ kJ/mol} \end{aligned}$$

    a. 1.5×10^6 c. 2.2×10^{12} e. 3.3×10^{13}
    b. 1.8×10^{147} d. 4.5×10^{-13} f. 7.2×10^{11}

22. Which process results in a decrease in entropy of the system?

    a. freezing a liquid   c. sublimation of a solid   e. making a solution
    b. warming a liquid   d. evaporation of a liquid   f. melting a solid

23. The enthalpy of vaporization of benzene is 30.8 kJ/mol at its boiling point of 80.0°C. Calculate the entropy change for the vaporization of benzene at 1.0 atm pressure.

    a. +87.2 J/K mol   c. +385 J/K mol   e. −87.2 kJ/K mol
    b. +0.385 J/K mol   d. +11.5 J/K mol   f. −11.5 kJ/K mol

24. Processes can be characterized by the sign of the change in enthalpy and the sign of the change in entropy as follows:

Process	Change in enthalpy ΔH	Change in entropy ΔS
1	+	+
2	+	−
3	−	+
4	−	−

Which process(es) must be spontaneous at constant temperature and pressure?

    a. only 1   c. only 3   e. 1 and 3   g. 2, and 4
    b. only 2   d. only 4   f. 2 and 3   h. all of them

25. If 1000 coulombs of electrical charge need to be delivered to an electrochemical cell, a current of 4.0 amp would have to flow for how many minutes?

    a. 250   c. 4000   e. 0.24
    b. 4.2   d. 67   f. 120

26. Calculate the standard cell potential for $Cu(s)|Cu^{2+}(aq) \| Ag^+(aq)|Ag(s)$ if the standard electrode reduction potentials are

$E° \; Ag^+(aq)|Ag(s) = +0.799 \text{ V}$
$E° \; Cu^{2+}(aq)|Cu(s) = +0.339 \text{ V}$

    a. 1.14 V   c. 3.14 V   e. 0.46 V
    b. 0.121 V   d. 2.39 V   f. 1.92 V

27. Based upon your answer to the previous question, calculate $\Delta G°$, the standard free energy change, for the reaction shown:

$$2Ag^+(aq) + Cu(s) \rightleftharpoons Cu^{2+}(aq) + 2Ag(s)$$

    a. −44.4 kJ   c. −320 kJ   e. −88.8 kJ
    b. −210 kJ   d. −178 kJ   f. −112 kJ

28. Calculate K_{sp}, the solubility product, for silver bromide, if

$$AgBr(s) + e^- \rightleftharpoons Ag(s) + Br^-(aq) \qquad E° = +0.0732 \text{ V}$$
$$Ag^+(aq) + e^- \rightleftharpoons Ag(s) \qquad E° = +0.799 \text{ V}$$

    a. 3.0×10^{-25} c. 2.6×10^{-12} e. 6.2×10^{-12} g. 5.5×10^{-13}
    b. 1.8×10^{-15} d. 3.4×10^{-30} f. 9.5×10^{-10} h. 3.1×10^{-14}

29. An electrode made of a metal that is more active is often used in the cathodic protection of a metal that is less active. Such an electrode is called

    a. a galvanized electrode
    b. a rust inhibitor
    c. a catalytic poison
    d. an inert electrode
    e. a sacrificial anode
    f. red lead

30. What current is required to deposit 0.500 g of chromium metal from a solution of chromium(III) sulfate in a period of one hour?

    a. 40.2 amp   c. 0.0129 amp   e. 0.773 amp   g. 46.3 amp
    b. 0.0112 amp   d. 0.257 amp   f. 2.32 amp   h. 2780 amp

CHAPTER 22

Introduction

The Periodic Table is divided into various blocks of elements. These blocks are characterized by the electronic subshell being filled. Thus there are the s, p, d, and f blocks. The s block (the two–column block on the left) and the p block (the six–column block on the right) contain the representative elements. In this chapter we will look at the properties of some of these elements.

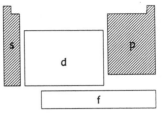

Contents

22.1   The Periodic Table—a Guide to the Elements

Periodic trends in the chemical and physical properties of the elements were examined in Chapter 8 (Section 8.6). Atomic and ionic size, ionization energy, electron affinity, and electronegativity all vary in a systematic way through the Periodic Table.

The underlying properties of the elements that are responsible for these trends are the nuclear charge and the electron configuration, especially the valence shell electron configuration. For example, when elements combine to form compounds they often do so in a way that enables them to achieve a noble gas electron configuration.

Elements on the right of the Periodic Table can achieve a noble gas configuration by adding electrons to form negative ions: Cl^-, O^{2-}, N^{3-}, etc. Elements on the left of the Periodic Table form positive ions: Na^+, Mg^{2+}, Al^{3+}, etc. Elements from the left and right combine to form ionic compounds. The metal (on the left) is oxidized and the nonmetal (on the right) is reduced.

Elements on the right side on the Periodic Table (above the diagonal line through the p block) combine to form covalent compounds; electrons are shared in the bonds.

Similarities between the elements in a group are expected because their valence shell electron configurations are the same. The same generic formulas are expected for their compounds. Incorrect formulas can be recognized fairly easily.

Read the Problem Solving Tips and Ideas 22.1 on page 1010 for hints on how to recognize formulas for nonexistent compounds.

22.2   Hydrogen

Hydrogen was first prepared by Robert Boyle in 1660 from the reaction between sulfuric acid and iron Henry Cavendish prepared a pure sample in 1766. The name means "water–former."

There are three isotopes of hydrogen:
    protium
    deuterium
    tritium

Hydrogen is the ninth most abundant element in the earth's crust. Used as a fuel, it is present in coal gas and water gas (a mixture of CO and H_2). About 300 billion liters are produced worldwide each year primarily from the reaction of methane gas and water at high temperatures (900–1000°C).

In the laboratory, hydrogen can be made by the reaction of a metal like zinc or iron with a strong acid, metals like sodium with water, or metals like aluminum with a strong base. Metal hydrides react with water to produce hydrogen.

Three types of hydrogen compounds are known: metal hydrides containing the hydride ion H^-, such as NaH and CaH_2; covalent hydrogen compounds such as diborane B_2H_6, ammonia NH_3, and water H_2O; and nonstoichiometric interstitial hydrides of some transition metals.

The largest use of hydrogen is in the production of ammonia by the Haber Process. A considerable quantity is also used to make methanol CH_3OH. Some is used in the partial hydrogenation of vegetable oils.

22.3   Sodium and Potassium

Sodium and potassium were first prepared by Sir Humphry Davy in 1807 by electrolysis of molten carbonates Na_2CO_3 and K_2CO_3.

The Downs cell used for the production of sodium (and chlorine) is illustrated in Figure 22.5 on page 1015 in the text.

The Group 1A elements are called the **alkali metals**. Sodium and potassium are the sixth and seventh most abundant elements in the earth's crust. Sodium is prepared by the electrolysis of molten sodium chloride and potassium is made by reacting sodium with potassium chloride.

Both metals are relatively soft with low melting points (Na: 93.5°C and K: 65.65°C). They are both highly reactive; potassium often dangerously so. The salts of alkali metals are usually soluble. Alkali metals form halides with the halogens. Sodium reacts with oxygen to form a peroxide Na_2O_2 and potassium forms a superoxide KO_2. Only lithium of the alkali metals forms a nitride.

Electrolysis of saturated brine is the basis for the chlor–alkali industry—the products are chlorine gas and sodium hydroxide. Sodium carbonate (washing soda) is also commercially important (glass manufacture; paper manufacture; cleaning materials). Sodium bicarbonate (baking soda) is another commonly used compound of sodium.

22.4   Calcium and Magnesium

Calcium and magnesium (and strontium and barium) were first prepared by Sir Humphry Davy in 1808 by the electrolysis of molten salts.

Group 2A elements are called the **alkaline earth metals**. Like the alkali metals, the elements are reactive. However, the ionic bonding in the salts of the alkaline earth metals is stronger than in the compounds of the alkali metals due to the 2+ charge (compared to 1+). So the salts generally have higher melting points and are less soluble.

Common minerals are limestone $CaCO_3$, gypsum $CaSO_4.2H_2O$, fluorspar CaF_2, magnesite $MgCO_3$, dolomite $CaCO_3.MgCO_3$, talc $3MgO.4SiO_2.H_2O$, and asbestos $3MgO.4SiO_2.2H_2O$. Marble is calcite (limestone) crystallized under high geologic pressure.

Magnesium metal is used in lightweight alloys (with aluminum). It is obtained from seawater by precipitation as the hydroxide. Conversion to the chloride, and its subsequent electrolysis, produces the magnesium metal.

Calcium is used extensively in a variety of forms. Fluorite, CaF_2, is used in the steel industry where it facilitates the separation of molten iron from the silicate slag. It is used as a source of fluorine used to make hydrofluoric acid which in turn is used to make cryolite (for aluminum production) and fluorocarbons such as teflon. Fluorapatite, $CaF_2.3Ca_3(PO_4)_2$, is used as a primary source of phosphoric acid H_3PO_4 used in turn in the fertilizer industry.

Limestone, when heated, forms lime (quicklime). Mixed with sand and water, lime forms a paste called mortar that has been used for centuries in securing stones in building. The lime forms slaked lime (calcium hydroxide) with the water and the slaked lime absorbs carbon dioxide from the air to reform the original calcium carbonate.

22.5   Aluminum

Aluminum is the third most abundant element in the earth's crust. In the pure state it is soft and weak but alloyed with other elements such as magnesium, manganese, and copper it is much stronger. Readily oxidized, it is in fact resistant to corrosion because the thin oxide film that forms on the surface protects the metal underneath.

Aluminum is found in nature in the form of many different aluminosilicates. It is also found as hydrated aluminum oxide (bauxite) which is used as the source of the metal. The aluminum oxide is purified by the Bayer process which relies on its amphoteric properties. The aluminum oxide is dissolved in cryolite $NaAlF_6$ and electrolyzed in the Hall–Héroult process.

Originally prepared in small quantities by the reduction of aluminum oxide using metals such as potassium, aluminum is now produced in vast quantities by the Hall–Héroult process.

22.6   Silicon

Silicon is the second most abundant element in the earth's crust. Clays, pottery, porcelain, bricks, and a lot of rocks are all aluminosilicates. Lime glass, pyrex, and lead crystal are all glasses made from silica. If silica is reduced at 3000°C with purified coke, reasonably pure silicon can be obtained. The extremely pure silicon required for semiconductors is prepared by distillation of silicon tetrachloride, reduction of the tetrachloride by magnesium to silicon, and then zone-refining of the silicon.

Quartz is one form of pure crystalline silica SiO_2. The crystals are used in electronics as frequency controllers. Silica dissolves in hot molten sodium hydroxide or sodium carbonate to form sodium silicate. If sodium silicate is treated with acid, an amorphous silica gel is obtained. When dry, this is a very porous material.

Silicates come in a wide variety of forms. All are based upon the tetrahedral SiO_4 unit. Ratios of silicon to oxygen range from 1:4 (for orthosilicates), 1:3.5 (for pyrosilicates), 1:3 (for cyclic metasilicates or linear asbestos or amphiboles), 1:2.5 (for sheet structures like talc or mica), to 1:2 (for silica or the 3-dimensional aluminosilicates). Zeolites are natural or specifically designed synthetic aluminosilicates that are often used as catalysts. Silicones are polymeric silicon-based ethers that are chemically inert and have a wide range of uses.

The zeolite structure is shown at the bottom of page 1026 in the text.

22.7   Nitrogen and Phosphorus

Nitrogen and phosphorus are essential to life on earth; both are present in biochemical molecules in the human body—proteins, nucleic acids, and others. Nitrogen N_2 is obtained by fractional distillation of liquid air. Its compounds, including ammonia, nitric acid, ammonium nitrate, and urea, play an important part in the economy. Phosphorus P_4 is obtained from the apatite minerals $3Ca_3(PO_4)_2.CaX_2$ where X = F, Cl, or OH. Its major use is in fertilizers.

Nitrogen is a colorless gas (b.pt. −196°C). It makes up 80% of the earth's atmosphere. Its reluctance to form compounds is due to the thermodynamic stability of the triple bond (bond energy 946 kJ mol^{-1}), its kinetic inertness, and its nonpolarity. Nitrogen gas cannot be used in biological systems until it is fixed, i.e. changed into nitrogen compounds such as ammonia. Some bacteria and organisms are able to do this, but most plants cannot. The fixed nitrogen must be provided by a fertilizer. The Haber process, the conversion of nitrogen and hydrogen into ammonia was the first successful method for the production of ammonia. As such it revolutionized the fertilizer industry and, with the Ostwald process, the explosives industry.

Another nitrogen-hydrogen compound is hydrazine N_2H_2. There are several nitrogen oxides, with nitrogen having oxidation numbers from +1 to +5: NO (nitric oxide), N_2O (nitrous oxide), N_2O_3 (dinitrogen trioxide), NO_2 (nitrogen dioxide and its dimer dinitrogen tetroxide N_2O_4), N_2O_5 (dinitrogen pentoxide—a molecule in the gas state but $NO_2^+NO_3^-$ in the solid state).

Nitric acid HNO_3 is manufactured from ammonia by the Ostwald process. 20% of ammonia produced in the Haber process is used in the subsequent manufacture of nitric acid. The Ostwald process is the oxidation of ammonia by air over a rhodium/platinum catalyst. The largest use of the nitric acid (80%) is in making ammonium nitrate. Nitric acid is not only a strong acid but an excellent oxidizing agent. Only four metals are not attacked by nitric acid: the noble metals gold, platinum, rhodium, and iridium. However, a mixture of nitric acid and hydrochloric acid (1:3 ratio) is aqua regia which will attack even the noble metals.

22.8   Oxygen and Sulfur

Oxygen O_2 is the most abundant element in the earth's crust. It also makes up about 20% of the earth's atmosphere. It is isolated from air by careful fractional distillation. Sulfur S_8 is also abundant in the earth's crust (15th). It is found in the elemental form but also as metal sulfides and sulfate minerals.

A second allotrope of oxygen is ozone O_3 found in the upper atmosphere where it absorbs UV solar radiation. Sulfur also exists as two common allotropes, rhombic and monoclinic, both based upon an eight-membered ring of sulfur atoms. It has other forms as well. Sulfur dioxide SO_2 is a by-product of the roasting of metal sulfide ores. It is also made directly by burning sulfur. Oxidation to sulfur trioxide SO_3 is done by the contact process using vanadium pentoxide as a catalyst. SO_3 is used to produce sulfuric acid—the chemical produced in the greatest quantity per annum. Much (70%) is used in the fertilizer industry, converting phosphate rock into soluble phosphates.

22.9   Chlorine

Chlorine is 11th in abundance in the earth's crust, largely as sodium chloride in sea water and solid deposits. It is manufactured by the electrolysis of concentrated brine. 70% is used in the production of organic chemicals—in particular vinyl chloride used to make PVC. The other major use is as a bleaching agent and disinfectant.

Hydrogen chloride HCl is a gas. It dissolves in water to produce the strong acid, hydrochloric acid. Oxoacids of chlorine and the corresponding oxoanions range from hypochlorous acid (HOCl) and hypochlorite (OCl^-) to perchloric acid ($HClO_4$) and perchlorate (ClO_4^-). All are strong oxidizing agents although the perchlorate ion is kinetically unreactive in dilute solution.

Liquid bleach, the bleach used in the home is the result of adding chlorine to cold sodium hydroxide solution:

$$Cl_2(g) + 2OH^-(aq) \rightleftharpoons OCl^-(aq) + Cl^-(aq) + H_2O(l)$$

The chlorine used in disinfecting swimming pools is essentially the same reaction using calcium hydroxide in place of sodium hydroxide.

More than half the sodium perchlorate manufactured is converted to ammonium perchlorate which is used as a rocket propellant—each shuttle launch uses about 750 tons.

Chlorine was first made by Karl Wilhelm Scheele in 1774 by the reaction of sulfuric acid on sodium chloride and manganese dioxide.

Review Questions

1. How is the valence shell electron configuration related to the Group number of an element?

2. How can the formula of an ionic compound be predicted?

3. Why are the stoichiometries of the compounds for all elements in a particular group often the same?

4. How does the character of the hydrogen compounds of the representative elements change across the Periodic Table?

5. Write the formula of a typical metal oxide. Is it acidic or basic in solution? Write the formula of a typical nonmetal oxide. Is this oxide acidic or basic in solution? What is an amphoteric oxide?

6. What is the difference between an allotrope and an isotope?

7. List, with name and symbol, the ten most abundant elements in the earth's crust.

8. Research, and list, the ten substances produced in the greatest quantity each year in the United States.

Answers to Review Questions

1. The valence shell electron configuration determines the group to which the element belongs. The group number is the number of electrons in the valence shell. For example, chlorine has the electron configuration $[Ne]\, 3s^2\, 3p^5$, it has seven valence electrons and is in Group 7A.

2. The charges on the positive and negative ions of the representative elements are predictable. The charges on the polyatomic ions are also known. Since in a compound the positive and negative charges must balance, the stoichiometry is easily established. For example, the formula for calcium phosphide must be Ca_3P_2 because the calcium ion always has a charge of +2 and the phosphide ion always has a charge of −3. And 3 × +2 balances 2 × −3.

3. The stoichiometries of the compounds for all elements in a particular group are often the same because the elements all have the same valence shell electron configuration. For example NH_3, PH_3 and AsH_3; CCl_4, $SiCl_4$, and $GeCl_4$; HF, HCl, HBr, and HI; LiCl, NaCl, KCl, and RbCl; MgO and CaO; etc.

4. With metallic elements on the left, hydrogen forms ionic (or saline) hydrides. In these compounds the hydrogen has a negative oxidation number −1. With some transition metals in the center, hydrogen forms nonstoichiometric hydrogen compounds: the hydrogen is in the atomic state. With the nonmetals on the right, hydrogen forms molecular compounds in which it has an oxidation number +1. The character changes from ionic hydride to covalent hydrogen compound.

5. For example magnesium oxide MgO. Metal oxides are basic in solution. For example sulfur trioxide SO_3. Nonmetal oxides are acidic in solution. An amphoteric oxide is an oxide like Al_2O_3 that is either acidic or basic depending upon its environment. Transition metals, with their variable oxidation states can often form a range of oxides. For example CrO_3 is acidic, Cr_2O_3 is amphoteric, and CrO is basic.

6. An allotrope is a different crystalline or structural form of an element. For example, diamond, graphite, and fullerene are three allotropes of carbon. Monoclinic and rhombic sulfur are two allotropes of molecular sulfur S_8. Isotopes are atoms of the same element having different numbers of neutrons and therefore different mass numbers.

All other elements together add up to the remaining approx 0.5%

7.

Oxygen	O	49.5%	Sodium	Na	2.6%
Silicon	Si	25.7%	Potassium	K	2.4%
Aluminum	Al	7.5%	Magnesium	Mg	1.9%
Iron	Fe	4.7%	Hydrogen	H	0.9%
Calcium	Ca	3.4%	Titanium	Ti	0.6%

8. This list is usually published in *Chemical & Engineering News* in j
    each year. The top ten occasionally switch places, but in general th
    is as follows. Steel and sodium chloride are not included in the list;
    hydrogen which is usually used immediately it is made:

1	sulfuric acid	45 billion kg	contact process	
2	nitrogen	31 billion kg	fractional distillation of ai
3	oxygen	25 billion kg	fractional distillation of ai.
4	ethylene	22 billion kg	thermal cracking of petrol
5	lime	19 billion kg	decomposition of limeston
6	ammonia	16 billion kg	Haber process
7	phosphoric acid	12 billion kg	from phosphate rocks
8	sodium hydroxide	12 billion kg	electrolysis of brine
9	propylene	12 billion kg	thermal cracking of petroleum
10	chlorine	12 billion kg	electrolysis of brine

335

Study Questions and Problems

What stuff 'tis made of, whereof it
is born, I am to learn.

William Shakespeare
(1564-1616)

1. Predict the formulas of the compounds that will result when the follow-
    ing elements react:

    a. $Na(s)$ and $P_4(s)$
    b. $Al(s)$ and $Br_2(l)$
    c. $Ca(s)$ and $O_2(g)$
    d. $Cl_2(g)$ and $Si(s)$

2. Write formulas for the following compounds based upon the formulas
    for elements in the same group that are shown.

    a. Water H_2O hydrogen selenide
    b. Ammonia NH_3 nitrogen triiodide
    c. Chlorate ClO_3^- iodate ion
    d. Triiodide I_3^- difluorochloride ion
    e. Bromine Br_2 iodine monochloride

3. Which formulas in the following series are incorrect?

    a. $CaCl_2$, CaS, KCl, KS
    b. NH_3, PH_3, CH_3, GeH_4
    c. C_{60}, S_8, P_4, O_3, Br, Xe

4. List the methods used for industrial production of the following ele-
    ments:

    a. hydrogen   g. nitrogen
    b. sodium   h. phosphorus
    c. potassium   i. oxygen
    d. magnesium   j. sulfur
    e. aluminum   k. chlorine
    f. silicon

5. List some of the major uses for the following elements or their compounds:

    a. hydrogen
    b. sodium
    c. magnesium
    d. aluminum
    e. silicon
    f. nitrogen
    g. phosphorus
    i. oxygen
    j. sulfur
    k. chlorine

6. Classify the following hydrides:

    SiH_4, CaH_2, HI, H_2Se

7. How does a peroxide or superoxide differ from an oxide ion? Describe the reaction between carbon dioxide and potassium superoxide.

8. Write formulas for the following compounds:

    sodium chloride   calcium sulfate
    potassium hypochlorite   nitric acid
    phosphine   methanol
    magnesium nitrate   dolomite
    dinitrogen tetroxide   limestone
    aluminum bromide   phosphoric acid
    sulfuric acid   baking soda

Answers to Study Questions and Problems

1. a. $Na(s)$ and $P_4(s)$ → Na_3P sodium phosphide
    b. $Al(s)$ and $Br_2(l)$ → Al_2Br_6 aluminum bromide
    c. $Ca(s)$ and $O_2(g)$ → CaO calcium oxide (quicklime)
    d. $Cl_2(g)$ and $Si(s)$ → $SiCl_4$ silicon tetrachloride

2. a. Water H_2O hydrogen selenide H_2Se
    b. Ammonia NH_3 nitrogen triiodide NI_3
    c. Chlorate ClO_3^- iodate ion IO_3^-
    d. Triiodide I_3^- difluorochloride ion ClF_2^-
    e. Bromine Br_2 iodine monochloride ICl

3. a. $CaCl_2$, CaS, KCl, KS KS should be K_2S
    b. NH_3, PH_3, CH_3, GeH_4 CH_3 should be CH_4
    c. C_{60}, S_8, P_4, O_3, Br, Xe Br should be Br_2

4.   Production methods:

    a. hydrogen   catalytic steam reformation of methane and the water gas shift reaction

    b. sodium   electrolysis of fused sodium chloride

    c. potassium   reaction of sodium gas with molten KCl

    d. magnesium   precipitation from seawater as hydroxide, conversion to chloride, and electrolysis of the fused chloride

    e. aluminum   purification of aluminum oxide by the Bayer process and subsequent electrolysis in cryolite by the Hall–Héroult process

    f. silicon   reduction of silica with coke at 3000°C

    g. nitrogen   fractional distillation of air

    h. phosphorus   reduction of phosphates using carbon and silica

    i. oxygen   fractional distillation of air

    j. sulfur   indirectly from sulfide ores, or mined directly using the Frasch process

    k. chlorine   electrolysis of brine

5.   Major uses:

    a. hydrogen   production of ammonia (Haber process) and methanol

    b. sodium   coolant in nuclear reactors

         sodium hydroxide   industrial base, paper, aluminum processing

         sodium carbonate   manufacture of glass, water treatment, paper manufacture, cleaning materials

    c. magnesium   lightweight alloys

    d. aluminum   lightweight alloys and structural components

    e. silicon   electronics industry

         sodium silicate   industrial detergents, pigments, silica gel

         silicones   oils, greases, resins, lubricants, polymers

    f. nitrogen   production of ammonia, oil recovery, inert atmosphere

         ammonia   production of nitric acid, fertilizer (80%)

         nitric acid   manufacture of nitrates, fertilizers (65%), explosives (25%)

    g. phosphorus   fertilizers

    h. oxygen   steel industry, plastics, waste water treatment

    i. sulfur   manufacture of sulfuric acid,

         sulfuric acid   production of phosphate fertilizers, chemical manufacture (dyes, drugs, explosives, paper)

    j. chlorine   production of organic chemicals, PVC, bleaching agent

6. SiH_4 silane   a covalent hydrogen compound

       CaH_2 calcium hydride   a saline or ionic hydride

       HI   hydrogen iodide   a covalent hydrogen compound

       H_2Se hydrogen selenide   a covalent hydrogen compound

7.   An oxide ion is a single oxygen atom with a charge of 2–. Both the peroxide ion and the superoxide ions are ions of the oxygen molecule. The peroxide ion is O_2^{2-} and the superoxide ion is O_2^-. The peroxide link or group is easily recognized by the two oxygen atoms joined together. The reaction between carbon dioxide and potassium superoxide produces 3 moles of oxygen for every 2 moles of carbon dioxide:

$$4KO_2 \ + \ 2CO_2 \ \rightarrow \ 2K_2CO_3 \ + \ 3O_2$$

8.
sodium chloride	NaCl
potassium hypochlorite	KClO
phosphine	PH_3
magnesium nitrate	$Mg(NO_3)_2$
dinitrogen tetroxide	N_2O_4
aluminum bromide	Al_2Br_6
sulfuric acid	H_2SO_4
calcium sulfate	$CaSO_4$
nitric acid	HNO_3
methanol	CH_3OH
dolomite	$CaCO_3.MgCO_3$
limestone	$CaCO_3$
phosphoric acid	H_3PO_4
baking soda	$NaHCO_3$

CHAPTER 23

Introduction

Between the s block and the p block in the Periodic Table there is a ten-column block of metals. Some of these metals are familiar, like copper and nickel (the coinage metals), or zinc and iron. Some are precious, like silver, palladium, platinum and gold. Others are not very familiar, like hafnium and rhenium. One, mercury, is a liquid. Some are required for life, like iron and cobalt; others, like cadmium and mercury, are highly toxic. One, technetium, does not occur naturally on earth. This block of metals is the d block, often referred to as the transition elements or the transition metals. We will examine them in this chapter.

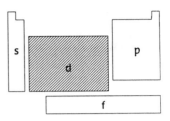

Tc-99 is radioactive and is used in medical imaging.

Contents

23.1   Properties of the Transition Elements

Most transition metals are expected to have high melting points, to shine when polished, to conduct an electrical current, and to be malleable and ductile; in other words, to have the typical properties of metals. And in general they do; however, the properties do vary. Some metals are easier to oxidize than others and their oxidation states vary considerably. The melting points vary dramatically from mercury at −38.9°C to tungsten at 3410°C.

The electron configurations of the metals vary from d^1 to d^{10}; their ions are formed by losing the s electrons first, followed by varying numbers of the d electrons. Unlike the representative elements, the transition metal ions usually do not have noble gas configurations. Their compounds are often paramagnetic—indicating the presence of unpaired electrons.

Oxidation numbers vary; +2 and +3 are common in ionic compounds; in complex ions involving covalent bonding the oxidation number may be as high as +6 or +7 (for example in CrO_4^{2-} and MnO_4^-) and as low as −3 at the center of the series. The stability of the higher oxidation states increases in the 2nd and 3rd series. For example, CrO_4^{2-} is quite easily reduced but WO_4^{2-} is quite stable.

The sizes (atomic radii) of the transition metals decrease initially from left to right and then increase slightly toward the end. The increased nuclear charge pulls the s electrons closer initially, making the atom smaller. When the d orbital set is almost full, the increased electronic repulsion causes the slight in-

There are three series of transition metals:
the first—the 3d
the second—the 4d
and the third—the 5d.

The fourth series consists of elements not found in nature. The series has only recently been completed. In some cases, only two or three atoms of the new element are made.

See Figure 8.11 on page 356 of the text.

crease. The second and third series are almost identical in size. This is due to the lanthanide contraction—adding 14+ to the nuclear charge while placing electrons in the very poorly shielding f orbitals. Thus, for example, tantalum and niobium, or silver and gold, have the same size. The platinum group metals Ru, Os, Rh, Ir, Pd, and Pt all have very similar properties and the same size. The lanthanide contraction is responsible for the high densities of the 3rd series elements.

Osmium has a density
= 22.49 g cm^{-3}.

See Figure 23.7 on page 1053
of the text.

The melting points increase toward the center of each series where the d orbital subshell is about half-filled—the largest number of unpaired electrons.

23.2   Commercial Production of Transition Metals

Some metals, copper, silver, and gold, for example, are found in their native state (as the element) in nature. Most metals, however, have to be obtained from ores—the oxides, sulfides, carbonates, or other salts. Most ores are mixtures and the first task is to isolate the desired ore. Then the metal has to be extracted.

A diagram of a blast furnace is
shown in Figure 23.10 on page
1055 of the text.

Iron is produced from its oxide by reduction with coke (carbon) in a blast furnace. The blast is carefully controlled air blown in at the bottom of the furnace. Molten iron is tapped off at the bottom of the furnace and forms cast iron or pig iron. The impurities form a slag which floats on the surface in the furnace and can be easily removed.

This is a major use of oxygen
prepared by the fractional
distillation of air.

The pig iron is purified in a basic oxygen furnace. The oxygen oxidizes nonmetal impurities (S, most of the C, and P). The result is carbon steel. Other elements can be added to modify the properties of the steel.

Copper occurs in sulfide ores and often with iron. These ores are concentrated by flotation and then roasted to oxidize the iron. Heated with limestone and sand, the iron forms a silicate slag which can be removed. Sulfur in the ore reduces the copper(II) sulfide to copper(I) sulfide which melts and flows to the bottom of the furnace. Blown with air, this copper matte is converted to copper and sulfur dioxide. The crude copper is purified by electrolysis.

23.3   Coordination Compounds

When a nickel(II) salt, like $NiCl_2$, dissolves in water, the nickel(II) ion is surrounded by water molecules. The energy of the ion–solvent interaction drives the reaction. When ammonia is added to the solution, ammonia molecules replace the water molecules around the nickel(II) to form another **complex ion** or **coordination compound**:

The ammonia molecule is a
stronger ligand than the water
molecule and therefore replaces
the water in the complex.

The [Ni(NH$_3$)$_6$]$^{2+}$ complex ion is
illustrated in Figure 23.16 on
page 1060 of the text.

$$[Ni(H_2O)_6]^{2+} + 2Cl^-(aq) + 6NH_3 \rightarrow [Ni(NH_3)_6]^{2+} + 2Cl^-(aq) + 6H_2O$$

In this reaction the ammonia is called a **ligand**. It forms a coordinate bond with the metal ion, supplying both electrons for the bond. It is an electron pair donor—i.e. a Lewis base. The nickel(II) ion is the Lewis acid.

The number of ligands attached to the metal is called the **coordination number**. In the nickel–ammonia complex the coordination number is 6. The six ligands are arranged at the corners of an octahedron around the nickel. A ligand like ammonia, with one lone pair to donate, is called monodentate. Other ligands may have more than one lone pair of electrons, often on different donor atoms, and are polydentate. Ethylenediamine $H_2NCH_2CH_2NH_2$ is a bidentate ligand; ethylenediaminetetraacetate (EDTA) is a hexadentate ligand. If a polydentate ligand can form coordinate bonds to the same metal ion, it can form a ring of atoms. Such a ring is called a **chelate ring** and the ligand is referred to as a chelating ligand.

Naming coordination compounds follows a set of rules just like the naming of ordinary compounds you learned earlier. These rules are listed and discussed on pages 1064 and 1065 in the text and in the answers to the Review Questions at the end of this study guide.

23.4 Structures of Coordination Compounds and Isomers

The geometry of a coordination compound is the arrangement of donor atoms of the ligands around the metal ion. Common coordination numbers are 2, 4, and 6 and the associated geometries are linear, tetrahedral, and octahedral. Examples of linear complexes are $[CuCl_2]^-$ and $[Ag(NH_3)_2]^+$. Examples of tetrahedral complexes are $[CoCl_4]^{2-}$, $[Ni(CO)_4]$ and $[Zn(NH_3)_4]^{2+}$. Some nickel complexes with a coordination number of 4 are square planar rather than tetrahedral. For example $[Ni(CN)_4]^{2-}$ and $[Ni(dmg)_2]$. Examples of octahedral complexes are $[Fe(H_2O)_6]^{2+}$ and $[Co(NH_3)_6]^{3+}$.

There are two main divisions of isomerism: structural and stereoisomerism. **Structural isomerism** involves different bonds. An example of structural isomerism in a coordination compound is the complex ion $[Co(NH_3)_5Cl]Cl_2.H_2O$ and its isomer $[Co(NH_3)_5H_2O]Cl_3$. The two isomers involve different ligands coordinated to the metal. Another example is $[Co(NH_3)_5NO_2]^{2+}$ in which the nitrite ion can coordinate through its nitrogen atom, or, in another structural isomer, through its oxygen atom.

Stereoisomerism involves the same bonds but a different arrangement. Stereoisomerism can be divided into **geometrical isomerism** and **optical isomerism**.

Square planar complexes can exhibit geometrical cis–trans isomerism. For example, in the complex $[Pt(NH_3)_2Cl_2]$, the two ammonia molecules can be next to each other, or opposite one another—the bonds are the same but the arrangement is different. Cis–trans isomerism is also possible in octahedral complexes but not in tetrahedral complexes. Another geometrical isomerism possible in octahedral complexes is fac–mer isomerism (facial–meridional). In a complex such as $[Co(NH_3)_3(NO_2)_3]$, the three ammonia ligands can be together on a face (the facial isomer) or they can be all in line (the meridional isomer).

A left hand and a right hand are mirror images of each other, but they are not the same—they are **nonsuperimposable**. Some molecules have nonsuperimposable mirror images and are said to be **chiral**. The mirror images are called **enantiomers**. They are optically active—the two enantiomers rotate the plane of plane polarized light in opposite directions. This is the basis of **optical isomerism**. The trisethylenediamine complex of cobalt(III) $[Co(en)_3]^{3+}$ is an example of an optically active octahedral complex.

A picture of an EDTA complex is shown in Figure 23.18 on page 1062. Note how the ligand encapsulates the metal ion.

The word chelate derives from the Greek meaning claw.

Not all polydentate ligands can chelate.

Notice that the charges on the complex ions may be + or −, or the complex may be neutral.

The ligand dmg is dimethyl-glyoxime, a bidentate chelating ligand often used in the gravimetric determination of nickel.

Recall the structural isomers butane and 2-methyl-propane.

Recall the geometrical isomers cis- and trans- 2-butene.

The easiest way to see this is to build some 3-dimensional models. Some are shown in Figures 23.20 and 23.21 on page 1068 of the text.

Molecules that have a plane of symmetry running through the center of the molecule, or molecules that have a center of symmetry, *cannot* have nonsuperimposable mirror images—they *cannot* be chiral—they are called achiral.

(Technically, the presence of *any* S_n axis prohibits chirality.)

23.5   Bonding in Coordination Compounds

The bonding between the ligand and the metal was described using **valence bond theory**. The formation of a coordinate bond is a Lewis acid-base reaction. Whereas the VB picture is satisfactory for the ground state of the complex it says nothing about the possible excited states of the complex and cannot explain the colors that characterize almost all transition metal complexes. **Crystal field theory** is an electrostatic approach to the bonding that does explain the spectra of the complexes. However, this theory assumes that there is no covalent bonding (sharing of electrons) between the metal and the ligand at all. A third theory, **molecular orbital theory**, incorporates the best features of both valence bond theory and crystal field theory.

> Crystal field theory is usually called ligand field theory when applied to transition metal complexes.

In an isolated atom or ion, all the d orbitals are equal in energy (they are said to be degenerate). However, when the atom or ion is approached by a set of ligands, repulsion between the electrons on the ligands and the electrons of the metal atom or ion causes those orbitals pointing toward the ligands to be pushed higher in energy. The d orbitals lose their degeneracy—they are split in energy—and how they split depends upon the symmetry of the ligand field imposed.

> The six ligands do not have to approach along the three cartesian axes—they can approach from any direction. The axes are defined as the directions in which they do approach.

For example, in an octahedral complex, the six ligands approach along the three Cartesian axes, x, y, and z. Therefore those orbitals that lie on these axes are pushed up in energy. These are the d_{z^2} and the $d_{x^2-y^2}$ orbitals. The other three d orbitals, d_{xy}, d_{xz}, d_{yz}, are raised in energy, but not as much.

> Recall that electrons are lost from the 4s before the 3d.

The d orbital splitting accounts for observed magnetic and spectroscopic properties (color) of metal complexes. For example, an Mn^{2+} ion has a d^5 electron configuration. In a free isolated Mn^{2+} ion, these five electrons occupy the five d orbitals, one in each orbital. As an octahedral ligand field approaches, and gets stronger and stronger, the d orbitals are split more and more in energy and eventually a point is reached when the system is more stable with the electrons paired in the lower set of three d_{xy}, d_{xz}, d_{yz} orbitals. The energy of stabilization (referred to as the ligand field stabilization energy) is enough to overcome the energy required to pair the electrons.

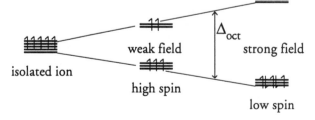

23.6   The Colors of Coordination Compounds

> Traces of transition metals in gemstones give the stones their colors.

One of the attractive properties of transition metal compounds is their color. They are quite unlike the majority of the representative metal compounds that are colorless. The visible spectrum lies in the range 400 nm(violet) to 700 nm(red). The reason why a solution of a transition metal complex appears one color is that the complex absorbs the complementary color. For example, the hexaaquanickel(II) nitrate appears green because the complex absorbs red light.

> Recall the colors of the visible spectrum ROYGBIV.
>
> Complementary colors are shown in Figure 23.33 on page 1079 of the text.

Transition metal complexes absorb light because electrons move from a low energy level to a higher level. The energy difference corresponds to a frequency in the visible portion of the electromagnetic spectrum, and therefore the complexes appear colored. It is the electron transitions between the d orbitals that are largely responsible for the colors. If the splitting of the d orbitals is changed, then the color is changed. Ligands can be ordered in their ability to split the d orbitals in energy, i.e. how strong the ligands are. Such an ordering of ligands is called the **spectrochemical series**.

This is essentially the same as the atomic absorption spectra mentioned in Chapter 7.

Some intense colors, such as that of the purple potassium permanganate and yellow cadmium sulfide are due to another type of transition called charge-transfer.

Review Questions

1.   What is a transition element? Are all transition elements metals?

2.   Review the magnetic properties paramagnetism, diamagnetism, and ferromagnetism.

3.   What is the lanthanide contraction?

4.   Review and summarize the rules for naming coordination compounds.

5.   What is a common size for a chelate ring (how many atoms in the ring)?

6.   Is there such a thing as a polydentate ligand that cannot chelate?

7.   Is it possible for a single atom or monatomic ion to be bidentate?

8.   Draw a flow chart indicating the various types of isomerism possible in coordination compounds. Give examples of each type.

9.   Describe differences between valence bond theory and crystal field theory. How does molecular orbital theory fit into the picture?

10. Describe the significance of the color wheel. What are complementary colors?

11. Describe how a change in the strength of a ligand field can change the magnetic properties of a complex compound.

12. Define the following terms:
     a.   chelate
     b.   ligand
     c.   polydentate
     d.   coordination compound
     e.   spectrochemical series
     f.   pairing energy
     g.   degeneracy

Answers to Review Questions

1. A transition element is an element in the ten-column block of elements between the s block and the p block. They are characterized by valence shell electron configurations d^1s^2 through $d^{10}s^2$. All are metals.

2. There are three types of magnetism: Diamagnetic substances are repelled by a magnetic field; they have no unpaired electrons. Paramagnetic substances are drawn into a magnetic field; they have at least one unpaired electron. The more unpaired electrons per atom, the more paramagnetic the substance. Ferromagnetic substances possess a strong intrinsic magnetism; these substances have all the unpaired electron spins lined up in the same direction.

3. Progressing through the lanthanide elements, 14 protons are added to the nucleus and 14 electrons are added to the 4f orbitals. This large increase in the nuclear charge pulls in the valence s electrons responsible for the size of the atom. The s electrons feel the large nuclear charge because the shielding provided by the 4f orbitals is very poor. As a result the second and third transition series are coincidentally almost identical in size. Tantalum and niobium have the same size; silver and gold have the same size, etc. Another result of the reduced size of the third series transition elements is their very high densities.

4. Rules for naming coordination compounds:

    > 1. When the compound is a salt, name the cation first.
    >
    > 2. Name the ligands in alphabetical order.
    >
    > 3. If the ligand is an anion, change the ending to –o
    > | –ite | to | –ito |
    > |------|----|------|
    > | –ate | to | –ato |
    > | –ide | to | –o |
    >
    > 4. If the ligand is neutral, its ordinary name is used except for
    > | water: | called aqua |
    > |--------|-------------|
    > | ammonia: | called ammine |
    > | carbon monoxide: | called carbonyl |
    >
    > 5. Indicate the number of ligands by the prefixes:
    > | two: | di | or | bis |
    > |------|----|----|-----|
    > | three: | tri | | tris |
    > | four: | tetra | | tetrakis |
    > | five: | penta | | pentakis |
    > | six: | hexa | | hexakis |
    >
    > 6. If the complex is anionic, add the suffix -ate to the metal.
    >
    > 7. Denote the oxidation state of the metal by Roman numerals in parentheses following the name of the metal.

Use the prefixes in the second column, with the ligand name in parentheses, if confusion would result from the use of the simpler prefixes.

5.   The common size for a chelate ring is five or six atoms in the ring—although there are many examples of other ring sizes.

6.   If there are two donor atoms in a molecule pointing away from each other, then the ligand cannot chelate using these donor atoms. Simple examples are the cyanide ion CN^- and the thiocyanate ion SCN^-, in which the lone pairs on the terminal atoms point 180° in opposite directions. Such ligands can bridge two metal atoms.

7.   If a monatomic ion, like a chloride ion for example, has more than one lone pair of electrons, then it can donate pairs to different metal atoms (in different directions). The atom or ion acts as a bridging ligand and would be bidentate.

8.
Isomerism
- Structural (several varieties all involving a different set of metal-ligand bonds in the complex)
- Stereo
  - Geometrical (cis-trans, fac-mer)
  - Optical

9.   In valence bond theory the bond between the ligand and the metal is considered to be a covalent bond formed by the donation of a pair of electrons from the ligand to the metal. It is a Lewis base–Lewis acid interaction. At a fairly simple level it is a satisfactory picture of the ground state (lowest energy state) of the coordination compound. Hybridization of the metal orbitals can be used to rationalize the shape of the complex. Crystal field theory is an electrostatic approach to the bonding that is far more successful than valence bond theory in explaining the magnetic and spectroscopic properties of the metal complexes. It considers the effect of the electrostatic field of the ligands upon the energies of the electrons in the d orbitals of the metal. However, this theory assumes that there is no covalent bonding (sharing of electrons) between the metal and the ligand. The third theory, molecular orbital theory, incorporates the best features of both valence bond theory and crystal field theory.

10.   A color wheel illustrates the relationship between the color absorbed and the color transmitted by a substance. If a color is extracted from the visible spectrum, then what remains is the complement of that color. Complementary colors are opposite one another on the color wheel.

See Figure 23.33 on page 1079 in the text.

11.   When a metal ion is approached by a set of ligands, repulsion between the electrons on the ligands and the electrons of the metal atom or ion causes those orbitals pointing toward the ligands to be pushed higher in energy. The d orbitals are split in energy. In an octahedral complex, when the six ligands approach along the three cartesian axes, x, y, and z, the d_{z^2} and $d_{x^2-y^2}$ orbitals are pushed higher relative to the d_{xy}, d_{xz}, and d_{yz} orbit-

als. As an octahedral ligand field gets stronger, the d orbitals are split more in energy and eventually the system is more stable with the electrons paired in the lower set of three d_{xy}, d_{xz}, d_{yz} orbitals. The crystal field stabilization energy is then sufficient to overcome the energy required to pair the electrons. The paramagnetism of the complex will decrease (and may become diamagnetic).

12. a. When a polydentate ligand forms coordinate bonds to the same metal ion, it forms a ring of atoms called a chelate ring.

  b. A ligand forms a coordinate bond with a metal ion, supplying both electrons for the bond. It is an electron pair donor (a Lewis base).

  c. A ligand that possesses more than one lone pair of electrons, often on different donor atoms is called a polydentate ligand.

  d. When a group of ligands form coordinate bonds to a metal ion (or atom), the result is a coordination compound.

  e. The spectrochemical series is a series of ligands ordered by their ability to split the d orbitals in energy. The relative ordering is derived from the analysis of the spectra of the complexes—hence the name spectrochemical.

  f. The pairing energy is the energy required to pair electrons in a single orbital. According to Hund's rule of maximum spin multiplicity, the lowest energy state is one in which the electrons are spread out among the orbitals in a degenerate set as much as possible. If the crystal field splitting is sufficiently large so that the stabilization energy exceeds the pairing energy, the low spin state becomes the more stable.

  g. Degenerate means equal in energy; degeneracy indicates the number of orbitals of equal energy. For example, the degeneracy of a set of d orbitals in the absence of a ligand field is 5—they are all equal in energy. In the presence of an octahedral ligand field, this degeneracy is "lost".

Study Questions and Problems

Iron rusts from disuse; stagnant water loses its purity and in cold weather becomes frozen; even so does inaction sap the vigor of the mind.

Leonardo da Vinci
(1452-1519)

1. Name the following coordination compounds:

  a. $Na_2[Pt(CN)_4]$

  b. $K_3[Fe(C_2O_4)_3] \cdot 3H_2O$

  c. $[Co(NH_3)_5Cl]Cl_2$

  d. $[Fe(OH)(H_2O)_5]Cl_2$

  e. $[Cr(en)_3]Br_3$

  f. $NH_4[PtCl_3(NH_3)]$

2. Write formulas for the following coordination compounds:

  a. hexaaquanickel(II) sulfate

  b. sodium tetrachlorocuprate(II)

  c. hexaamminechromium(III) nitrate

  d. pentacarbonylmanganese(0)

3. Suppose you prepared two octahedral cobalt(III) complexes, one containing one dimethylamine ligand and five ammonia ligands and the other containing two methylamine ligands and four ammonia ligands. How would you name them?

4. How many geometrical isomers of $[Co(NH_3)_3Cl_2Br]$ are there? Draw them.

5. A series of nickel complexes vary in color as follows:

   $[Ni(H_2O)_6]^{2+}$ green

   $[Ni(NH_3)_6]^{2+}$ blue

   $[Ni(en)_3]^{2+}$ violet   en = ethylenediamine

   $[Ni(dmg)_2]$ red   dmg = dimethylglyoxime anion

   $[Ni(CN)_4]^{2-}$ yellow

   Even though the geometry of the complex changes from octahedral to square-planar, this series represents quite well the magnitude of the splitting of the d orbitals by the various ligands and the color reflects the relative strengths of the ligands. Using a color wheel, determine the approximate wavelength being absorbed, and hence the order of these ligands in the spectrochemical series.

6. Why, if the crystal field is sufficiently strong, is the complex of a cobalt(III) ion diamagnetic, whereas if the field is weaker, the complex is paramagnetic?

7. A four-coordinate complex of nickel(II) $[NiX_2Y_2]^{2-}$, where X and Y are two different monodentate anions, can be prepared as two different stereoisomers. Is the complex diamagnetic or paramagnetic? Explain.

Answers to Study Questions and Problems

1.   a. $Na_2[Pt(CN)_4]$ sodium tetracyanoplatinate(II)

      b. $K_3[Fe(C_2O_4)_3] \cdot 3H_2O$ potassium tris(oxalato)ferrate(III) trihydrate

      c. $[Co(NH_3)_5Cl]Cl_2$ pentaamminechlorocobalt(III) chloride

      d. $[Fe(OH)(H_2O)_5](NO_3)_2$ pentaaquahydroxoiron(III) nitrate

      e. $[Cr(en)_3]Br_3$ tris(ethylenediamine)chromium(III) bromide

      f. $NH_4[PtCl_3(NH_3)]$ ammonium amminetrichloroplatinate(II)

2.   a.   hexaaquanickel(II) sulfate $[Ni(H_2O)_6]SO_4$

      b.   sodium tetrachlorocuprate(II) $Na_2[CuCl_4]$

      c.   hexaamminechromium(III) nitrate $[Cr(NH_3)_6](NO_3)_3$

      d.   pentacarbonylmanganese(0) $[Mn(CO)_5]$

3. $[Co(NH_3)_5((CH_3)_2NH)]^{3+}$ pentaammine(dimethylamine)cobalt(III)

      $[Co(NH_3)_4(CH_3NH_2)_2]^{3+}$ tetraamminebis(methylamine)cobalt(III)

This illustrates why there is the second series of prefixes for denoting the number of ligands. The use of the prefix di- in the name for the second complex would cause confusion between the ligand methylamine and the ligand dimethylamine in the first complex.

4.   Geometrical isomers of $[Co(NH_3)_3Cl_2Br]$:

Make some models of these isomers and check the answer.

The three ammonia ligands can be fac– or mer–; the two chloro– ligands can be cis– or trans–. However, if the ammonia ligands are fac–, then the chloro– ligands cannot be trans–, which limits the number of possible isomers to three.

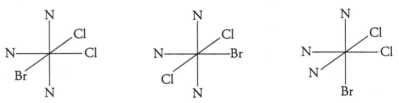

5. $[Ni(H_2O)_6]^{2+}$ green   complement: red

      $[Ni(NH_3)_6]^{2+}$ blue (cyan)   orange

      $[Ni(en)_3]^{2+}$ violet   yellow   increasing

      $[Ni(dmg)_2]$ red   green   frequency

      $[Ni(CN)_4]^{2-}$ yellow   violet

Spectrochemical series:

$$CN^- \quad > \quad dmg \quad > \quad en \quad > \quad NH_3 \quad > \quad H_2O$$

Note that ligands in the spectrochemical series are arranged in order of the Group number of their donor atoms: C > N > O > halide

6.   In a strong octahedral ligand field there is a large separation between the two sets of d orbitals (the splitting is large). A considerable amount of energy is gained if the electrons are in the lower set instead of being spread out over all five orbitals. However, some energy is required to pair the electrons. So there is a competition between the ligand field stabilization energy and the pairing energy. If the crystal field is sufficiently strong the stabilization energy is sufficient to overcome the pairing energy and the cobalt(III) is diamagnetic, whereas if the field is weaker, there is insufficient energy to pair the electrons and the complex is paramagnetic.

7.   If two different stereoisomers exist, then the complex must be square planar. Cis-trans geometrical isomerism is impossible in a tetrahedral geometry. If the metal is nickel(II), then the complex must be diamagnetic. All square planar complexes of nickel (and platinum and palladium) are diamagnetic (all the d^8 electrons are paired), whereas tetrahedral complexes of nickel are paramagnetic.

Build some models and prove that this is true.

CHAPTER 24

Introduction

Chemistry is most often concerned with the behavior of the electrons, particularly the valence electrons, of an atom. It is the electrons of an atom that are responsible for the chemical properties of the element. However, somewhere between the realms of chemistry and physics lies what is called nuclear chemistry—the study of the nuclei of elements. This study includes the transmutation of elements (changing one element into another), the radioactive decay of elements, preparing artificial isotopes of elements for tracer studies and medical imaging experiments, and the preparation of new elements beyond the 112 now known.

Contents

24.1   The Nature of Radioactivity

There are three types of nuclear radiation: alpha(α), beta(β), and gamma(γ). **Alpha** radiation is composed of helium nuclei and **beta** radiation is composed of electrons; both are ejected at high energies from radioactive nuclei. **Gamma** radiation is electromagnetic radiation at the high frequency end of the spectrum. Alpha(α) radiation is stopped relatively easily. Beta(β) radiation, also particulate in nature, is stopped by thin sheets of metal. Gamma(γ) radiation, on the other hand, is nonparticulate and is the most penetrating, requiring thick layers of lead or concrete for effective shielding.

24.2   Nuclear Reactions

Radioactivity is the result of the change of the isotope of one element into the isotope of another element. This is the **transmutation** of an element—a goal sought by alchemists since the beginning of chemistry. The particles in the nucleus—the neutrons and protons— are called **nucleons**. In a nuclear reaction the total number of nuclear particles (nucleons) is conserved. In a nuclear reaction, there can be no change in the total mass number (the sum of the mass numbers of all the particles); nor, in order for the charge to be conserved, can

Alpha(α) radiation was discovered and characterized by Thomson and Rutherford.
Beta(β) radiation was shown to be high energy electrons by Becquerel.
Gamma(γ) radiation was discovered by Villard.

Other particles may be emitted or captured by nuclei. For example, many artificial isotopes are prepared by neutron capture.

Gamma(γ) radiation results when an excited nucleus decays to its ground state.

Transmutation by alpha(α) emission was first proposed by Rutherford and Soddy in 1903. It was Soddy who coined the word "isotope".

there be any change in the total atomic number. For example consider the decay of uranium–234 to thorium–230:

$$^{234}_{92}\text{U} \quad \rightarrow \quad ^{230}_{90}\text{Th} + ^{4}_{2}\text{He}$$

The sum of the superscripts (the mass numbers) must be the same on both sides of the equation 234 = 230 + 4, and the sum of the subscripts (the atomic numbers) must be the same on both sides of the equation 92 = 90 + 2.

Often, in a nuclear decay reaction, the daughter isotope formed is also radioactive and decays in its turn. This will continue until a stable isotope is reached. The series of decay reactions is called a **decay series**. These decay series can be illustrated clearly in a plot of mass number *vs.* atomic number.

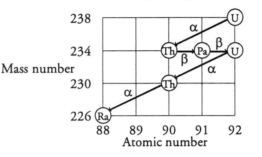

The beginning of the uranium–238 decay series is shown.

Alpha(α) emission reduces the mass number by 4 and the atomic number by 2.

Beta(β) emission changes a neutron into a proton, so it leaves the mass number unchanged and increases the atomic number by 1.

Other nuclear reactions are positron emission, electron capture, neutron or proton capture or emission, etc. However, in natural decay processes for the heavy elements the two nuclear reactions are α and β emission.

24.3   Stability of Atomic Nuclei

Some nuclei are stable, some are unstable; how are they different? There are relatively few stable nuclei—relatively few stable combinations of protons and neutrons in the nucleus. The presence of the neutrons is necessary to bind the nucleus together into a stable unit.

For light elements up to calcium, the stable isotopes have equal numbers of protons and neutrons, sometimes one additional neutron. Beyond calcium, the neutron-proton ratio increases and the **band of stability** deviates more and more from the straight (atomic number Z = mass number A) line. All elements beyond bismuth are unstable and are radioactive—no matter how many neutrons are present. The rate of disintegration increases as the nucleus becomes more massive. It is interesting that nuclei with even atomic numbers are more stable than those with odd atomic numbers. Those with an even number of protons and neutrons are the most abundant. Only five stable isotopes have an odd number of both protons and neutrons.

Heavy radioactive nuclei can move into the zone of stability from both directions by α emission (from the right) and β emission (from the left). For the lighter radioactive isotopes with too few neutrons, positron emission or electron capture moves the nuclei from the right into the zone of stability.

The force that binds the protons (and neutrons) together in the nucleus must be very powerful to overcome the electrostatic repulsion between the protons. The energy involved in the binding together of the nucleus can be calcu-

See Figure 24.2 on page 1093 of the text.

There are three naturally occurring radioactive series:

Uranium-238 ending at Pb-206
Uranium-235 ending at Pb-207
Thorium-232 ending at Pb-208

The positron is a particle having the same mass as an electron but a charge of +1. It was discovered by Anderson in 1932.

See, for example, the chart in Figure 24.4 on page 1097 of the text.

There are (magic) numbers for the protons and neutrons in particularly stable nuclei: 2, 8, 20, 28, 50, 82, and then 114 (for protons), and 126 and 184 (for neutrons). These numbers are reminiscent of the occupancy of the electron shells outside the nucleus.

lated by adding together the masses of the individual nucleons and comparing the total to the actual mass of the nucleus. The **mass defect** (difference) is the mass that has been converted into the **binding energy**. It is Einstein's equation $E=mc^2$ that indicates the relation between mass and energy. In this equation c is the velocity of electromagnetic radiation equal to 3.00×10^8 ms^{-1} and its magnitude illustrates that a little mass is equivalent to a lot of energy.

The greater the binding energy per nucleon, the greater the stability of the nucleus. The nucleus with the greatest stability is iron–56. Heavier nuclei undergo **fission** to move toward iron–56; lighter nuclei undergo **fusion** to move toward iron–56.

See Figure 24.5 on page 1101 of the text.

24.4 Rates of Disintegration Reactions

The rates of the disintegration of radioactive nuclei vary widely, from fractions of a second to billions of years. Because radioactive decay is a first-order process, the stability of an isotope is usually expressed as a half-life.

Recall from Chapter 15 that the half-life for a first order process is independent of the quantity of substance initially present.

The rate of nuclear decay is often described in terms of the activity (A) of the sample—it is the number of disintegrations per unit time and therefore depends upon the number of radioactive nuclei in the sample. If the activity is measured at time 0, and is found to be A_0, and the activity is measured again at time t and is found to be A, then the ratio of A/A_0 must equal the ratio of N/N_0, where N and N_0 are the numbers of radioactive nuclei at time 0 and at time t.

The first order integrated rate law $\ln(N/N_0) = -kt$ relates the change in the number of radioactive nuclei to the time t. Measuring the activity of a sample at different times allows calculation of the rate constant k and the half-life of the isotope.

The age of some object can sometimes be determined based upon the decay of a radioactive isotope incorporated into the material. Carbon–14 dating is the most well known method. The concentration of carbon–14 in the atmosphere is relatively constant (formed in the upper atmosphere as a result of cosmic radiation). Growing plants absorb CO_2 containing carbon–14, but when the plant dies, the absorption ceases and the carbon–14 slowly decays. By measuring the counts per gram of carbon, the age of the material can be calculated.

The half-life of carbon–14 is 5730 years.

24.5 Artificial Transmutations

The first transmutation of an element occurred in 1919 when Rutherford bombarded nitrogen with α particles to produce an isotope of oxygen and a proton:

$$^4_2\text{He} + ^{14}_7\text{N} \rightarrow ^{17}_8\text{O} + ^1_1\text{H}$$

Because α particles are charged they are repelled by positive nuclei and experiments involving bombardment by α particles were not very successful. Subsequent experiments involving neutrons (that have no charge) proved much more successful. Virtually all nuclei undergo neutron capture with subsequent emission of γ radiation as the nucleus returns to its ground state. This reaction is termed an (n,γ) reaction and is the method used to prepare most radioiso-

The neutron bombardment is readily available in nuclear reactors.

Cyclotrons and linear accelerators are used to accelerate charged particles to sufficient energies to overcome the electrostatic repulsion of nuclei.

Neutron-absorbing control rods are made from boron or cadmium.

Rods made from graphite are used to moderate the reaction by slowing down the neutrons. Water is also used as a moderator. It was the graphite rods that caught fire at Chernobyl.

Slowing down the neutrons is beneficial. Slow neutrons do not cause the fission of fissionable isotopes such as uranium–238, they are more easily absorbed by the fissile uranium–235, and they allow the control rods to work more efficiently because the neutrons are absorbed more easily.

Weapon-quality uaranium has the uranium–235 concentrated to over 90%.

topes used in medicine and chemistry. Many of the first examples of the transuranic elements were prepared by initial (n,γ) reactions in nuclear reactors.

24.6   Nuclear Fission

When a uranium–235 nucleus absorbs a neutron, it splits into two smaller nuclei. In doing so, two important things happen. First, a lot of energy is released corresponding to the mass loss, and second, three neutrons are ejected. This second characteristic means that an accelerating chain reaction is possible as these three neutrons are captured by three other nuclei which in turn eject nine neutrons, and so on...

$$^{235}_{92}U + ^{1}_{0}n \rightarrow ^{236}_{92}U \rightarrow ^{141}_{56}Ba + ^{92}_{36}Kr + 3\,^{1}_{0}n \quad \Delta E = -2 \times 10^{10}\,kJ\,mol^{-1}$$

In an uncontrolled fission reaction, when the concentration of uranium–235 is sufficiently high and the size of the sample exceeds the critical mass, an explosion results. However, if the uranium–235 is not so concentrated and the fission is dampened by inserting neutron-absorbing rods into the reactor, the process can be controlled. Uranium–235 and plutonium–239 are two isotopes for which fission can be induced by slow neutrons. Naturally occurring uranium contains only 0.72% of the fissile U–235 isotope, and this must be enriched (concentrated) for use in nuclear power plants or nuclear weapons.

Nuclear fission power plants supply electrical energy without exhausting valuable fossil fuels and without polluting the atmosphere with smoke, ash, and acidic oxides. They do however pose a serious problem in what is to be done with the highly radioactive nuclear waste materials. And accidents do occur in nuclear reactors; the Chernobyl reactor, when it caught fire, contaminated surrounding land with long-lived radioactive isotopes of plutonium and cesium. This land will be uninhabitable for thousands of years.

24.7   Nuclear Fusion

Movement toward iron–56 can be achieved by the fission of heavy nuclei or the fusion of light nuclei. Both release energy. For example the fusion of two hydrogen nuclei to produce one helium nucleus:

$$4\,^{1}_{1}H \quad \rightarrow \quad ^{4}_{2}He + 2\,^{0}_{1}\beta \quad \Delta E = -2.5 \times 10^{9}\,kJ$$

High temperatures (~100,000,000°C) are required to initiate a fusion reaction. For example, to initiate a fusion bomb (a hydrogen bomb), a fission explosion is set off first. It is not yet possible to control a fusion process to harness the energy for peaceful uses. The plasma must be contained and it must be heated to the necessary (very high) temperature. Fusion processes have great promise in the abundance of the starting material (hydrogen) and in the greatly reduced radiation hazard of the products.

24.8 Radiation Effects and Units of Radiation

Several units have been developed to describe quantities of radiation. The **röntgen** is a measure of radiation exposure characterized by the amount of radiation produced in air by x-rays and γ-rays. The **rad** measures the radiation dose to tissue rather than air but is similar in magnitude. One rad represents a dose of 1×10^{-5} J per gram of tissue. Different types of radiation have different biological effects, so the situation is not so simple. To account for these differences, the unit **rem** is used. This is the dose in rads multiplied by a factor that is proportional to how lethal the radiation is (the so-called biological effectiveness): 1 for β and γ radiation, 5 for low energy neutrons and protons, 10 to 20 for α particles and high energy neutrons and protons. The **curie** (Ci) is a unit of radioactivity and corresponds to 3.7×10^{10} decay events (disintegrations) per second.

There is always some background radiation; it is estimated to be about 200 millirem/year and comes form a variety of sources.

rad stands for radiation absorbed dose.

rem stands for röntgen equivalent man.

There are other units, specifically the SI units:

becquerel (Bq) = 1 dps so 1 Ci = 3.7×10^{10} Bq

gray (Gy) = 1 J kg^{-1} so 1 Gy = 100 rad

sievert (Sv) = 100 rem

Sources and amounts of background radiation are listed in Table 24.2 on page 1120 of the text.

Review Questions

1. Describe the following types of radiation and their penetrating power: alpha(α), beta(β), and gamma(γ).

2. Describe the following nuclear reactions; specify the change in the nucleus, the change in the mass number and the change in the atomic number: alpha(α) emission, beta(β) emission, neutron emission, neutron capture (n,γ), positron emission, electron capture, gamma(γ) emission.

3. What is the transmutation of an element?

4. In any of the three natural decay series, how can you easily calculate the number of alpha(α) particles and beta(β) particles emitted from beginning to end. Consider for example the actinium series from uranium–235 to lead–207, or the uranium series from uranium–238 to lead–206.

5. Knowing the magic numbers, 2, 8, 20, 28, 50, and 82 for both protons and neutrons, deduce some particularly stable nuclei.

6. Describe the principles involved in writing an equation for a nuclear reaction.

7. Describe what happens to the mass number and the atomic number of a nucleus when a) an electron is emitted and b) a positron is emitted.

8. What is the nuclear binding energy and what is its relationship to the mass defect?

9. Describe the principles of carbon–14 dating. What are its limitations?

10. What are the requirements for a nuclear fission explosion?

11. What is a PET scan?

12. Tabulate various sources of background radiation. What is the most abundant source of radiation?

Answers to Review Questions

1. Alpha(α) (helium nuclei) radiation is stopped relatively easily—by a layer of clothing and skin or a few sheets of paper for example. Beta(β) radiation (high-speed electrons) is more penetrating and requires a thin sheet of metal to stop it. Gamma(γ) radiation is the most penetrating, passing easily through the body, and it requires thick layers of lead or concrete to provided effective shielding.

2.

Nuclear reaction	change in the nucleus	mass number	atomic number
alpha(α) emission	loss of $^4_2$He $(+\gamma)$	-4	-2
beta(β) emission	loss of $^{\;\;0}_{-1}$e $(+\gamma)$	none	$+1$
neutron emission	loss of $^1_0$n $(+\gamma)$	-1	none
neutron capture (n,γ)	gain of $^1_0$n $(+\gamma)$	$+1$	none
positron emission	change proton into neutron	none	-1
electron capture	change proton into neutron	none	-1
gamma(γ) emission	excited state to ground state	none	none

3. The transmutation of an element is the conversion of one element into another. It requires a change in the number of protons in the nucleus by nuclear reaction.

There is no chemical means to change one element into another.

4. A decrease in mass only occurs with alpha(α) emission. So the number of alpha(α) particles can be calculated from the change in mass number. The number of beta(β) particles emitted can be calculated from the change in atomic number required after 2 units have been subtracted for each alpha(α) particle. For example:

actinium series: $^{235}_{92}$U to $^{207}_{82}$Pb — mass number change = 28
therefore 7 α particles
therefore -14 protons
actual decrease = -10 ($92 \rightarrow 82$)
therefore 4 β particles

actinium series: $^{238}_{92}$U to $^{206}_{82}$Pb — mass number change = 32
therefore 8 α particles
therefore -16 protons
actual decrease = -10 ($92 \rightarrow 82$)
therefore 6 β particles

5. Some examples:
   2 protons, 2 neutrons: alpha(α) particle ${}^{4}_{2}\text{He}$
   8 protons, 8 neutrons: oxygen–16
   20 protons, 20 neutrons: calcium–40
   50 protons: tin
   82 protons: lead

6. In a nuclear reaction, there can be no change in the total mass number or the total atomic number. When the superscript (the mass number) and the subscript (the atomic number) are included in the symbols for all the participants in a nuclear reaction, then the balancing of the reaction is straightforward: the sum of the mass numbers on one side must equal the sum of the mass numbers on the other side, and the sum of the atomic numbers on one side must equal the sum of the atomic numbers on the other side.

7. When an electron is emitted, the mass number stays the same, but the atomic number increases by 1. Effectively, a neutron converts to a proton. When an positron is emitted, the mass number again stays the same, but the atomic number decreases by 1. For example:

$$ {}^{207}_{84}\text{Po} \quad \rightarrow \quad {}^{207}_{83}\text{Bi} \;+\; {}^{0}_{1}\beta $$

8. The nuclear binding energy is the energy theoretically released when all the subatomic particles making up a nucleus come together to form that nucleus. It is the energy that would have to be supplied to break up the nucleus into the individual particles. It is often expressed as the binding energy per nucleon.
   The mass defect is the difference in the mass between the mass of the nucleus and the sum of the masses of the individual nucleons making up that nucleus. It equals the mass lost when the nucleus is formed from its constituent nucleons.
   The relation between the binding energy and the mass defect is given by Einstein's equation $E = mc^2$.

9. The age of an object can be determined by the measurement of the relative quantity of carbon–14 present. The concentration of carbon–14 in the atmosphere is relatively constant—it is formed in the upper atmosphere as a result of cosmic radiation and is incorporated into CO_2 molecules. Growing plants absorb CO_2 containing the carbon–14 on a continuous basis, but when the plant dies, the absorption ceases. At this point no further carbon–14 is absorbed and the carbon–14 present in the plant slowly decays. By measuring the activity per gram of carbon, and comparing the activity with a live plant, the age of the object can be calculated. The half-life of carbon–14 is 5730 years and the method works well of objects of approximately that age. The limit at the far end is about 40,000 years because the activity after this length of time has decreased to inaccurate levels. At the near end the limit is about 100 years, since in 100 years not much decay has taken place.

10. The nuclear reaction must produce more neutrons than it absorbs; this leads to the potential for a chain reaction. However, the neutrons that are produced must be captured by other nuclei; so the concentration of the fissile material must be high enough and the sample size must be large enough. The size of the sample when the chain reaction becomes self-sustaining is called the critical mass.

11. Positron emission tomography (PET) is a technique in which a positron emitter like carbon–11, oxygen–15, nitrogen–13, or fluorine–17 is used to image human tissue with a resolution not possible with x–rays. For example, fluorine–17 is incorporated into a compound that is preferentially absorbed by the part of the body that needs to be examined. (For example a tumor.) The positrons that are given off are quickly annihilated by electrons and two γ–rays are produced. These γ–rays are emitted 180° apart and are detected by scanners. The position of the annihilation must lie on a straight line between the scanners and very precise imaging is possible. The positron emitter has a short half-life and the compound must be freshly prepared. The half-life of fluorine–17, for example, is 110 minutes.

12.

Cosmic radiation (depends upon altitude)	50 mrem
Medical x–rays	50 mrem
The earth	47 mrem
Elements in your body	21 mrem
Radiotherapy	10 mrem
Inhaled (for example radon)	5 mrem
Fallout from nuclear testing	4 mrem
Stone, rock, or concrete house construction	3 mrem
Industrial waste, watches, TV tubes	2 mrem
Internal medical diagnosis	1 mrem
Nuclear power industry	1 mrem

I am become death, the shatterer of worlds.

J. Robert Oppenheimer (1904-1967)

quoted from the Hindu text , the Bhagavad-Gita, when the first atomic bomb exploded at Trinity on July 16, 1945.

Study Questions and Problems

1. Write equations for the following reactions:

   a. The production of ^{56}Mn by neutron bombardment of ^{59}Co.

   b. The production of the new element dubnium by bombarding atoms of $^{249}_{98}$Cf with $^{15}_{7}$N nuclei.

   c. The bombardment of $^{27}_{13}$Al with α particles to produce $^{30}_{15}$P and the subsequent decay of $^{30}_{15}$P to silicon by positron emission.

   d. The conversion of potassium–40 to argon–40 by electron capture.

   e. The production of carbon–14 in the upper atmosphere by neutron bombardment of $^{14}_{7}$N.

2. Why are radioactive isotopes of intermediate half-lives more hazardous than radioisotopes with long half-lives or very short half-lives?

3. Predict the radioactive decay of the following radioactive isotopes based upon their position relative to the band of stability:
   a. carbon–14
   b. krypton–87
   c. thorium–230
   d. phosphorus–29
   e. europium–145

4. Calculate the quantity of energy released when one atom of uranium–235 is split by the impact of a neutron into barium–142 and krypton–92 releasing one additional neutron. Then calculate the quantity of energy released when one gram of uranium–235 undergoes fission in the same way.

$$\ce{^{235}_{92}U} + \ce{^{1}_{0}n} \rightarrow \ce{^{142}_{56}Ba} + \ce{^{92}_{36}Kr} + 2\,\ce{^{1}_{0}n}$$

   Atomic masses:

   $\ce{^{235}_{92}U}$ 235.4 amu

   $\ce{^{142}_{56}Ba}$ 141.92 amu

   $\ce{^{92}_{36}Kr}$ 91.92 amu

   $\ce{^{1}_{0}n}$ 1.0087 amu

5. Charcoal retrieved from the site of Stonehenge in England has a carbon–14 activity 62.0% that of carbon–14 in living plants. Assuming that the abundance of carbon–14 in the atmosphere has remained moreorless constant for past few thousand years, how old is the charcoal? The half-life of carbon–14 is 5730 years.

6. Strontium–90 is a hazardous isotope present in the fallout from nuclear explosions. If 1.00 gram of strontium–90 diminishes to 0.786 gram in 10 years, as measured by its activity, what is the half-life of strontium–90?

7. Technicium–99 is prepared for medical imaging experiments by neutron bombardment of molybdenum–98. The unstable molybdenum–99 produced by this bombardment decays by β emission to an excited technicium–99, which in turn relaxes to its ground state by γ emission. The technicium–99 is itself radioactive and decays by β emission. Write equations for this sequence of reactions.

8. If a radioactive isotope lies above the band of stability, which decay process would lead it toward the band, that is, form a more stable isotope?

9. Iron–56 is the isotope with the highest binding energy per nucleon. Calculate the binding energy per nucleon from the following data:

    mass of proton = 1.007275 amu
    mass of neutron = 1.008666 amu
    mass of electron = 0.0005486 amu
    atomic mass of iron–56 = 55.9349 amu

    Express your answer in units of J/nucleon, in MeV/nucleon, and in kJ/mol nucleon.

10. a. The fission of an americium–244 isotope produces iodine–134 and molybdenum–107. How many neutrons are also produced in each fission event?

    b. The fission of a californium–252 nucleus produces one barium–142 nucleus and one molybdenum–106 nucleus. How many neutrons are produced in this reaction?

11. Moderator rods and control rods in a nuclear fission reactor serve different functions. What are their functions and how are they different?

Answers to Study Questions and Problems

1. a. $^{59}_{27}Co + ^{1}_{0}n \rightarrow ^{56}_{25}Mn + ^{4}_{2}He$

    b. $^{249}_{98}Cf + ^{15}_{7}N \rightarrow ^{260}_{105}Db + 4 ^{1}_{0}n$

    c. $^{27}_{13}Al + ^{4}_{2}He \rightarrow ^{30}_{15}P + ^{1}_{0}n$ and $^{30}_{15}P \rightarrow ^{30}_{14}Si + ^{0}_{1}\beta$

    d. $^{40}_{19}K + ^{0}_{-1}e \rightarrow ^{40}_{18}Ar$

    e. $^{14}_{7}N + ^{1}_{0}n \rightarrow ^{14}_{6}C + ^{1}_{1}H$

2. Radioactive isotopes of intermediate half-lives are more hazardous because neither do they decay so slowly that the radiation is very low nor do they decay so quickly that they are around for a very short time.

3. Some idea of the position of the band of stability can be obtained from the atomic masses of the elements (since the atomic masses represent the average masses of the naturally occurring isotopes). Remember also that the neutron to proton ratio increases to greater than one above calcium.

    a. carbon–14   n/p ratio too high   emit $^{0}_{-1}\beta$
    b. krypton–87   n/p ratio too high   emit $^{0}_{-1}\beta$ or $^{1}_{0}n$
    c. thorium–230   naturally radioactive emit alpha(α)
    d. phosphorus–29   n/p ratio too low   emit $^{0}_{1}\beta$ or electron capture
    e. europium–145   n/p ratio too low   emit $^{0}_{1}\beta$ or electron capture

4. Mass before disintegration:

   $^{235}_{92}U + ^{1}_{0}n$: $235.04 + 1.0087 = 236.05$ amu

   Mass after disintegration:

   $^{142}_{56}Ba + ^{2}_{36}Kr + 2\,^{1}_{0}n = 141.92 + 91.92 + 2 \times 1.0087 = 235.86$ amu

   Mass converted into energy $= 0.19$ amu

   $\qquad\qquad\qquad\qquad\qquad = 0.19 \times 1.66 \times 10^{-27} = 3.15 \times 10^{-28}$ kg

   Equivalent energy = mc^2 $\quad = 3.15 \times 10^{-28}$ kg $\times (3.0 \times 10^8\ ms^{-1})^2$

   $\qquad\qquad\qquad\qquad\qquad = 2.84 \times 10^{-11}$ J for one event

   For the fission of one gram of uranium–235:

   Energy $= 2.84 \times 10^{-11}$ J $\times (6.022 \times 10^{23}$ mol^{-1} / 235.04 g mol$^{-1})$

   $\qquad\quad = 7.27 \times 10^{10}$ J g^{-1}

   $\qquad\quad = 7.27 \times 10^{7}$ kJ g^{-1}

5. The half-life for a first order decay $t_{1/2} = 0.693/k$, or $k = 0.693/t_{1/2}$.

   $t_{1/2} = 5730$ years

   so $\quad k = 1.210 \times 10^{-4}$.

   $\ln(N_0/N) = kt$

   so $\quad \ln(100/62.0) = 1.210 \times 10^{-4} \times t$

   $t = 3950$ years.

6. $\ln(N_0/N) = kt$

   $\ln(1.00/0.786) = kt$

   $t = 10$ years

   so $\quad \ln(1.00/0.786) = k \times 10$

   $k = 0.0241$

   $t_{1/2} = 0.693/k$

   so $\quad t_{1/2} = 28.8$ years.

7. $^{98}_{42}Mo + ^{1}_{0}n \rightarrow ^{99}_{42}Mo$

   $^{99}_{42}Mo \rightarrow ^{99}_{43}Tc^* + ^{0}_{-1}\beta$ \qquad *indicates an excited state

   $^{99}_{43}Tc^* \rightarrow ^{99}_{43}Tc + \gamma$

   $^{99}_{43}Tc \rightarrow ^{99}_{44}Ru + ^{0}_{-1}\beta$

8. The neutron to proton ratio is too high; so any decay process that decreases the neutron count, or increases the proton count, or changes a neutron into a proton, will make the isotope more stable. A good candidate is beta (β) emission.

9. Atomic mass of iron–56 = 55.9349 amu

   Mass of the iron–56 nucleus = $55.9349 - 26 \times 0.0005486$ amu

   $\qquad\qquad\qquad\qquad\qquad = 55.9206$ amu

Sum of the individual masses of the nucleons:
Protons: 26 × 1.007275 amu = 26.1892 amu
Neutrons: 30 × 1.008666 amu = 30.2600 amu
Total: 56.4491 amu

Mass defect = 56.4491 − 55.9206 amu = 0.5285 amu
          = 0.5285 amu × (1.6605 × 10^{-27} kg/1 amu) = 8.777 × 10^{-28} kg

Equivalent energy = mc^2 = 8.777 × 10^{-28} kg × (2.998 × 10^8 ms^{-1})^2
          = 7.888 × 10^{-11} J for 56 nucleons

For one nucleon:
          = 7.888 × 10^{-11} J/56 nucleons = 1.409 × 10^{-12} J/nucleon
          = 1.41 × 10^{-12} J/nucleon /1.6022 × 10^{-19} J/eV
          = 8.79 × 10^6 eV/nucleon
          = 8.79 MeV/nucleon.

For one mole of nucleons:
          = 8.48 × 10^8 kJ/mol of nucleons

10. a. Americium has an atomic number = 95, which is the sum the atomic numbers of iodine (53) and molybdenum (42). However, the mass number of americium–244 exceeds the sum of the mass numbers of iodine–134 and molybdenum–107 by 3, so three neutrons must be emitted in each event.

     b. Californium has an atomic number = 98, which is the sum the atomic numbers of barium (56) and molybdenum (42). However, the mass number of californium–252 exceeds the sum of the mass numbers of barium–142 and molybdenum–106 by 4, so four neutrons must be emitted in each event.

11. Moderator rods slow down the neutrons so that they are more readily absorbed by the uranium–235. The control rods are made from neutron absorbers and reduce the rate of fission—they control the reaction. The moderator rods (for example graphite) improve the efficiency of the control rods because slowing down the neutrons makes them easier to absorb by the material of the control rods.

EXAMINATION 5

Introduction

This last examination tests your knowledge and understanding of the descriptive chemistry in Chapters 22 through 24 of Kotz & Treichel. The first set of questions are true–false questions and the second set are multiple choice questions. Try the questions before looking at the answers provided at the end of this study guide.

True–false questions

1. Aluminum is the most abundant metal in the earth's crust.

2. The maximum oxidation number of a representative element equals its group number.

3. Only three of the known elements form no compounds.

4. Nitrogen dioxide is paramagnetic.

5. A cyclic metasulfuric acid can be made by the condensation of sulfuric acid.

6. Sulfurous acid is a diprotic acid.

7. The oxidation states of the halogens are always negative.

8. Oxygen O_2, nitrogen N_2, and fluorine F_2, are all paramagnetic.

9. Mirror images are optically active.

10. Optically active complexes are those complexes with intense colors.

11. If the formation constant for the formation of a transition metal complex is higher, then the complex is more stable.

12. Four coordinate complexes are always tetrahedral.

13. Diamagnetic complexes always involve strong ligands.

14. Yellow and blue are complementary colors.

15. A bidentate ligand is a ligand that forms a chelate ring with a metal ion.

16. Alpha (α) radiation is the most penetrating form of natural radiation.

17. Only heavy nuclei are naturally radioactive.

18. It is possible, in a nuclear reaction, to change one element into another.

19. Loss of a β particle does not change the mass number of a nucleus.

20. Stable nuclei always have a number of neutrons at least equal to the number of protons.

21. In a natural decay series, alpha (α) emission increases the neutron/proton ratio.

22. Because the nucleus is positively charged, only positively charged particles are ejected in radioactive decay.

23. In the decay of a radioactive nucleus, the half-life gets progressively longer as the quantity of radioactive isotope is depleted.

24. Uranium-238 decays spontaneously.

25. Fusion nuclear reactions have not yet been achieved.

Multiple choice questions

1. Most industrial hydrogen is used to

   a. synthesize methanol and other alcohols for ultimate conversion to gasolines
   b. produce ammonia in the Haber process
   c. combine with oxygen to produce water in fuel cells
   d. reduce metal oxides in the production of pure metals
   e. produce hot flames for welding
   f. fuel rockets

2. Which of the alkali metals is the least expensive to produce?

   a. Li   b. Na   c. K   d. Rb   e. Cs

3. Sodium hydroxide is produced industrially by the

   a. reaction of sodium with water
   b. hydrolysis of the salts of weak acids such as carbonic acid
   c. electrolysis of fused sodium chloride
   d. electrolysis of concentrated aqueous sodium chloride (brine)
   e. roasting of sodium carbonate (soda ash) followed by reaction with water

4. Quicklime is

   a. calcium fluoride   c. calcium sulfate   e. calcium hydroxide
   b. calcium chloride   d. calcium carbonate   f. calcium oxide

5. The general formula for the nitride of an alkaline earth metal is

   a. MN b. M_3N_2 c. M_2N_3 d. MN_3 e. M_3N

6. Most industrial oxygen is used in

    a. the synthesis of plastics
    b. the oxidation of fuels such as hydrazine in rocket engines
    c. fuel cells
    d. the steel industry to remove nonmetal impurities by oxidation
    e. the production of hot flames for welding
    f. medical treatment

7. What is the shape of the ozone molecule?

    a. linear
    b. trigonal
    c. V-shaped
    d. tetrahedral
    e. trigonal pyramidal
    f. square planar

8. The yellow-brown color of nitric acid is due to the formation of

    a. NH_3
    b. NO
    c. NO_2
    d. HNO_2
    e. N_2H_4

9. Most ammonia is used in

    a. the plastics industry
    b. manufacturing explosives
    c. household cleaners
    d. medical anesthesia
    e. the fertilizer industry

10. Which halogen is the most powerful oxidizing agent?

    a. fluorine
    b. chlorine
    c. bromine
    d. iodine
    e. astatine

11. Which element in Group 6A has the highest electronegativity?

    a. O
    b. S
    c. Se
    d. Te
    e. Po

12. Most sulfur mined by the Frasch process is used in industry for

    a. the synthesis of pharmaceuticals
    b. making gunpowder
    c. the manufacture of sulfuric acid
    d. a fuel in the steel industry
    e. the manufacture of fertilizer
    f. the vulcanization of rubber

13. The most important industrial use of silicon is

    a. in the production of semiconducting devices
    b. in the manufacture of glass
    c. in the production of synthetic aluminosilicate zeolite catalysts
    d. in the production of silicon carbide (carborundum)
    e. in the manufacture of synthetic gems such as zircon and beryl

14. Gangue is

     a. the overburden to be removed before the metal-bearing ore can be mined
     b. an ultimately fatal disease caused by exposure to heavy metals
     c. rock and soil which accompanies a metal-bearing mineral
     d. an ore from which iron can be profitably extracted
     e. the silicate slag produced in the manufacture of iron

15. The process used to convert a sulfide mineral to the easily reduced oxide is called

     a. smelting   c. blasting   e. leaching
     b. flotation   d. reduction   f. roasting

16. Most gold in the US is obtained from

     a. anode sludge produced in the electrolytic purification of copper
     b. cyanide leaching of gold bearing rock formations
     c. residues from the extraction of magnesium from seawater
     d. as a byproduct from the mining of silver
     e. from rivers in the Pacific Northwest

17. The reason why aluminum is not oxidized upon contact with air is because it

     a. forms an invisible oxide film that protects the underlying metal
     b. is too low in the activity series
     c. reacts with nitrogen to form a layer of nitride that protects the metal from oxygen
     d. is mixed with other metals to reduce its activity (and increase its strength)
     e. forms relatively weak bonds with oxygen

18. The most common oxidation states of chromium are

     a. 2   c. 2 and 3   e. 3   g. 4
     b. 3 and 4   d. 4 and 6   f. 2, 3, and 4   h. 2, 3, and 6

19. In the first transition series, the variation in atomic radius is

     a. a steady decrease from scandium to zinc
     b. a steady increase from scandium to zinc
     c. an increase to the center of the series and then a gradual decrease to the end
     d. an initial decrease in size to the center of the series and then a slight increase at the end
     e. a constant radius throughout the series

20. What is the ground state electron configuration for iron?

     a. $[Ar]\ 3d^5\ 4s^2$ c. $[Ar]\ 3d^6\ 4s^2$ e. $[Ar]\ 3d^5\ 4s^1$
     b. $[Ar]\ 3d^7\ 4s^1$ d. $[Ar]\ 3d^6\ 4s^1$ f. $[Ar]\ 3d^8$

21. The name for ammonia when it is used as a ligand in a transition metal complex is

     a. ammino   c. ammonio   e. amino
     b. amido   d. ammine   f. pepto

22. What is the name of the transition metal complex $K_2[NiCl_4]$?

    a. potassium tetrachloronickelate(II)
    b. potassium nickel(II) tetrachloride
    c. dipotassium tetrachloronickel(II)
    d. dipotassium nickel chloride(II)
    e. potassium tetrachloronickelate(III)

23. When one ethylenediamine molecule, two chloride ions Cl^-, and two ammonia molecules coordinate to one cobalt(III) ion in an octahedral complex, how many stereoisomers are possible?

    a. 1   c. 3   e. 5   g. 7
    b. 2   d. 4   f. 6   h. 8

24. The octahedral complex ion $[CoF_6]^{3-}$ is a high-spin or weak-field complex. How many unpaired electrons are there in the complex?

    a. 0   c. 2   e. 4   g. 6
    b. 1   d. 3   f. 5   h. 7

25. In a square planar complex, which orbital(s) is(are) highest in energy?

    a. $d_{x^2-y^2}$ b. d_{z^2} c. d_{xz} and d_{yz} d. d_{xy}

26. Which of the following elementary particles has an atomic number of 2?

    a. alpha particle   c. positron   e. deuteron
    b. proton   d. beta particle   f. neutron

27. In the three natural series of radioactive decay, each successive product is radioactive and decays to another nucleus until a final element is reached. What is this element?

    a. uranium   c. bismuth   e. iron
    b. actinium   d. thorium   f. lead

25. The symbol for a nuclide of the element lead is ^{214}Pb. How many neutrons are there in the nucleus of this isotope?

    a. 214   b. 82   c. 296   d. 132   e. 207

26. Complete the reaction: $^{218}_{84}Po \rightarrow ^{4}_{2}He + ?$

    a. $^{214}_{82}Pb$ b. $^{214}_{83}Bi$ c. $^{222}_{86}Rn$ d. $^{214}_{82}Rn$ e. $^{210}_{84}Po$

27. Which type of nuclear decay results in an increase in the atomic number of the nucleus?

    a. alpha emission   c. positron emission   e. gamma emission
    b. beta emission   d. electron capture   f. proton emission

28. When one radium-226 nucleus decays through a series of stages to polonium-214, what is the total number of alpha particles emitted?

    a. 1   c. 3   e. 5   g. 7
    b. 2   d. 4   f. 6   h. 8

29. The first artificial radioactive isotope was produced in 1933 by Irene and Frederick Curie when they bombarded aluminum with alpha particles. The isotope was produced with the emission of one neutron for each event. The radioactive nuclide decayed with the emission of a positron, producing which element?

    a. magnesium   c. silicon   e. sulfur
    b. aluminum   d. phosphorus   f. chlorine

30. A sample of charcoal from an ancient fireplace was found to have a disintegration rate of 13.6 emissions/min per gram of carbon due to the carbon-14 content. In living wood the disintegration rate is 15.3 emissions/min per gram of carbon. The half-life of carbon-14 is 5730 years. Approximately how old is the charcoal?

    a. 468 years   c. 974 years   e. 1630 years   g. 4320 years
    b. 675 years   d. 1340 years   f. 2990 years   h. 6230 years

31. Most stable isotopes have in their nuclei

    a. an even number of protons and an even number of neutrons
    b. an even number of protons and an odd number of neutrons
    c. an odd number of protons and an even number of neutrons
    d. an odd number of protons and an odd number of neutrons

32. Calculate the quantity of energy released per mole in the reaction of deuterium with helium-3 to produce one alpha particle and one proton.

    Mass of a deuterium nucleus = 2.01345 amu
    Mass of a proton = 1.007275 amu
    Mass of an alpha particle = 4.00150 amu
    Mass of a helium-3 nucleus = 3.01493 amu

    a. 1.76×10^{12} J   c. 1.76×10^{9} J   e. 5.88×10^{4} J   g. 9.03×10^{19} J
    b. 2.24×10^{11} J   d. 4.64×10^{12} J   f. 3.22×10^{10} J   h. 4.21×10^{11} J

SOLUTIONS

Contents

1 Challenges

1.1 Chapter 2 challenge:

Suppose the atomic mass unit had been defined as 1/10th of the mass of an atom of phosphorus. What would the atomic mass of carbon be on this scale?

This question was asked in another popular general chemistry text. It is intended to make you think—it does!

The answer was given:

Regardless of the starting definition of a mass scale , we should find that the relative masses remain constant from one scale to another. The following conversion factor therefore gives us the ratio of mass of a carbon atom to a phosphorus atom:

$$\frac{12.00 \text{ g C}}{30.97 \text{ g P}} = 0.3875 \frac{\text{g C}}{\text{g P}}$$

This conversion factor is correct; the mass of a carbon atom is indeed 0.3875 of the mass of a phosphorus atom.

If we start with 1/10th of the value assigned to phosphorus, namely 3.097 g, then the mass of carbon is obtained by using the above conversion factor:

$$3.097 \text{ g P} \times 0.3875 \frac{\text{g C}}{\text{g P}} = 1.20 \text{ g carbon}$$

The 3.097 g for phosphorus, and therefore the 1.20 g for carbon, is irrelevant. 3.097 g is merely 1/10th of a mole of phosphorus and therefore 1.20 g is merely 1/10th of a mole of carbon—which you probably already know.

The question asked for the atomic mass of carbon on a new scale based upon the assignment of 1/10th the mass of a phosphorus atom as the base unit.

Let this new atomic mass unit be the nmu:

The mass of a phosphorus atom (on average) would be 10 nmu by definition.

The mass of a carbon–12 atom would be 10 nmu × 0.3875 = 3.87 nmu. Avogadro's number would become the number of ^{12}C atoms in 3.87 grams of carbon–12, or the number of phosphorus atoms in 10 grams of phosphorus.

Note therefore that Avogadro's number would change. It would be less by a factor of (3.87/12) or (10/30.97).

1.2 Chapter 4 challenge:

$$4NH_3(g) + 5O_2(g) \rightarrow 4NO(g) + 6H_2O(g) \qquad 90\%$$

$$2NO(g) + O_2(g) \rightarrow 2NO_2(g) \qquad 90\%$$

$$3NO_2(g) + H_2O(l) \rightarrow 2HNO_3(aq) + NO(g) \qquad 90\%$$

It would seem, at a first glance, that only 2 moles of nitric acid are produced from 4 moles of ammonia. But the equations are not balanced with one another. For example, 4 moles of NO (nitric oxide) are produced in the first step, but only two are used in the next step. In addition, the nitric oxide produced in the third step is recycled back to the second step. A clearer picture is obtained from the overall reaction:

$$NH_3(g) + 2O_2(g) \rightarrow HNO_3(aq) + H_2O(l)$$

There is a 90% yield in each of three steps $NH_3 \rightarrow NO \rightarrow NO_2 \rightarrow HNO_3$. Notice the one mole of N in each step.
The overall yield is $(90\%)^3 = 73\%$ or 2700 grams.

See if you can obtain this equation by adjusting the coefficients in the three equations above, adding them together, and cancelling the NO and NO_2 from both sides.

1.3 Chapter 7 challenge:

Mass of an α particle = 4.003 g/mol or $4.003/6.022 \times 10^{23}$ g/particle.

Kinetic energy = $1/2\ mv^2$, so the velocity
$= [2 \times 6.0 \times 10^{11} \text{ J mol}^{-1}/\ 4.003 \times 10^{-3} \text{ kg mol}^{-1}]^{1/2}$
$= 1.73 \times 10^7 \text{ ms}^{-1}$

Use kg instead of g to be consistent with SI.

J = kg m² s⁻²

Using deBroglie's equation ($\lambda = h/mv$):
The wavelength $\lambda =$

$$\frac{6.626 \times 10^{-34} \text{ Js}}{(4.003 \times 10^{-3} \text{ kg mol}^{-1}/\ 6.022 \times 10^{23} \text{ mol}^{-1}) \times 1.73 \times 10^7 \text{ ms}^{-1}}$$

Check that the units cancel correctly in this expression.

$= 5.76 \times 10^{-15}$ m
$= 5.76$ fm

1.4 Chapter 8 challenge:

a. $0s^3\ 0p^{15}\ 1s^3\ 1p^{15}\ 1d^{21}$

⊞ represents a full (3 electrons) orbital.

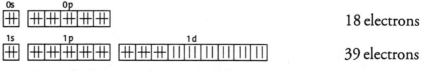

18 electrons

39 electrons

b. Number of orbitals within a subshell = $4\ell + 1$

The pattern, once you see it, is 2×3, 3×5, 4×7, 5×9, 6×11...

c. Number of orbitals in a principal quantum level = $(n + 2)(2n + 3)$
Number of electrons in a principal quantum level = $3(n + 2)(2n + 3)$

1.5 Chapter 11 challenge:

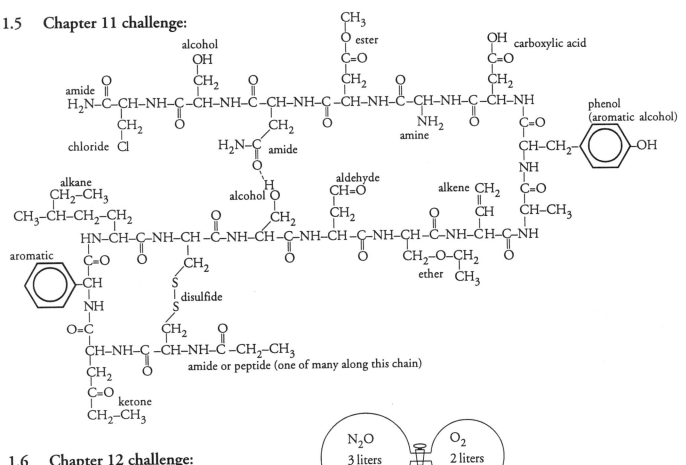

1.6 Chapter 12 challenge:

$$2N_2O(g) \ + \ 3O_2(g) \ \rightarrow \ 4NO_2(g)$$

When the tap is opened, the gases mix and both occupy the total space available. Assuming that no reaction occurs (yet):
Initial partial pressure of nitrous oxide = 3 atm × (3/5) = 9/5 atm.
Initial partial pressure of oxygen = 5 atm × (2/5) = 10/5 atm.

> Recall that the partial pressure of a gas in a mixture is the pressure it would exert if it were the only gas present.

The volume and temperature are the same for both gases, so the partial pressure is directly proportional to n, the number of moles, and the partial pressures can be used in the stoichiometric calculations just as n would be used.
The mole ratio required by the equation is 2 N_2O to 3 O_2, so oxygen is the limiting reactant.

The amount of N_2O used = 10/5 × (2/3) atm = 4/3 atm.
The amount of N_2O remaining at the end = 9/5 − 4/3 atm = 7/15 atm.
The amount of oxygen remaining at the end = zero (it's all used up).
The amount of NO_2 produced = 10/5 × (4/3) atm = 8/3 atm.

> 10/5 atm is the amount of the limiting reactant and 2/3 is the stoichiometric ratio.

> 4/3 is the stoichiometric ratio.

Total pressure = partial pressure of N_2O + partial pressure of NO_2
= 7/15 + 8/3 atm = 3.13 atm

1.7   Chapter 13 challenge:

A question in the Survey of Natural Sciences section on a recent sample Dental Admission Test read as follows:

> Potassium metal crystallizes in a body-centered cubic lattice. It has a density of 0.86 g/cc. What would the density be if potassium metal crystallized in a simple cubic lattice instead?
>
> a.   0.43   b.   0.57   c.   0.69   d.   0.76
>
> The correct answer was listed as a. (0.43 g/cc).

If r = radius of the atom:

volume of a simple cubic cell = $8r^3$

volume of a body-centered cell = $(64/3^{\frac{3}{2}}) r^3$
= $(64/5.2) r^3$
= $12.32 r^3$

a.   The "correct answer" was obtained using the fact that the body-centered cubic unit cell contains two potassium atoms whereas the simple cubic unit cell contains only one atom. The density of the simple cubic unit cell was therefore said to be one-half the density (i.e. 0.43 g/cc).

b.   The incorrect assumption is that the unit cell stays the same size. It doesn't. The volume of the body-centered unit cell is 1.54 times the volume of the simple cubic unit cell.

c.   The density is indeed less, but only by a factor of 1.54/2, i.e. 0.77. New density = 0.86 g/cc × 0.77 = 0.66 g/cc.

1.8   Chapter 14 challenge:

40 grams of NH_4NO_3 = 0.50 mol in 100 g water so molality = 5.0 m
If *i*, the van't Hoff factor, is assumed to be 1.6, the depression in the freezing point of the water = 1.86 × 5.0 × 1.6 = 15°C.

In fact, if you do the experiment, you will find that the lowest temperature you can reach is about −14.5°C (it depends upon how accurate your thermometer is).

Adding 40 grams of NH_4NO_3 to 100 g water at room temperature causes a temperature drop of about 22°C. This is in good agreement with the stated heat of solution of ammonium nitrate:

This assumes that the specific heat of the solution is the same as the specific heat of water.

heat   = specific heat × mass × temperature change
= 4.184 $JK^{-1}g^{-1}$ × 140 g × 22K   = 12.9 kJ for 0.5 mol
compared to $\Delta H_{solution}$ (NH_4NO_3) = 25.7 kJ mol^{-1}

However, starting from 0°C, the freezing point will be reached before the temperature has fallen 22°C; it can only fall 15°C before the water starts to freeze. Which, if you do the experiment, you will see that it does.

You should take into account the heat capacity of the solid ammonium nitrate if it's not at 0°C when it's added (perhaps about 1 kJ).

The amount of ice that forms is the amount necessary to supply the additional heat for the solution process—about 9 grams.
Heat of fusion of ice = +333 J g^{-1} so 3100J / 333J g^{-1} = 9.3 grams of ice

1.9 Chapter 15 challenge:

The rate equation for the reaction:

$$I^- + ClO^- \rightarrow Cl^- + IO^-$$

in basic aqueous solution is

$$\text{Rate} = k \frac{[I^-][ClO^-]}{[OH^-]}$$

The rate equation clearly indicates that the reaction is not a simply bimolecular reaction between the iodide ion and the hypochlorite ion. The appearance of hydroxide ion in the rate equation suggests a more complicated mechanism.

The inverse dependence on hydroxide suggests that hydroxide ion might be involved in an equilibrium that is established quickly at the beginning of the reaction.

There are four principal species initially in solution: I^-, ClO^-, OH^-, and water H_2O. Assuming the first step is a bimolecular reaction, which two species might interact? All are negatively charged except water, so perhaps either:

$$I^- + H_2O \rightleftharpoons HI + OH^-$$

or:

$$ClO^- + H_2O \rightleftharpoons HClO + OH^-$$

HI is a strong acid, so the first is unlikely in aqueous solution, but the second is a hydrolysis reaction that readily occurs.

See Chapter 18 for a discussion of hydrolysis.

The hypochlorous acid might then collide and react with an iodide ion:

$$HClO + I^- \rightarrow HIO + Cl^-$$

Reaction with I^- is the only reasonable route; considering what's in the solution:

The hypoiodous acid is in equilibrium with its anion, the hypoiodite anion:

$$HIO + OH^- \rightleftharpoons IO^- + H_2O$$

Reaction with OH^- would simply reverse the previous step (this will indeed happen).

Reaction with ClO^- will result in no change (just a proton transfer from ClO^- to ClO^-).

In the presence of base this equilibrium would be pulled strongly to the right. The rate determining step is the second step, the actual rearrangement:

$$\text{Rate} = k\,[HClO][I^-]$$

If the initial equilibrium is established quickly, so that the hypochlorous acid achieves a steady equilibrium concentration, then:

$$K = \frac{[OH^-][HClO]}{[ClO^-]} \quad \text{or} \quad [HClO] = K\,[ClO^-]/[OH^-]$$

This is called the steady state approximation.

Substituting this expression for [HClO] into the rate equation and combining the constants:

$$\text{Rate} = k\,\frac{[I^-][ClO^-]}{[OH^-]} \quad \text{which is the observed rate equation.}$$

1.10   Chapter 16 challenge:

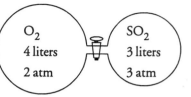

Using partial pressures:

		$2SO_2(g)$	$+$ $O_2(g)$	$\rightleftharpoons$ $2SO_3(g)$	
	initial volume	3	4		liters
	initial pressure	3	2		atm
ICE box	initial partial pressure	$3 \times 3/7$	$2 \times 4/7$		atm
	change	$-2x$	$-x$	$+2x$	atm
	equilibrium	$(9/7 - 2x)$	$(8/7 - x)$	$2x$	atm
	after solving for x	0.513	0.756	0.773	atm
	total pressure = sum of the partial pressures = 0.513 + 0.756 + 0.773 = 2.04 atm				

It is quite easy to solve expressions like this using an excel or lotus spreadsheet.

Enter the expression in one cell, referring to another cell that contains the value of x.

Then in the second cell that contains the value of x, try some sensible numbers (e.g. 0.2 or 0.3, etc.) and monitor how the value of the expression changes.

Be aware that it is possible to satisfy the requirement that the expression = 3.0 using unrealistic values for x (e.g. negative concentrations)!

Try it.

Solving the equilibrium constant expression for x requires solution of the equation:

$$K_p = \frac{P_{SO3}^2}{P_{SO2}^2 \, P_{O2}} = \frac{(2x)^2}{(9/7 - 2x)^2(8/7 - x)} = 3.00$$

$$x = 0.3864$$

Alternatively, if you really want to use concentrations instead of partial pressures:

		$2SO_2(g)$	$+$ $O_2(g)$	$\rightleftharpoons$ $2SO_3(g)$	
	initial volume	3	4		liters
	initial pressure	3	2		atm
	temperature	1000	1000		K
	no of moles	0.1097	0.0975	0	mol
	initial concentration	0.01567	0.01393	0	mol L^{-1}
ICE box	change	$-2x$	$-x$	$+2x$	
	equilibrium	$0.01567 - 2x$	$0.01393 - x$	$2x$	
	after solving for x	0.00625	0.00922	0.00942	mol L^{-1}
	total no of moles in 1 L = 0.00625 + 0.00922 + 0.00942 = 0.0249 mol				
	Pressure = (n/V) RT = 0.0249 × 0.08206 × 1000 = 2.04 atm (same answer)				

$$K_c = K_p(RT)^{\Delta n} = 246.18$$

1.11   Chapter 20 challenge:

ΔH_{fusion} (ice) = 333 Jg^{-1}
Heat capacity of water = 4.184 JK^{-1}g^{-1}

a.   The temperature of the ice-water mixture at the end of the experiment must be 0°C (the temperature of an ice–water mixture at equilibrium at 1 atm pressure).

b.   The heat required to raise the temperature of the water from −10°C to 0°C:
Heat   = 4.184 JK^{-1}g^{-1} × 100 g × 10K
         = 4184 J
Amount of water changed to ice as a result:
Heat   = 333 Jg^{-1} × mass
4184 J = 333 Jg^{-1} × mass
Mass of ice formed = 12.6 grams

c.   The system is completely isolated, so there is no change in the surroundings: ΔS_{surr} = 0.

d.   Indeed the entropy change must be positive, the reaction is spontaneous. The negative change as a result of the water freezing is compensated by the increase in the temperature of the water from −10°C to 0°C.

The entropy change for the freezing of water at 0°C :

$\Delta S = \Delta H_f/T$ = −333 Jg^{-1}/273.15 = −1.22 JK^{-1}g^{-1}

for 12.56 grams: ΔS = 1.22 × 12.56 = −15.31 JK^{-1}

The entropy change involved in warming the water from −10°C to 0°C must be more positive than 15.31 JK^{-1} . (It is, it's +15.61 JK^{-1}).

2   Puzzles & Crosswords

28 Elements
—solution
(page 26)

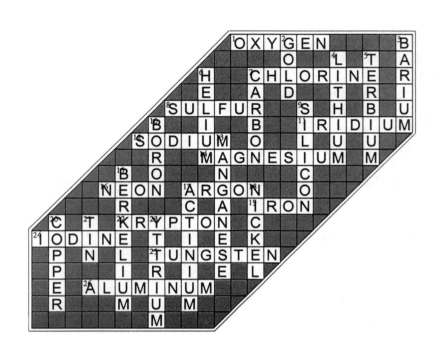

Across:

1.   Acid former discovered by Joseph Priestley.
7.   Halogen used in water purification.
8.   Brimstone.
11.   Produces iridescent colors.
12.   Below 4 down, its lamps are yellow.
14.   The milk of its hydroxide settles the stomach.
16.   A colored sign and below it...
17.   ...another noble gas.
19.   A ferrous metal first cast as pigs.
22.   Noble origin of the man of steel?
24.   Its tincture is antiseptic.
25.   But there's no W anywhere in the name!
26.   The most abundant metal in the earth's crust.

Down:

2.   The goal of the alchemists.
3.   An alkaline earth metal, bachelor of arts?
4.   Named for greek stone.
5.   With or without the T, it's found where 23 down is.
6.   Rutherford fired its nuclei at 2 down.
7.   Basis of compounds in living things and below it...
9.   ...a semiconductor from the valley?
10.   From a mineral hauled by 20 mules.
13.   A transition metal sometimes mistaken for 14 across.
15.   For the site of Seaborg's laboratory.
17.   Is it the first of the actinides?
18.   Named for the old devil himself.
20.   A penny from Cyprus...
21.   ...with this it's bronze.
23.   Source and symbol initially Ytterby in Sweden.

Numbers
—solution
(page 42)

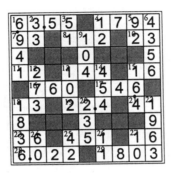

Across:

1.   Molar mass of copper.
4.   The year Antoine Lavoisier lost his head.
7   Neptune's number.
8.   The most recently detected element (Feb 96).
10. The exponent of Avogadro's number.
11. The number of inches in one foot...
13. ...and the number of sq. in in one sq.ft.
15. The molar mass of oxygen atoms.
16. One atmosphere pressure in torr.
17. 273°C in K.
18. The mass of the empirical formula for 9 down.
19. The volume of one mole of a gas at 0°C and 1 atm.
21. Niobe's number.
23. Three moles of carbon.
25. The ignition temperature of paper in °F.
27. The molar mass of methane CH_4.
28. The first four digits of Avogadro's number.
29. The year John Dalton announced his atomic theory.

Down:

1.   Lithium's mass.
2.   The atomic number of arsenic...
3.   ...and that of the element under it.
4.   The isotope of carbon upon which the amu is based.
5.   The atomic number of 20 down.
6.   The group numbers of silicon, aluminum, nitrogen, and sulfur, respectively.
9.   The mass of four moles of acetylene C_2H_2.
12. It's freezing in K.
13. A number for the prize-giver.
14. The number of grams in one pound.
15. The number of cubic cm in one cubic inch.
18. The mass of a proton divided by the mass of an electron.
20. The isotope of uranium used in reactors.
22. The year Linus Pauling won his second Nobel Prize.
24. The number of carbon atoms in a common fullerine.
25. The number of neutrons in 2 down's isotope–75.
26. The number of electrons in a sodium atom...
27. ...and after it has lost one to form the sodium ion.

People

—solution
(page 110)

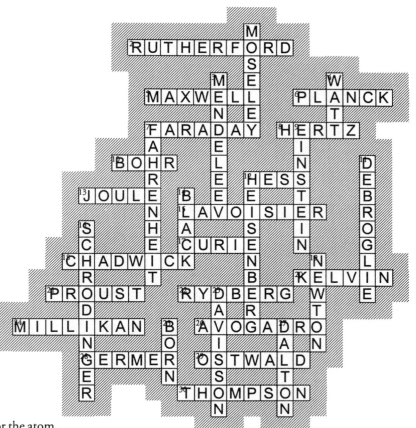

Across:

2. Proposed the nuclear model for the atom.
5. Described radiation in terms of electric and magnetic waves.
6. Developed the quantum theory in 1900.
7. Wrote the laws of electrolysis in 1833.
8. Proved the existence of electromagnetic radiation.
10. Incorporated ideas of quantization of energy into a model for the electronic structure of hydrogen.
12. Developed the law of constant heat summation.
13. A man of energy, a student of 27 down.
15. Proposed the law of conservation of matter.
17. Discovered polonium, named after her native land.
18. Discovered the neutron in 1932.
20. Proposed a temperature scale in 1848 for which the zero point was −273°C.
21. Proposed the law of constant composition.
22. Developed an equation used to predict the lines in the emission spectrum of hydrogen.
24. Measured the charge on a single electron.
26. An Italian whose ideas were a long time in being accepted.
28. See 23 down.
29. Introduced the term "mole" for an amount of substance.
30. Suggested the presence of electrons in all matter.

Down:

1. Determined that elements in the Periodic Table should be arranged by atomic number, not mass.
3. Discovered periodicity in the properties of elements and designed the first Periodic Table.
4. What was his name, the unit for power? A student of 14 down.
7. His degree is smaller than that of Celsius.
9. He used Planck's new quantum theory to explain the photoelectric effect in 1905.
11. Proposed wave–particle duality.
12. A man of some uncertainty.
14. Distinguished heat, temperature, and heat capacity.
16. Used 11 down's assertion to write a wave equation for the electron in a hydrogen atom.
19. The SI unit for force is named after him.
23. With 28 across established the wave–particle duality of an electron beam.
25. Suggested that φ^2 should be interpreted as the probability of finding the electron at any point about the nucleus.
27. Forcefully revived the idea of atoms in 1803.

People of this chapter and their States
—solution
(page 186)

Across:

1. William Thomson's new temperature scale.
4. Formulated the law of effusion.
6. A sceptical chemist, his was the first gas law.
8. The total pressure is the sum of the partial pressures.
10. Country of 7 down.
12. A balloonist of some renown.
14. Country of 4 across.
15. Remembered by his number.
16. Country of 12 across, 3 down, and 11 down.
17. Country of 15 across and 13 down.

Down:

2. Country of 8 across.
3. Formulated the law of combining volumes.
5. Country of 9 down.
7. He was concerned with probabilities and distributions.
9. Developed a gas law for real gases.
11. Mathematician for whom the SI unit of pressure is named.
13. Invented the mercury barometer in 1643.

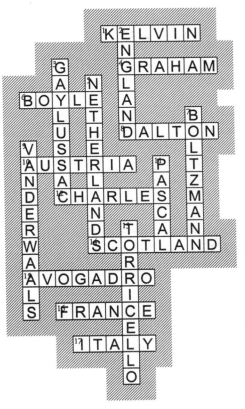

Kinetics

—solution
(page 234)

Across:

2. What the square brackets [ ] represent.
4. The inverse of rate.
5. One of the platinum group metals used as a catalyst.
6. Required by a system for the reaction to go.
8. Must be correct for an effective collision.
9. Slope.
10. The order that is sum of the exponents in the rate equation.
11. A transient species produced and then consumed in a reaction sequence.
12. The sequence of bond-making and bond-breaking that is the way the reaction happens.
14. *and 19 across:* The largest growth in catalyst use is in this area.
19. *See 14 across.*
20. One molecular event in a reaction sequence.
21. His equation relates *1 down*, *3 down*, and *6 across*.
22. The time taken for a reactant concentration to be reduced by 50%.
25. The slowest step in a *12 across*.
28. The common one is the exponent of 10.
31. The instantaneous rate is the *9 across* of this.
33. Only for an *20 across* is the order equal to this.
36. These must do it with sufficient *35 down* and the correct *8 across* if they are to react.
37. A *17 down* in a biological system.
38. *See 29 down.*

Down:

1. The k in the rate equation.
3. Increase this and the kinetic energy of the molecules increases.
5. The relationship between *2 across* and *4 across*.
7. The exponent of a concentration term in the rate equation.
12. The number of reacting species that come together in an elementary reaction step.
13. The order when the *16 down* is 1.
15. Its magnitude is responsible for dust explosions.
16. The power to which a term is raised.

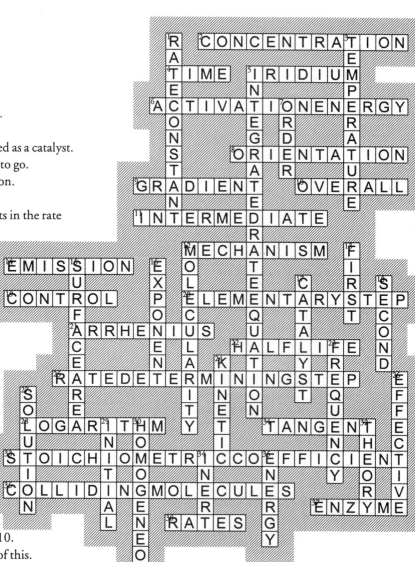

17. It changes the route of a reaction.
18. The order when the *16 down* is 2.
23. The factor A in the equation of *21 across*.
24. This effect rather than a thermodynamic effect determines the rate of a reaction.
26. Only these collisions result in reaction.
27. Reactions are often done in this because the molecules are free to move about.
29. *and 38 across:* A method used to determine reaction order.
30. A *17 down* in the same phase as the reactants.
32. An explanation.
34. Kinetically unreactive.
35. Used when work is done.

Equilibria

—solution
(page 266)

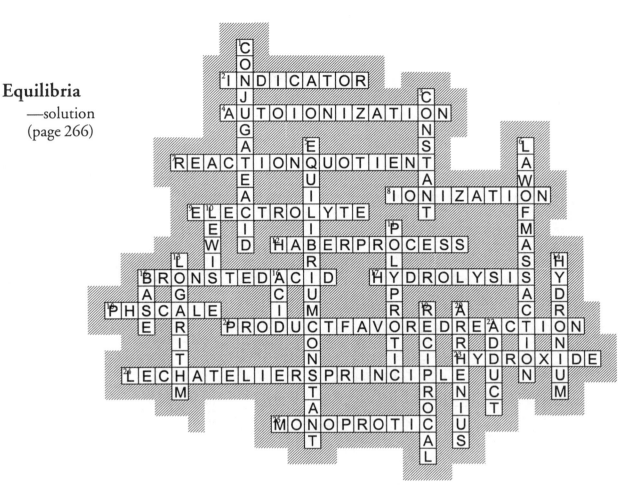

Across:

2. In a titration, it lets you know when you have reached the equivalence point.
4. When water produces *14 down* and *23 across* ions.
7. Product concentrations divided by reactant concentrations, not necessarily at equilibrium.
8. It's assumed to be complete for a strong *9 across* in solution.
9. Weak or strong, it conducts an electrical current in solution.
12. Industrial manufacture of ammonia.
15. It transfers a hydrogen ion to form its conjugate base.
17. What happens to the salts of weak acids or weak bases in solution.
18. Normally from 0 to 14, an indication of hydronium ion concentration.
21. A process for which *5 down* is large.
23. The conjugate base of water; produced by Arrhenius bases in aqueous solution.
24. The idea that a system in equilibrium will adjust to accommodate any alteration in conditions.
25. An acid with only one hydrogen atom per molecule.

Down:

1. The partner of a base.
3. Unchanging.
5. Q at equilibrium.
6. A statement of the fact that the *5 down* is *3 down* regardless of how the masses of the components present are changed—developed by Guldberg & Waage.
10. He classified an acid as an electron pair acceptor and a base as an electron pair donor.
11. Sulfuric acid, oxalic acid, phosphoric acid, carbonic acid, for example.
13. Negative this of the concentration of *14 down* is pH.
14. Ion produced by the protonation of a water molecule.
15. This increases the hydroxide ion concentration...
16. ...whereas this increases the hydronium ion concentration in aqueous solution.
19. If the reaction is reversed, then the new equilibrium constant is this of the original.
20. His was the first set of definitions describing an acid and base in aqueous solution.
22. One name for what is produced when a Lewis acid and a Lewis base react, forming a coordinate bond.

Thermodynamics

—solution
(page 302)

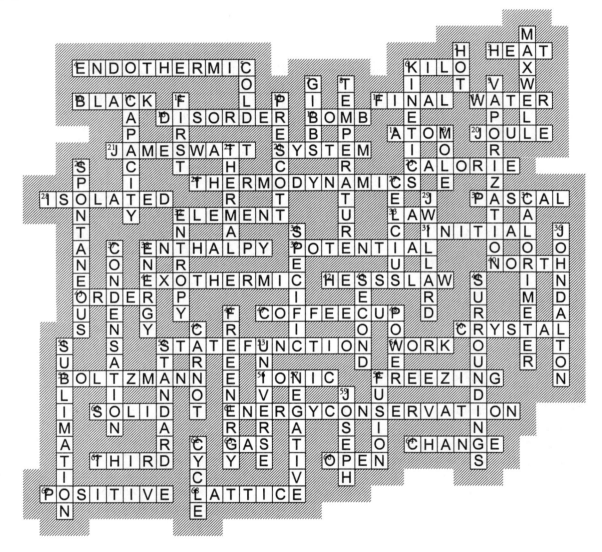

Across:

3.   *and 54 across:* q and w—two ways to increase the internal energy of a system.
4.   A process in which heat is absorbed.
6.   Prefix for 1000 ×.
10. *see 59 down.*
14. The state of the system at the end.
15. At its freezing point, its solid state is less dense than its liquid state.
16. Entropy.
17. A constant volume calorimeter.
18. The smallest particle of an element.
20. SI unit for energy.
21. The unit for *48 down* was named after him.
23. The part of the *53 down* under examination.
25. An older unit of energy.
26. The study of heat and work.
28. A system to (or from) which neither energy nor matter can be added (or removed).
30. Newtons per square meter.
32. In its standard state has an enthalpy of formation equal to zero.
33. A statement of regularity in natural occurrence or phenomena.
35. The state of the system at the beginning.
38. Sometimes described as the "heat within."
39. Energy due to position.
40. Where the needle points.
41. A process in which heat is released.
42. The enthalpy change is independent of the route.
45. To decrease the entropy.
47. A common laboratory constant pressure calorimeter.
50. A perfect one has zero entropy at zero K.
52. A property of a system that is independent of its previous history.
54. *see 3 across.*
55. S = k logW is engraved on his tombstone in Vienna.
56. A bond resulting from the transfer of electron(s).
58. Changing state from liquid to solid.
60. The state of lowest entropy and...
61. The first law.
63. ...the state of highest entropy.
64. Δ.
65. *see 12 down.*
66. A system to (or from) which both energy and matter can be added (or removed).
67. The sign of ΔH for an endothermic process.
68. The energy required to break up a *50 across*.

Down:

1.   Regarded by many as the greatest physicist between Newton and Einstein.
2.   Heat travels from this...
5.   ...to this.
6.   The rate and mechanism of reactions.
7.   *see 29 down.*
8.   The measure of how hot an object is.
9.   Conversion from liquid to gas.
11. The ability to absorb heat per degree rise in temperature.
12. *and 43 down, 65 across:* the three laws of *26 across*.
13. Middle name of *20 across*.
19. SI unit for an amount of a substance.
22. Energy in the form of heat.
24. A process for which the entropy of the universe increases.
27. In magnitude a degree the same as a K.
29. *and 7 down:* His is *46 down*.
31. A device for determining heat change by measuring temperature change.
32. S.
34. The heat required to raise the temperature of one gram by one degree K.
36. His were the laws of multiple proportions and partial pressures.
37. The change from vapor to liquid.
38. It's conserved so the first law says.
43. *see 12 down.*
44. Everything but the system.
46. Available for useful work.
48. Energy divided by time.
49. He derived an expression for the efficiency of heat engines.
51. Conversion from solid to vapor.
52. This state of a substance is its most stable state at 1 bar pressure and a specified temperature (usually 25°C).
53. *23 across* plus *44 down*.
57. The sign of the free energy change for a spontaneous process.
58. Its heat melts the solid state.
59. *and 10 across:* The first to distinguish temperature and heat capacity.
62. Like *49 down*'s, or Born-Haber, a sequence of processes that start and end at the same state.

3 Answer keys for the examinations

3.1 Examination 1

True–false questions

1. F All atoms of an element are not identical (Dalton was wrong). Isotopes exist with different numbers of neutrons in the nucleus.

2. F Iodine is a halogen (Group 7A or 17).

3. T This is the law of conservation of mass. A chemical reaction is just a rearrangement of atoms; atoms are neither created nor destroyed.

4. T Heat capacity (JK^{-1}) = specific heat capacity ($JK^{-1}g^{-1}$) × mass (g).

5. T A Celsius degree interval is equal in magnitude to a Kelvin degree interval.

6. F The base SI unit for mass is the kilogram (kg).

7. T ΔH is an extensive property, if heat is liberated in one direction, heat will be absorbed in the reverse direction. For example, breaking bonds is endothermic, making bonds is exothermic.

8. T The prefix milli– means ×10^{-3}.

9. T A compound must consist of two or more different elements—otherwise it wouldn't be a compound.

10. T Neon–22 has 12 neutrons in its nucleus (mass number 22, atomic number 10; difference 12 neutrons).

11. T An empirical formula is the simplest possible integer ratio.

12. T Energy = power × time (J = Ws) or Power = energy/time (W = Js^{-1}).

13. T Volume, pressure, and temperature are all state functions.

14. T Perchloric acid $HClO_4$ is one of the strong acids. Chlorous acid $HClO_2$ is not.

15. F Some molecules are elements. For example H_2, O_2 and O_3, Br_2, P_4, S_8, C_{60}.

16. F The pH of a solution at the equivalence point depends upon the acid and base involved. For example, a strong acid and a weak base produces an acidic solution.

17. T Bromine in its standard state is a liquid, so $\Delta H_f^{\circ}(Br_2(l))$ = zero but $\Delta H_f^{\circ}(Br_2(g))$ does not = zero; energy is required to vaporize the liquid.

18.   **F**   The expression is incorrect; the correct expression is # moles = mass / molar mass.

19.   **T**   Melting always requires energy.

20.   **T**   The oxidizing agent is an oxidizing agent because it accepts electrons.

21.   **F**   An exothermic reaction is often product-favored, but not always. Some spontaneous reactions are endothermic.

22.   **F**   Elements in the Periodic Table are arranged in order of their atomic numbers not masses.

23.   **F**   An allotrope is one form of an element. For example, the three forms of carbon: diamond, graphite, and the fullerenes.

24.   **T**   Lead sulfate is one of three insoluble sulfates (lead, barium, and strontium).

25.   **T**   Non-metal oxides produce acidic solutions when dissolved in water. For example, sulfur dioxide dissolves in water to produce sulfurous acid H_2SO_3.

Multiple choice questions

1.   e.   force = mass acceleration = $kg\ m\ s^{-2}$

2.   b.   feet

3.   e.   the number of electrons equals the number of protons.

4.   g.   27 protons = atomic number 27, therefore Co; 24 electrons means a charge of 3+.

5.   a.   the anion having the fewer oxygens has a name ending in –ite.

6.   b.   Rutherford's gold foil experiment indicated that most of the mass of the atom is concentrated at a nucleus.

7.   e.   Set up a table to organize your calculations:

	$2N_2O$	$3O_2$ →	$4NO_2$
moles initially	6 moles	8 moles	?
		limiting reactant	?
moles used or made	8 × (2/3)	all of it	8 × (4/3)
	= 16/3 moles		= 32/3 moles
amount left	2/3 mole	zero	32/3 moles
	remaining	none left	produced

The 2/3 and 4/3 are the stoichiometric ratios from the equation.

8.   c. $Ca(CO_3)_2$ is not calcium carbonate. The carbonate ion has a charge of 2–, not 1–.

9.   f.   256 gram sample of sulfur = 8 moles

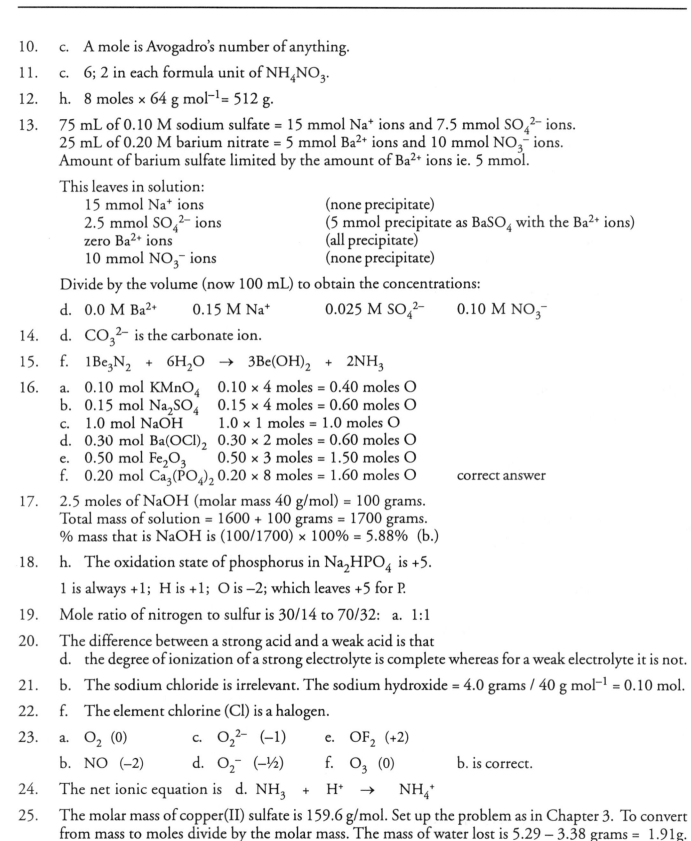

10. c. A mole is Avogadro's number of anything.

11. c. 6; 2 in each formula unit of NH_4NO_3.

12. h. 8 moles × 64 g mol^{-1}= 512 g.

13. 75 mL of 0.10 M sodium sulfate = 15 mmol Na$^+$ ions and 7.5 mmol SO$_4^{2-}$ ions.
     25 mL of 0.20 M barium nitrate = 5 mmol Ba^{2+} ions and 10 mmol NO$_3^-$ ions.
     Amount of barium sulfate limited by the amount of Ba^{2+} ions ie. 5 mmol.

     This leaves in solution:
15 mmol Na$^+$ ions	(none precipitate)
2.5 mmol SO$_4^{2-}$ ions	(5 mmol precipitate as BaSO$_4$ with the Ba^{2+} ions)
zero Ba^{2+} ions	(all precipitate)
10 mmol NO$_3^-$ ions	(none precipitate)

     Divide by the volume (now 100 mL) to obtain the concentrations:

     d. 0.0 M Ba^{2+} 0.15 M Na$^+$ 0.025 M SO$_4^{2-}$ 0.10 M NO$_3^-$

14. d. CO$_3^{2-}$ is the carbonate ion.

15. f. 1Be$_3$N$_2$ + 6H$_2$O → 3Be(OH)$_2$ + 2NH$_3$

16. a. 0.10 mol KMnO$_4$ 0.10 × 4 moles = 0.40 moles O
     b. 0.15 mol Na$_2$SO$_4$ 0.15 × 4 moles = 0.60 moles O
     c. 1.0 mol NaOH   1.0 × 1 moles = 1.0 moles O
     d. 0.30 mol Ba(OCl)$_2$ 0.30 × 2 moles = 0.60 moles O
     e. 0.50 mol Fe$_2$O$_3$ 0.50 × 3 moles = 1.50 moles O
     f. 0.20 mol Ca$_3$(PO$_4$)$_2$ 0.20 × 8 moles = 1.60 moles O   correct answer

17. 2.5 moles of NaOH (molar mass 40 g/mol) = 100 grams.
     Total mass of solution = 1600 + 100 grams = 1700 grams.
     % mass that is NaOH is (100/1700) × 100% = 5.88% (b.)

18. h. The oxidation state of phosphorus in Na$_2$HPO$_4$ is +5.

     1 is always +1; H is +1; O is −2; which leaves +5 for P.

19. Mole ratio of nitrogen to sulfur is 30/14 to 70/32: a. 1:1

20. The difference between a strong acid and a weak acid is that
     d. the degree of ionization of a strong electrolyte is complete whereas for a weak electrolyte it is not.

21. b. The sodium chloride is irrelevant. The sodium hydroxide = 4.0 grams / 40 g mol^{-1} = 0.10 mol.

22. f. The element chlorine (Cl) is a halogen.

23. a. O$_2$ (0)   c. O$_2^{2-}$ (−1)   e. OF$_2$ (+2)

     b. NO (−2)   d. O$_2^-$ (−½)   f. O$_3$ (0)   b. is correct.

24. The net ionic equation is   d. NH$_3$ + H$^+$ →   NH$_4^+$

25. The molar mass of copper(II) sulfate is 159.6 g/mol. Set up the problem as in Chapter 3. To convert from mass to moles divide by the molar mass. The mass of water lost is 5.29 − 3.38 grams = 1.91g.

	CuSO$_4$	H$_2$O
mass ratio:	3.38 g	1.91 g
divide by molar mass:	/159.6	/18.02
	= 0.0212	= 0.106
convert to integers by dividing by the smaller:	= 1	= 5

The formula of the hydrate is CuSO$_4$.5H$_2$O.

26. Reverse the second equation and add: $\Delta H° = -84.7$ kJ $- 52.3$ kJ $= -137$ kJ (for one mole).
The quantity of ethylene used is 56 grams $= 56$ g / 28 g mol^{-1} $= 2.0$ mol.
Total enthalpy change $= -274$ kJ. (e.)

27. Heat (J) = specific heat capacity (JK^{-1}g^{-1}) × mass (g) × temp change (K)

Heat lost by lead $= 0.128$ JK^{-1}g^{-1} × 30.0 g × 68.4 K
                         $= 262.66$ J
Heat gained by water $= 4.184$ JK^{-1}g^{-1} × mass g × 1.6 K
                         $= 6.69$ × mass J

Heat lost = heat gained: 262.66 J $= 6.69$ × mass J
Mass = 39 grams. (e.)

28. $\Delta H = q + w = 100$kJ $- 60$ kJ $= 40$ kJ. (b.)

29. Aluminum nitrate is Al(NO$_3$)$_3$. There are three moles of nitrate ions in each mole of sluminum nitrate. Moles of aluminum nitrate = concentration (M) × volume (L) $= 0.020 × 0.200 = 0.0040$ mol.
Number of moles of nitrate ions $= 0.0040 × 3 = 0.012$ mol. (e.)

30. The data provided is mass composition data. To convert from mass to moles divide by the molar mass. So the mass percent for each element is divided by its molar mass as follows:

	carbon	hydrogen	oxygen
mass % ratio:	34.6%	3.90%	61.5%
divide by molar mass:	/12.011	/1.008	/16.00
	= 2.88	= 3.87	= 3.84
convert to integers by dividing by the smallest:	= 1	= 1.34	= 1.33

The empirical formula is C$_3$H$_4$O$_4$. (b.)

31. Write a balanced equation for the combustion:

$$CH_4(g) + 2O_2(g) \rightarrow CO_2(g) + 2H_2O(l)$$

9.0 grams of water (molar mass 18 g/mol) is 0.50 mol of water.
The stoichiometry of the equation indicates that 0.25 mol of methane was burned.
Heat released $= 0.25$ mol × 890 kJ/mol $= 222$ kJ. (c.)

3.2 Examination 2

True–false questions

1.   T   Red light has a lower frequency (less energy per photon).

2.   T   Allowed values for m_ℓ run from $+\ell$ to $-\ell$.

3.   F   Paramagnetic materials are drawn into a magnetic field.

4.   F   Fluorine is the smallest of the 2nd Period elements (except perhaps Ne) due to the large effective nuclear charge.

5.   T   There are $(2\ell+1)$ orbitals in a subshell, therefore $2(2\ell+1)$ electrons.

6.   F   The most stable ion of sodium is the Na^+ ion (although Na^- does exist).

7.   T   A lone pair of electrons is a nonbonding pair of electrons.

8.   T   This is how the bond order is determined in valence bond theory.

9.   T   Electronegativity is a property of an atom *in a compound*.

10.   F   This would be a metallic bond.

11.   T   The see-saw molecular geometry is derived from the trigonal bipyramidal arrangement where there is one lone pair of electrons.

12.   F   Bonding pairs of electrons repel each other less than nonbonding pairs of electrons.

13.   F   A triple bond indicates that three pairs of electrons are shared in the bond.

14.   T   There's always some overlap of orbitals and some covalent contribution to the bond.

15.   F   Many ions are nonpolar (they are symmetrical in structure).

16.   T   The geometry of a set of sp hybrid orbitals is linear.

17.   T   The number of hybrid orbitals always equals the number of atomic orbitals used.

18.   F   Cis-trans isomerism is impossible about a triple bond; and it is impossible about a double bond in which the two groups at one end of the bond are the same.

19.   T   An antibonding orbital is always higher in energy than the corresponding bonding orbital.

20.   T   It depends upon their orientation.

21. **F** The three π bonding pairs of electrons are shared by all carbon atoms in the ring. The electrons are delocalized around the ring. The bond order of each bond is 1.5.

22. **T** Resonance is required in VB theory for a satisfactory picture of the bonding.

23. **T** A single bond is a σ bond.

24. **T** Polyamides, like nylon, result from the condensation of difunctional amines and difunctional acids.

25. **T** Copolymers result from the polymerization of two or more monomers. An example is SBR (styrene–butadiene rubber).

Multiple choice questions

1. b. 3.3×10^{-2} cm; wavelength × frequency = velocity. Make sure the units are compatible. Change the velocity of 3.0×10^8 m s^{-1} into 3.0×10^{10} cm s^{-1} for example.

2. a. 3.97×10^{-18} J; $E = h\nu$ and $\lambda \times \nu = c$. So $E = hc/\lambda$. Again, make sure the units cancel correctly.

3. b. 2390 kJ mol^{-1}; to change the units from J photon^{-1} to kJ (mol of photons)$^{-1}$, multiply by Avogadro's number; and then divide by 1000 to convert to kJ.

4. b. The angular momentum quantum number ℓ designates the type of orbital (s, p, d, or f).

5. e. When $n = 4$, the secondary quantum number ℓ must be no greater than 3.

6. a. If the quantum number $\ell = 0$, then the orbital is s and only 2 electrons can be accommodated.

7. b. Aluminum $1s^2\, 2s^2\, 2p^6\, 3s^2\, 3p^1$

8. e. The 3d is filled after the 4s sublevel.

9. e. The configuration is $4s^2\, 3d^6$, so four electrons are unpaired.

10. c. Hund's rule of maximum multiplicity.

11. f. Se^{2-}; should be larger than 184 pm and between 195 and 230 pm.

12. c. Covalent (but polar).

13. b. 5 valence electrons; P.

14. d. XeF_2

15. a. +1; a half share in 8 electrons = 4; compare to 5 for the isolated atom.

16. e. Upper right

17. f. Octahedral

18. c. Tetrahedral

19. d. PCl_3

20. b. A set of sp^3 hybrid orbitals is formed by combination of one s and three p orbitals.

21. a. Only one σ bond in a single, double, or triple bond.

22. c. p_z and p_z; the orbitals must be perpendicular to the bond axis, but they must also line up with one another.

23. g. write the equation for the formation of methane:

$$C(s) \ + \ 2H_2(g) \ \rightarrow \ CH_4(g) \qquad\qquad \Delta H_f^\circ \ ?$$

The process involves breaking up the carbon solid, breaking up two hydrogen molecules, and forming four C–H bonds. The total energy involved is

$716 + (2 \times 436) - (4 \times 413)$ kJ $= -64$ kJ

24. f. 6 pairs, therefore octahedral

25. b. 6; these are illustrated on page 164 of this study guide.

26. b. Orbital hybridization is the combination of atomic orbitals to form a new set of orbitals with directional properties more appropriate for bonding

27. c. $(\sigma_{1s})^2(\sigma_{1s}^*)^2(\sigma_{2s})^2(\sigma_{2s}^*)^2(\pi_{2px})^2(\pi_{2py})^2(\sigma_{2pz})^2$; a triple bond (same as N_2)

28. d. 4 p_π orbitals, therefore the same number of molecular orbitals.

29. a. 2,3–dimethylbutane

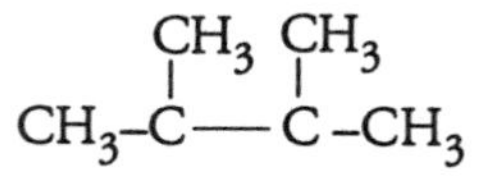

$$\begin{array}{cc} CH_3 & CH_3 \\ | & | \\ CH_3\text{--}C\!\!-\!\!-\!\!C\text{--}CH_3 \end{array}$$

30. c. A cyclic alkane has the general formula C_nH_{2n} (an alkane less 2H).

31. b. A –C=C– double bond

32. f. methane $\rightarrow$ methanol $\rightarrow$ formaldehyde HCHO $\rightarrow$ formic acid $\rightarrow$ carbon dioxide

33. d. Esters are formed as a result of the condensation of an acid and an alcohol

3.3 Examination 3

True–false questions

1.   T   All gases are real; although most behave ideally at normal temperature and pressures.

2.   T   Volume is directly porportional to temperature.

3.   T   A useful thing to know in gas law calculations.

4.   F   The kinetic energy must be the same, so a larger molecule moves more slowly.

5.   F   There are exactly 100 kPa in one bar.

6.   T   The principal reason for deviation from ideal behavior is intermolecular attraction.

7.   F   The number of moles of gases decreases (from 4 to 2 moles) so the pressure decreases.

8.   T   This is the significance of the rms speed.

9.   F   Some gases are colored (for example Cl_2, Br_2, NO_2).

10.   T   Both are condensed states of matter.

11.   T   The strongest intermolecular force is the force of attraction between ions and polar molecules.

12.   F   Even for polar molecules, dispersion forces often contribute more to the intermolecular attraction.

13.   T   Doubly–charged metal ions are more strongly solvated than singly–charged metal ions.

14.   F   Ethanol, like all alcohols, contains the –O–H functional group.

15.   F   HF has the highest boiling point due to hydrogen bonding.

16.   F   Very few are; water is the most common example.

17.   T   The heat of vaporization parallels the boiling point; in fact the ratio of the two for many compounds is a constant number called Trouton's constant.

18.   T   True, the liquid and vapor cannot be distinguished beyond the critical point.

19.   F   The surface tension is due to intermolecular attraction.

20.   T   There are only seven crystal systems.

21.   F   One system has four associated crystal lattices, some only one.

22.　**F**　There are only 2 structural units within a bcc unit cell and 4 within a fcc unit cell. Molecules at the edge of a unit cell are shared with adjacent unit cells.

23.　**T**　Molality is a temperature-independent concentration unit; molarity changes with temperature.

24.　**F**　Supersaturated solutions exist; they are unstable.

25.　**T**　Gases usually dissolve exothermically.

26.　**F**　Solution processes are usually entropy-driven; many have a positive ΔH.

27.　**T**　According to Raoult's law, the vapor pressure of the solvent depends on the mole fraction of the solvent in the solution. Addition of any solute reduces that mole fraction.

28.　**F**　The freezing point depression has nothing to do with the reduction in vapor pressure. Both are due to the increase in disorder of the liquid phase when a solute is added.

29.　**T**　The van't Hoff factor i is an integer only in an extremely-dilute solution; association between ions occurs in concentrated solutions.

30.　**F**　Immiscible means that the liquids don't mix.

Multiple choice questions

1.　　e.　The SI derived unit for pressure is the pascal Pa.

2.　　b.　An ideal gas law problem (just make sure the units are correct).
P = 0.50 atm; V = 2.5 L, R = 0.08206 L atm K–1 mol–1; T = 283.15K, so n = 0.054 mol.

3.　　b.　The pressure increases from 760 torr to 1013 torr (an increase of one-third or by a factor of 4/3). The volume must decrease in the same ratio; ie. to 3/4 of the initial volume.

4.　　c.　　$C_3H_8(g) + 5O_2(g) \rightarrow 3CO_2(g) + 4H_2O(l)$

One mole of any gas at STP has a volume of 22.4 liters. The molar mass of propane is 44 g, so 2.20 grams of propane is 2.2/44 moles = 0.050 moles. According to the stoichiometry of the reaction one mole of propane requires five moles of oxygen; so 0.25 moles of oxygen are required in this reaction.

5.　　b.　Partial pressure of hydrogen = 1 atm pressure × (2.0/6.0) = 1/3 atm
　　　　　Partial pressure of nitrogen = 1.5 atm pressure × (4.0/6.0) = 1 atm
　　Therefore the total pressure = 1/3 + 1 atm = 1.33 atm

6.　　e.　Relative rates of diffusion are proportional to the inverse of the square roots of the molar masses (Graham's law). The molar mass of nitrogen gas is 28 g mol^{-1}. (Note that the answer has to be greater than 2, which limits your choice.)

7.　　a. b. c. (any phase change that requires breaking bonds is endothermic): melting, vaporization, sublimation.

8.   d.   The lines represent the conditions under which two phases coexist in equilibrium.

9.   d.   some molecules in a liquid have sufficient energy to escape intermolecular attraction.

10.   e.   In a cubic lattice, a structural unit at the corner of a unit cell contributes 1/8.

11.   c.   Volume of one unit cell = $(362)^3$ pm^3 = 4.74×10^{-23} cm^3
Mass of one unit cell = volume × density = 4.24×10^{-22} g
Mass of one copper atom = 63.546 g / 6.022×10^{23} = 1.055×10^{-22} g
This is 1/4 of the mass of one unit cell; so there are 4 atoms per unit cell.

12.   b.   The metal ion most strongly solvated is the one with the largest charge and the smallest size.

13.   a.   It triples; Henry's law states that the solubility of a gas depends upon the partial pressure of that gas above the solvent.

14.   b.   the partial vapor pressure of acetone = 0.50 × 176 torr = 88 torr
the partial vapor pressure of ethanol = 0.50 × 50 torr = 25 torr
total vapor pressure = 113 torr.
According to Dalton's law of partial pressures, the mole fraction of ethanol = 25/113 = 0.22
(Ethanol is the less volatile of the two solvents.)

15.   d.   Molality is the number of moles / 1000 g solvent, so in 200 g of solvent, we need 0.03 moles of NaOH. The molar mass is 40 g mol^{-1}.

16.   d.   20% methanol means 20 g of methanol in 100 g solution, or 20 g methanol in 80 g ethanol. The concentration of methanol is therefore 250 g /kg ethanol.
The molar mass of methanol is 32 g mol^{-1}. Number of moles = 250 g / 32 g mol^{-1} = 7.8 mol.

17.   a. $\Delta T = k_b \times m \times i = -3.0 \times 2.0 \times 1 = -6.0°C$, so the boiling point of pure acetic acid is 118°C.

18.   c. $\Pi = cRT$; remember to use the correct units.
$\Pi = 9.8 / 760$ atm; R = 0.08206 L atm K^{-1} mol^{-1}; T = 293.15K; so c = 5.36×10^{-4} mol/L.
Number of moles in 25 mL = 1.34×10^{-5} mol; this is the number of moles in the 50 mg sample.
Molar mass = 0.050 g / 1.34×10^{-5} mol = 3730 g mol^{-1}.

19.   a.   An aerosol is a colloidal dispersion of a liquid or solid in a gas.

20.   f. $\Delta T = k_b \times m \times i$. Look for the highest product $m \times i$. Colligative properties depend upon the concentration of particles, not their identity. For Na_2SO_4, $m \times i = 0.040 \times 3 = 0.120$.

3.4 Examination 4

True–false questions

1. F   The rate of a reaction is a kinetic property; it does not matter whether the reaction is endothermic or exothermic. However, exothermic reactions will invariably increase the temperature which will increase the rate.

2. F   It depends upon whether the reaction is exothermic or endothermic.

3. T   The exponents equal the stoichiometric coefficients only for an elementary step.

4. F   The rate constant k depends upon the temperature.

5. F   Homogeneous in this context means that the catalyst is in the same phase as the reactants.

6. T   The half-life of any first-order process is independent of the amount initially present.

7. F   Only for an elementary step (*cf.* question 3).

8. F   It depends upon whether the reaction is exothermic or endothermic.

9. T   This is a common definition of equilibrium.

10. F   While this is true for a gaseous equilibrium, in general substances in a difference phase from that in which the equilibrium exists are omitted from the expression. So, liquids and solids are omitted from an equilibrium constant expression for the gas phase, but solids and gases are omitted from the expression for an equilibrium in the solution (liquid) state.

11. T   When $\Delta n=0$, $K_p = K_c$.

12. T   K is the ratio of product concentrations to reactant concentrations at equilibrium.

13. T   Equilibrium can be reached from either side.

14. F $\Delta G°$ influences the position of equilibrium: if it is positive the reaction will be reactant-favored but the reaction will still go until the equilibrium position is reached.

15. T   Both describe a system not at equilibrium: K not equal to Q and ΔG not equal to zero.

16. F   It depends upon whether the reaction is exothermic or endothermic.

17. F   When a reaction and its equation are reversed, the K for the new reaction is the reciprocal of the K for the original reaction.

18. F   The words strong and weak describe the degree of ionization of the acid.

19. **F** They are unrelated.

20. **F** The higher the pH, the lower the hydronium ion concentration. pH is a function of the concentration of the hydronium ion, and is only indirectly related to the strength of the acid.

21. **T** This is one way to define what a strong acid is.

22. **T** The Lewis acid definition is the broader definition.

23. **T** The acid-base reaction involves the transfer of a hydrogen ion.

24. **F** But very often true. If the salt is derived from a weak acid *and* a weak base, and the weak base is weaker than the acid, then the solution of the salt is acidic (e.g. ammonium formate). The other exception occurs when the acid is polyprotic. For example sodium dihydrogen phosphate is acidic because the salt can ionize further.

25. **T** K_a for an acid $\times$ K_b for its conjugate base = K_w.

26. **F** The pH of pure water decreases as the temperature increases; there are more hydronium ions (and hydroxide ions) in solution—the solution is still neutral.

27. **T** The product of the two is constant so if one increases, the other must decrease.

28. **T** Due primarily to the very low value for K_w.

29. **T** The ideal buffer point: $pH = pK_a$.

30. **T** When a salt solution is saturated, the concentration of the salt equals its solubility.

31. **F** It depends upon the stoichiometries of the salts.

32. **F** Q_{sp} must exceed K_{sp} for precipitation.

33. **F** The overall equilibrium constant is the product of the two: $K_{sum} = K_1 \times K_2$.

34. **T** This is the common ion effect.

35. **F** Sodium salts are soluble.

36. **F** Spontaneous reactions sometimes happen very slowly—kinetics *vs.* thermodynamics again.

37. **T** Entropy increases as the temperature increases.

38. **F** The entropy of the universe is always increasing; it is the energy that is constant.

39. **T** If $\Delta H_{sys} < 0$, and $\Delta S_{sys} > 0$, then ΔG_{sys} must be < 0.

40.   **T**   At equilibrium, $\Delta G = 0$, and $\Delta G° = -RT\ln K$.

41.   **T**   Vaporization requires energy, so it cannot be enthalpy driven. Vapor is more disordered than liquid.

42.   **T** $\Delta G° = \Delta H° - T\Delta S°$, and depends upon T.

43.   **F**   E° is intensive; increasing the size of the cell increases the current and capacity to do work.

44.   **T**   An anode is where the oxidation takes place, providing the electrons.

45.   **T**   A faraday is the charge on one mole of electrons = 96485C.

46.   **F**   A battery is a collection of cells in series (primary or rechargeable).

47.   **T**   Reduction always occurs at the cathode in a voltaic or electrolysis cell.

48.   **F**   One coulomb is the quantity of charge passed when one amp flows for one second.

49.   **T**   All are equal to energy: 1J = 1 watt.second = 1 volt.coulomb = 1 volt.amp.sec. T

50.   **T**   If the standard reduction potential for a metal is more negative the metal is a better reducing agent.

Multiple choice questions

1. $N_2H_4(g) + 3O_2(g) \rightarrow 2NO_2(g) + 2H_2O(g)$ e.   63 mol L^{-1}s^{-1}
     Check the stoichiometry: oxygen is used 3/2 times faster than water is formed.

2.   b.   It must halve three times; the half-life $t_{1/2}$ is ten days; so the time required is 30 days.

3.

	Initial [NO] mol liter^{-1}	Initial [H$_2$] mol liter^{-1}	Initial rate mol liter^{-1} s^{-1}
Experiment 1:	0.5×10^{-3}	0.5×10^{-3}	2.5×10^{-5}
Experiment 2:	0.5×10^{-3}	1.0×10^{-3}	5.0×10^{-5}
Experiment 3:	1.0×10^{-3}	1.0×10^{-3}	2.0×10^{-4}

Compare experiments 1 and 2: [NO] is the same [H$_2$] is doubled; the rate doubles, so the order with respect to [H$_2$] is 1.
Compare experiments 2 and 3: [NO] is doubled [H$_2$] is the same; the rate quadruples, so the order with respect to [NO] is 2 (since $2^2 = 4$).
The rate equation for the reaction is   e. Rate = k [NO]^2 [H$_2$]

4.   The Arrhenius equation that relates k and T is: k (reaction rate constant) = $A\,e^{-\frac{E_a}{RT}}$
     A useful form of this equation for
     comparing k at two temperatures is: $\ln(\frac{k_2}{k_1}) = -\frac{E_a}{R}(\frac{1}{T_2} - \frac{1}{T_1})$
     Use consistent units: degrees K (not °C) and J (not kJ).
     c. 2.8×10^6 s^{-1}

5. Write the expression for the equilibrium constant and substitute the equilibrium concentrations:

$$Kc = \frac{[NH_3]^2}{[N_2][H_2]^3} = 4^2 / 1 \times 3^3 = 0.59 \text{ (b.)}$$

6. $$CS_2(g) + 3Cl_2(g) \rightleftharpoons S_2Cl_2(g) + CCl_4(g)$$

According to the stoichiometry of the reaction, to form 1 mol CCl_4 you have to use 3 mol Cl_2. So, to form the 0.30 mol of CCl_4, 0.90 mol of Cl_2 must have been used (mole ratio of 3:1). Initially there were 4.0 mol Cl_2 so at equilibrium they must be (4.0 – 0.90 mol) remaining. f. 3.1 mol.

7. Establish an ICE box: $2NO + O_2 \rightleftharpoons 2NO_2$

I	3.00	2.00	
C	–2.00	–1.00	+2.00
E	1.00	1.00	2.00

$$Kc = \frac{[NO_2]^2}{[NO]^2[O_2]} = 4^2 / 1^2 \times 1 = 4 \text{ (h.)}$$

8. $$CaCO_3(s) \rightleftharpoons CaO(s) + CO_2(g) \qquad \Delta H = \text{positive – endothermic}$$

Increase CO_2 at equilibrium by: a. heating the system
(Adding solid $CaCO_3$ or CaO has no effect)

9. $HSO_4^- + OH^- \rightleftharpoons H_2O + SO_4^{2-}$, the HSO_4^- ion is the conjugate acid of SO_4^{2-} (c.)

10. The strongest acid that can exist in liquid ammonia is the protonated solvent molecule f. NH_4^+. This is true of any protonic solvent.

11. $pH = -\log_{10}[H^+] = 1.70$ (c.)

12. A solution made by dissolving NaCN in water will be b. basic because CN^- hydrolyzes.

13. The ammonium ion hydrolyzes: $NH_4^+ + H_2O \rightleftharpoons NH_3 + H_3O^+$ to produce an acidic solution.

$$K_a = \frac{[NH_3][H_3O^+]}{[NH_4^+]} = \frac{[H_3O^+]^2}{[NH_4^+]} = \frac{[H_3O^+]^2}{[0.50]}$$

$K_a = K_w / K_b$ and K_b for $NH_3 = 1.8 \times 10^{-5}$, so $[H_3O^+] = 1.67 \times 10^{-5}$; pH = 4.78 (a.)

All the choices that have a pH > 7.0 can be discarded immediately (the solution must be acidic).

14. This is a buffer solution problem; the solution contains a weak acid and its conjugate base. Use the Henderson-Hasselbalch equation, or simply write the equation for the ionization of the weak acid.

$$pH = pK_a + \log\left(\frac{[\text{conjugate base}]}{[\text{acid}]}\right) = 3.70 + \log(0.10/0.50) = 3.00 \text{ (c.)}$$

15. Concentration of "Bi_2S_3" = 2.0×10^{-12} mole per liter.

Concentration of Bi^{3+} = 4.0×10^{-12} mole per liter.

Concentration of S^{2-} = 6.0×10^{-12} mole per liter.

K_{sp}, for bismuth sulfide = $[Bi^{3+}]^2 \times [S^{2-}]^3$ = $(4.0 \times 10^{-12})^2 \times (6.0 \times 10^{-12})^3$ = 3.5×10^{-57} (d.)

16. In the presence of a common ion, the solubility is suppressed. The little amount of Br^- from the lead bromide is negligible.

The solubility product K_{sp} for $PbBr_2 = [Pb^{2+}][Br^-]^2$ = is $[Pb^{2+}][0.30]^2 = 9.0 \times 10^{-6}$.

$[Pb^{2+}] = 1.00 \times 10^{-4}$ —this is the solubility of $PbBr_2$ (a.)

17. The solubility product for SrF_2 must be exceeded. $K_{sp} = 8.0 \times 10^{-10}$.

a. $Q_{sp} = (1.0 \times 10^{-2}) (1.0 \times 10^{-3})^2 = 1.0 \times 10^{-8}$ $> K_{sp}$ so precipitation (a.)
b. $Q_{sp} = (1.0 \times 10^{-5}) (1.0 \times 10^{-4})^2 = 1.0 \times 10^{-13}$
c. $Q_{sp} = (1.0 \times 10^{-6}) (1.0 \times 10^{-6})^2 = 1.0 \times 10^{-18}$

18. The formation of the cyanide complex sequesters the Ag^+ and reduces the concentration of free Ag^+ ions in solution. Note that silver nitrate is very soluble in water.

$$Ag^+ (aq) \ + \ 2CN^- (aq) \ \rightleftharpoons \ [Ag(CN)_2]^- \qquad K_f = 1.0 \times 10^{21}$$

Assume that all the silver forms the cyanide complex, so that the concentration of the complex is 0.10M. Only a very small fraction of the complex ion will dissociate to form free silver ions.

$$1/K_f = \frac{[Ag^+][CN^-]^2}{[Ag(CN)_2{}^-]} = \frac{[Ag^+][0.8]^2}{[0.10]} \qquad [Ag^+] = 1.6 \times 10^{-22}$$

19. The maximum amount of useful work = $\Delta G° = -257.2$ kJ/mol. The enthalpy change $\Delta H°$ for the combustion of carbon monoxide is -283.0 kJ/mol. The difference is the heat liberated in the reaction. For the combustion of 2.0 moles of carbon monoxide, the difference is two time the difference.
b. 51.6 kJ

20. $\Delta G°_{reaction} = \Sigma(\Delta G°_f \text{ (products)} - \Delta G°_f \text{ (reactants)}) = 2 \times -273.2$ kJ/mol $= -546.4$ kJ (c.)

21. $\Delta G°_{reaction} = \Sigma(\Delta G°_f \text{ (products)} - \Delta G°_f \text{ (reactants)}) = 2 \times +51.31 - 2 \times +86.55$ kJ/mol $= -70.48$ kJ

$\Delta G° = -RT\ln K$

-70.48 kJ $\times 1000$ J/kJ $= -8.314$ JK^{-1}mol$^{-1} \times 298.15$ K $\times \ln K_p$

$\ln K_p = +28.43$

$K_p = 2.23 \times 10^{12}$ (c.)

22. Decrease in entropy of the system: a. freezing a liquid

23. At the boiling point the system is in equilibrium and ΔG is zero. Therefore $\Delta H_{vap} - T\Delta S_{vap}$ must equal zero or, in other words, ΔH_{vap} must equal $T\Delta S_{vap}$.

$\Delta S_{vap} = \Delta H_{vap}/T = 30{,}800/383.15 = +87.2$ JK^{-1}mol^{-1}

24. If the change in enthalpy ΔH is − and the change in entropy ΔS is +, then ΔG must be − and the reaction must be spontaneous. (c.)

25. One coulomb of charge passes when a current of 1 amp flows for one second.

    Time (seconds) = charge (coulomb)/current (amp) = 1000 coulombs / 4.0 amp = 250 seconds.

    = 2 minutes 10 seconds (b.)

26. Cu*(s)*|Cu^{2+}*(aq)* || Ag$^+$*(aq)*|Ag*(s)*

    E° Ag$^+$*(aq)*|Ag*(s)* = +0.799 V
    E° Cu^{2+}*(aq)*|Cu*(s)* = +0.339 V

    Change the sign for the anode oxidation and add: −0.339 + 0.799 V = 0.46V (e.)

27. $\Delta G^°_{rxn}$ = −nFE° = −2 × 96485 × 0.46 = 88,800 J = 88.8 kJ (e.)

28. K$_{sp}$, the solubility product for silver bromide = [Ag$^+$][Br$^-$]

    AgBr*(s)* + e$^-$ ⇌ Ag*(s)* + Br$^-$*(aq)*   E° = +0.0732 V
    Ag$^+$*(aq)* + e$^-$ ⇌ Ag*(s)*   E° = +0.799 V

    Reverse the second equation and add:

    AgBr*(s)* ⇌ Ag$^+$*(aq)* + Br$^-$*(aq)*   E° = +0.0732 − 0.799 = −0.726V

    E° = (RT/nF)lnK   or   lnK = nE°/0.0257V at 25°C;   K$_{sp}$ = 5.4 × 10^{-13} (g.)

29. e. a sacrificial anode

30. 0.500 g of chromium metal = 0.500/51.996 = 0.00962 mol.

    Number of electrons required = 3 × 0.00962 mol = 0.0288 faraday = 2783 C.

    Current (amp) = charge (coulombs) / time (seconds) = 2783 coulombs /3600 seconds.

    = 0.773 amp or 773 mA (e.)

3.5 Examination 5

True–false questions

1. T Aluminum is the most abundant metal in the earth's crust.

2. T The maximum oxidation number of a representative element equals its group number.

3. T Only three of the known elements from no compounds—helium, argon, and neon.

4. T Nitrogen dioxide has an unpaired electron.

5.　**F**　Metasulfuric acid does not exist; condensation beyond the pyrosulfuric acid produces the acid anhydride, sulfur trioxide.

6.　**T**　Sulfurous acid H_2SO_3 is a diprotic acid.

7.　**F**　The oxidation states of the all halogens except fluorine can be positive—for example in their compounds with fluorine or oxygen.

8.　**F**　Only oxygen O_2 is paramagnetic.

9.　**F**　Mirror images are optically active only if they are nonsuperimposable.

10.　**F**　Optical activity refers to the existence of optical isomers; whether the compound chiral or not.

11.　**T**　If the formation constant is higher then the complex is more stable.

12.　**F**　Four coordinate complexes can be square planar for example.

13.　**F**　Diamagnetic complexes do not always involve strong ligands. For example, complexes of zinc are always diamagnetic regardless of the strength of the ligand.

14.　**T**　Yellow and blue are complementary colors; they are opposite on the color wheel.

15.　**F**　A bidentate ligand is a ligand with two pairs of electrons to donate; it does not have to chelate.

16.　**F**　Gamma (γ) radiation is the most penetrating form of natural radiation.

17.　**F**　There are many hundreds of light radioactive isotopes (for example, tritium).

18.　**T**　It is possible, in a nuclear reaction, to change one element into another—transmutation.

19.　**T**　Loss of a β particle changes a neutron into a proton, so the atomic number changes but the mass number does not.

20.　**T**　The more massive the nucleus, the more neutrons are required (relative to the number of protons).

21.　**T**　α emission increases the neutron/proton ratio because there are fewer protons than neutrons.

22.　**F**　Both positively charged and negatively charged are ejected in radioactive decay (e.g. α and β).

23.　**F**　For a first order decay, the half-life is independent of the quantity present.

24.　**T**　Uranium–238 decays spontaneously by alpha (α) emission to thorium–234.

25.　**F**　Nuclear fusion reactions have been achieved in nuclear explosions. Controlled fusion reactions for power generation have not yet been achieved.

Multiple choice questions

1. b. Most industrial hydrogen is used to produce ammonia in the Haber process.

2. b. Sodium is the least expensive to produce.

3. d. Sodium hydroxide is produced by the electrolysis of concentrated aqueous sodium chloride (brine).

4. f. Quicklime is calcium oxide.

5. b. The general formula for the nitride of an alkaline earth metal is M_3N_2.

6. d. Most industrial oxygen is used in the steel industry to remove nonmetal impurities by oxidation.

7. c. V-shaped

8. c. The yellow-brown color of nitric acid is due to the formation of NO_2.

9. e. Most ammonia is used in the fertilizer industry.

10. a. Fluorine is the most powerful oxidizing agent.

11. a. Oxygen has the highest electronegativity.

12. c. Most sulfur mined by the Frasch process is used in industry for the manufacture of sulfuric acid.

13. a. The most important industrial use of silicon is in the production of semiconducting devices.

14. c. Gangue is rock and soil which accompanies a metal-bearing mineral.

15. f. The process used to convert a sulfide mineral to the easily reduced oxide is called roasting.

16. a. Most gold is obtained from anode sludge produced in the electrolytic purification of copper.

17. a. Aluminum forms an invisible oxide film that protects the underlying metal.

18. h. The most common oxidation states of chromium are 2, 3, and 6.

19. d. In the first transition series, the variation in atomic radius is an initial decrease in size to the center of the series and then a slight increase at the end.

20. c. The ground state electron configuration for iron is $[Ar]\ 3d^6\ 4s^2$.

21. d. The name for ammonia when it is used as a ligand in a transition metal complex is ammine.

22. a. The name for $K_2[NiCl_4]$ is potassium tetrachloronickelate(II).

23. d. **Four:** Chloride ions cis, ammonia molecules trans, ethylenediamine has to be cis.
    Ammonia molecules cis, chloride ions trans.
    Both ammonia molecules and chloride ions cis—two optical isomers.

24. e. The d configuration is d^6, four electrons unpaired in the high spin configuration.

25. a. The $d_{x^2-y^2}$ orbital is highest in energy in a square planar complex.

26. a. An alpha (α) particle has an atomic number of 2.

27. f. Lead

25. d. The number of neutrons is the difference between the mass number and the atomic number;
    $214 - 82 = 132$.

26. a. Complete the reaction: $^{218}_{84}Po \rightarrow ^{4}_{2}He + ^{214}_{82}Pb$

    Note that the mass numbers and atomic numbers on each side must balance.

27. b. Beta emission (β) results in an increase in the atomic number of the nucleus; a neutron is changed
    to a proton.

28. b. The difference in the mass numbers $= 226 - 214 = 12$, so 3 alpha (α) particles are emitted. The
    decrease in the atomic number is 4, so the number of β particles emitted must be $6 - 4 = 2$.

29. c. Addition of an alpha (α) particle adds 2 to the atomic number: $13 + 2 = 15$. Emission of a neutron
    does not change the atomic number, and subsequent positron decay decreases the atomic number
    by one: $15 - 1 = 14$. The element produced is silicon.

30. c. 974 years. $\ln(15.3/13.6) = kt$ where $k = 0.693/5730$.

31. a. Most stable isotopes have an even number of protons and an even number of neutrons.

32. a. Mass difference $= (2.01345 + 3.01493) - (1.007275 + 4.00150) = 1.96 \times 10^{-2}$ amu.
    Mass in kg $= 1.96 \times 10^{-2}$ amu $\times 1.66054 \times 10^{-27}$ kg/amu $= 3.27 \times 10^{-29}$ kg.
    Equivalent energy $= 3.27 \times 10^{-29}$ kg $\times (3.0 \times 10^8)^2 = 2.94 \times 10^{-12}$ J.
    $= 2.94 \times 10^{-12}$ J $\times 6.02214 \times 10^{23}$ mol$^{-1} = 1.77 \times 10^{12}$ Jmol^{-1}

One of the ultimate advantages of
an education is simply coming to
the end of it.

B. F. Skinner
(1904-1990)